РОСТ КРИСТАЛЛОВ

ROST KRISTALLOV

GROWTH OF CRYSTALS

VOLUME 9

Growth of Crystals

Volume 9

Edited by

N. N. Sheftal' and E. I. Givargizov

Institute of Crystallography
Academy of Sciences of the USSR, Moscow

Translated by
J. E. S. Bradley
Senior Lecturer in Physics
University of London

 CONSULTANTS BUREAU · NEW YORK AND LONDON

The original Russian text, published for the Institute of Crystallography of the
Academy of Sciences of the USSR by Nauka Press in Moscow in 1972 has been
corrected by the editors for this edition.

Proceedings of E. S. Fedorov All-Union Symposium on Crystal Growth,
Jubilee Meeting, May 21-24, 1969

ISBN 978-1-4684-1691-6   ISBN 978-1-4684-1689-3 (eBook)
DOI 10.1007/978-1-4684-1689-3

Library of Congress Catalog Card Number 58-1212
Softcover reprint of the hardcover 1st edition 1972

© 1975 Consultants Bureau, New York
A Division of Plenum Publishing Corporation
227 West 17th Street, New York, N.Y. 10011

United Kingdom edition published by Consultants Bureau, London
A Division of Plenum Publishing Company, Ltd.
Davis House (4th Floor), 8 Scrubs Lane, Harlesden, London, NW10 6SE, England

FOREWORD

For 50 years the Fedorov Institute of Crystallography, Mineralogy, and Petrography at Leningrad Mining Institute has held annual memorial meetings for E. S. Fedorov. Immediately after the jubilee meeting (May 21-24, 1969), the Fedorov All-Union Symposium on Crystal Growth was held, and the proceedings of that symposium constitute Volume 9 of Growth of Crystals.

The symposium surveyed the advances made in the USSR in those aspects of growth concerned mainly with morphology and structure in natural crystals or closely related artificial ones, work which confirmed their relation to E. S. Fedorov and to mineralogical crystallography.

Crystallography is one of the older branches of natural science but has recently undergone a striking rejuvenation on account of new methods and new concepts. Photogoniometric methods have been developed in goniometry, while crystal optics has found new lines of advance in electrooptics and techniques in the ultraviolet and far infrared regions. Morphologic studies now use a vast range of techniques, from the hand lens to the electron microscope or cinemicrography. X-ray analysis is steadily becoming more automatic, and fast computers are used with accelerated methods of structure interpretation.

Crystal growth is one of the younger divisions of crystallography; previously, it had been of interest only in experimental mineralogy, but now it is an important branch of science and technology with close relations to industry.

In Fedorov's time, crystal growth was a very small part of crystallography, but it attracted his attention, and he produced some major valuable results in this area.

It appears that Soviet crystallographers were among the first to appreciate the value of work on crystal growth and led research into this topic here, as well as providing the basis for progress in the subject generally.

The present symposium provides a further opportunity to demonstrate the strong connection between mineralogy and crystallography, a connection that we value greatly and consider to be largely responsible for our success.

We hope that publication of these papers from the symposium will prove of value to many specialists on crystal growth: crystallographers, mineralogists, physicists, and chemists, as well as students envisaging research in this area.

Academician N. V. Belov

CONTENTS

III. CRYSTAL GROWTH FROM MOLTEN SOLUTIONS

IV. CRYSTAL GROWTH FROM MELTS

VI. THEORY OF CRYSTAL GROWTH

APPENDIX

INTRODUCTORY ADDRESS FROM THE
ORGANIZING COMMITTEE

N. N. Sheftal'

This is the first symposium on crystal growth to be organized in Leningrad by the Institute of Crystallography. It is being held in one of the oldest colleges in the country, the Mining Institute, where Fedorov worked, and where his work on morphology and crystal growth has been continued and greatly extended by Shafranovskii, Grigor'ev, Mokievskii, and many others. These studies strengthen the link of crystallography with geology and mineralogy.

The Mining Institute is closely associated with Leningrad University, where major research on crystal growth is conducted by Petrov, Gendelev, Treivus, Kasatkin, and many others under the general direction of Tatarskii and Frank-Kamenetskii.

The symposium continues the series of conferences on crystal growth organized by the Institute of Crystallography: the Moscow conferences of 1956, 1959, and 1963, which had participants from the socialist countries, and the symposium at the Seventh Crystallography Congress (1966), which produced wider international links. As previously, the basic topics are real crystal formation and growth control.

Publication of the proceedings has facilitated exchanges with other countries, and the English translations have familiarized workers abroad with Soviet advances in crystal growth.

This work has also been carried forward by the international conferences organized since 1966, together with the founding of the International Union on Crystal Growth and of the journal "Crystal Growth."

It is now much more difficult to organize periodic conferences because the volume of work has expanded so greatly, and we are very much indebted to those, such as Sirotov, Ovsienko, Rzhanov, Aleksandrov, and Kuznetsov, who have undertaken the organization of the individual aspects of the topic.

This symposium is an All-Union Conference on Crystal Growth restricted principally to the mineralogical aspect; it is devoted to the features of crystallography and crystal growth most closely related to mineralogy, which is for many of us still our principal interest.

However, if the interests of crystallography did not encompass crystal growth generally, the subject could not have dominated the early years of the science of crystal growth in this country.

The expanding work on crystal growth continually poses new problems, of which the following are some of the most important:

1. Application of chemical crystallography to discover new crystals and to improve their properties.

1

2. Research on the effects of the structure and size of the building blocks on the morphology and perfection of the crystals.
3. Research on the scope for growing perfect single crystals at high supercoolings and high rates.
4. Application of chemical crystallography to modify the properties of crystals by suitable doping.
5. Research on the effects of magnetic, electric, and other fields on the perfection of growing crystals.
6. Synthesis and growth of bulk crystals, films, and whiskers under the microscope.
7. Growth control in whisker systems.
8. Development of the vapor—liquid—crystal method.
9. Research on methods of making single-crystal films on nonorienting substrates.
10. Research on mechanisms whereby doping improves crystal perfection and growth rate.
11. Production of large single crystals of the type now made hydrothermally by chemical transport reactions not involving pressure.
12. Research on the scope for growing single crystals in gels.

I consider that a central task for Soviet science in crystal growth is to found a theory of real crystal formation at the macroscopic level; as Shubnikov said, "We need a theory that will enable us to grow single crystals."

Soviet science has made many advances on the basis of the researches of Fedorov, Wulff, Shubnikov, and Belov, and further advances are to be looked for.

In conclusion, the organizing committee thanks the Mining Institute, especially the rector L. N. Kell, the staff of the Department of Crystallography, and the directors I. I. Shafranovskii, V. A. Mokievskii, and V. F. Alyardin.

Finally, the committee thanks Academician N. V. Belov, the senior Soviet crystallographer, for much assistance in organizing the symposium and publishing its proceedings.

E. S. FEDOROV'S PAPERS ON CRYSTAL GROWTH

V. B. Tatarskii

Evgraf Stepanovich Fedorov will be very long remembered because his unexampled labors had decisive effects on several branches of science. Books dealing with his life and activities give very little information on his researches on crystal genetics, and it may well be that these studies are unknown to most of those working on crystal genesis. A symposium dedicated to Fedorov on crystal growth gives opportunity to deal with this side of his life's work.

There is a certain break in the traditions in crystallography — in recent decades, where the number of crystallographers has increased very considerably, mainly by influx of physicists and some chemists. The old mineralogical traditions of crystallography have largely been forgotten, and with them have been lost from view certain lines of evidence formerly employed in crystallography. On the other hand, some experiments and observations of Federov's have a direct bearing on aspects of crystal growth that so far have not been solved and so may be of interest not merely in a historical respect.

Fedorov published over 400 papers, of which about 20 papers and notes dealt with crystal growth, these ranging in volume from half a page to two or three pages; papers of 10-12 pages are exceptional. However, Fedorov's notes contain may ideas and describe such a variety of experiments and observations that here we can consider only a third of all the evidence on crystal growth contained in his papers. The literature cited contains only papers directly quoted in the text; a more complete annotated list of Fedorov's papers on crystal growth is to be found in the Proceedings of the Fedorov Jubilee Meeting of 1969.

The first article on this topic [1] is a continuation of experiments of 1881-3.* This deals with experiments in which one sees the origin of the method of sphere crystallization; circular glass plates were glued on to ground rock salt plates, and then the material at the edges of these glass plates was dissolved away and the plates were transferred into saturated solution. After 5-6 hr, the circle had become a polygon, whose principal vertices corresponded to the vertices of the corresponding section of the cube. No growth was observed along the normals to the cube faces. If a solution contained carbamide (in which sodium chloride grows as octahedra), there was no growth at all in this time, so carbamide does not accelerate growth on cube faces but retards it on octahedron ones, which is due to adsorption: the particles of carbamide have special attachment to the NaCl crystal in the [111] direction, and they are retained there and do not allow the sodium chloride particles to be deposited.

An interesting experiment was performed with the pure solution. A cover glass with a circular hole was glued to a plate of rock salt; in an unsaturated solution, a circular depression

---

*See [2, 3] for notices of earlier studies.

3

was produced in the plate, and then in a saturated solution the circle became a polygon within a few days. In the section on the cube one could see traces of the (210), (110), and (120) faces, while on the section on the rhombododecahedron there were traces of (211), (111), and (122). Fedorov pointed out that this was the first time that a complex combination had been obtained on NaCl in pure aqueous solution. This method of Fedorov's was taken up by his pupil Artem'ev and applied as the method of sphere crystallization, which in particular allowed one to define the symmetry of a crystal and gives an objective criterion for the relative importance of the various faces. The method is still used and the scope for use of it is far from exhausted. Here we describe a modification of this method used by Fedorov himself.

In [1], much attention was given to skeletal figures or growth figures as Fedorov calls them. The following is a single observation: first of all one gets growth figures, then small nuclear crystals appear, which grow while the growth figures dissolve. This effect is well-known to anybody who has examined crystallization under the microscope, but it has never appeared in textbooks on crystallography. Fedorov drew the important conclusion from this as regards a saturated solution: a solution supersaturated in relation to crystals is very much unsaturated in relation to growth figures. Later in the article he mentions the higher solubility of corners and edges of crystals, the transition of material in a saturated solution from a convex corner to a concavity, and theoretically demonstrates the higher solubility of close-packed faces (ones with simple symbols in the regular setting of the crystal).

Fedorov in later papers devoted considerable attention to effects arising from solubility differences between faces; some of these papers may be mentioned, because they describe interesting experiments and solubility differences employed as a method of investigation.

In [4], which deals with solubility differences between faces, it is stated that all specialists accept that this is so, but direct experiment would solve it beyond doubt. The solubility is the greater the denser the face and the less the more complex the symbol, and consequently it is zero for an irrational face, and such a face should grow even from a very weak solution, but this lies outside the scope for experimental test, since growth of an irrational face causes the latter to be replaced at once by a set of rational ones, for which the solubility is not zero. One can grind a plane, and the lateral faces close to this may have complex symbols, and consequently low solubility.

Further there is an experiment. At a vertex of cleaved piece of rock salt, a plane was ground close to (621), which Fedorov placed on the cleaved face of another rock salt crystal with an intervening drop of saturated solution. After $2\frac{1}{2}$ minutes, the face was unaltered, whereas the polished plane had become slightly mat, i.e., there was deposition of material from the saturated solution on the irrational plane. The solution became very dilute, but the time was inadequate for the cube face underneath it to begin to dissolve. The experiment was repeated with a 10-min delay, after which the plane had become very mat and a hand lens revealed on it a rosette of faces. The face of the other crystal underneath has a triangular recess, whose depth increased from the center to the edge almost to 1 mm.

Further, various forms of experiments with other crystals are described, and the paper concludes with the following experiment: a slightly ground vicinal face of an octahedron in alum was used, and the roughness was smoothed off, so the resulting plane was close in position to an actual octahedron face; in this case again the ground face grew while the untreated face dissolved.

In November 1907 Fedorov presented his paper "Solubility differences between faces in the mineral kingdom" [5]. The paper begins with the following argument: a solution saturated for the most important faces of the crystal will be supersaturated for other faces, so (a paradoxical consequence), a finely ground powder placed in a saturated solution will cause the latter to become more dilute. Then he describes an experiment: rock salt powder was poured on to glass, and a circular cover glass was placed above and then above this was a large drop of saturated solution followed by a cleaved rock salt surface resting partly on the cover glass and partly on the powder. After 10 min, the cleaved face had a step with a circular shape, while above the powder the cube face had dissolved. Then he describes crystals of various minerals on which the most important faces have been replaced by recesses edged by narrow sets of secondary faces. The goniometer pattern showed that these were not growth skeletons but a result of etching of facetted crystals. The crystal had evidently been acted on by a solution saturated by the secondary faces but unsaturated for the closest-packed ones.

In November 1908 he wrote "Experiments on crystallization between two spheres" [6]. A hemisphere was cut from an alum crystal and attached to a piece of glass via an octahedron plane. In the octahedron face of another crystal he made a hemispherical recess. One hemisphere was faced in the other in a parallel position, and the gap was filled with saturated solution, with observation of the course of the regeneration. On the convex surface he observed the usual regeneration of a sphere, with production of areas corresponding to the most important faces, with a mat surface in between. No faces were visible on the concave hemisphere, but the goniometer revealed signals of the same form as for the convex one, but of a much lower quality. In another experiment, a convex hemisphere was turned through 15° around a vertical threefold symmetry axis, and in this case the crystal faces on it were obtained as uneven and giving poor signals, while the secondary faces gave good signals. After several such experiments, it was concluded that close spacing between close-packed face and an irrational surface is unfavorable to the growth of the former on account of solubility differences; in spite of diffusion, the solution was diluted.

There is also his paper of 1909 "Experimental solution of the origin of vicinal places" [7]. First of all he quotes the well-known fact that a crystal is not an ideal structure, so one has to accept that an ordinary crystalline material is not completely uniform but is distorted, and this is a logical necessity, and it is unclear why this obvious fact has been continuously ignored everywhere except in this country. Some experiments previously done are then mentioned: the vicinal faces on alum are in equilibrium with a saturated solution, and if an uneven surface is produced by grinding, the resulting octahedron space grows from a saturated solution and even from an unsaturated one, so the irregular ridges and recesses that we see on a freely crystallized octahedron face of alum represent a set of not completely parallel distorted octahedral faces.

Further follows an experiment on crystallization between spheres; if parts of the structure can deviate, the deviation on average is the larger the further the parts are apart. The deposition on a convex sphere occurs around the pole of the (111) face, while on a concave sphere this is impossible; particles are deposited on all parts of the hemisphere, and the planar areas are negligible in size. The (111) signal from the goniometer is then a set of signals from innumerable microfaces. If the crystal lattice were ideal, the signal quality would be as from a plane on the convex hemisphere, but in fact the difference is enormous. Fedorov considered this a decisive experiment for the demonstration of nonideal structure in a crystal. In fact, if the first experiment does not satisfy us, because we now know that the structure of a ground and polished face differs considerably from that of a regular face of the crystal, the second experiment gives the first direct proof of nonuniformity in a crystal structure, which now is universally accepted.

We consider again his paper of 1901. He observed skeletal growth in various crystals under the microscope, for which he concluded in a paper published in 1903 "One of the most general laws of crystallization" [8], which is devoted to his law of crystallographic limits, which does not have any relevance to the present topic, but which we mention here because in the course of the discussion he mentions another law, namely that markedly positive crystals are tabular and commonly show good cleavage, while neutral crystals are isometric, and markedly negative ones are in the form of needles.

Fedorov used the terms positive and negative for crystals that were elongated or flattened; in general, the crystal shape reflects the lattice type as well as the angular relationships and the parameter ratios for the unit cell. Crystallochemical analysis allows one to find the parallelohedron, as the above is called, for any substance, i.e., to determine the structure. The weak point in crystallochemical analysis is the need to employ Bravais's rule. This rule is inaccurate, so many such determinations have not subsequently been confirmed, but the elongation or flattening of the elementary cell is very often subject to this rule, because the result is derived from goniometric data, while Bravais's rule plays only a secondary part. This enabled Fedorov to establish the regularity of which he speaks in his papers of 1901 and 1903. The explanation is as follows [8]. The adhesion (bond strength) in a given direction is always such as to influence the relative separation of the particles (Fedorov crystalline particle is a moleculer or a group of molecules). However, the adhesion is affected also by the individual properties of the atoms and their disposition in a particle, so this is the limiting law, which makes itself felt statistically and especially strongly in extreme cases, i.e., if one axis of the parallelohedron is very much elongated, the crystal always takes the form of a thin plate; if the axis is very much shortened, the crystal always takes the form of a needle.

In the latter form, this rule has now been completely confirmed, but it is not given in textbooks and is not familiar to many crystallographers. In recent publications one sometimes found arguments based on the belief that the highest growth rate of a crystal will correspond to the direction of the largest parameter in the unit cell.

Fedorov used Bravais's rule but saw that the sequence of face development was far from often in accordance with it. In his paper of 1901, he offers a theoretical basis for the rule: the denser the packing on a face, the greater the solubility as a rule, and consequently the less the supersaturation and the slower the growth for a given solution concentration. Subsequently, Fedorov several times attempted to elucidate the reasons for deviations from Bravais's rule and took the view that they are related to the fine structure, which at that time was inaccessible. His last mention of this topic we find in two papers written in September 1916 "Discussion of crystallization via the disposition of atoms" [9] and "Determination of the density of the atoms on crystal faces" [10]. Several structures had been determined by this time and this enabled him to consider Bravais's rule in a new light.

Fedorov wrote that x-rays showed that each type of atom forms its own lattice, and from the earliest structural studies it became clear that the shape of a crystal is determined not so much by the lattice, which is composed of the centers of parallelohedra, as by the density in the atomic arrays. If we assume that the main importance attaches to planes with the highest atomic densities, one attains a conflict with the experimental evidence for many of the structures that have been examined; on the other hand, experiments show the particular importance of individual zones, which vary from one crystal to another. Certain phenomena force us to give preference not to the closest-packed plane but to the closest-packed series of nodes; then the main faces are ones that are determined by these rows as axes, and it is these faces in these zones that control not only the structure but also the crystallization conditions.

Further he considers the question which zones and which places in them are most important in different types of structure; it is clear from this analysis that by 1916 Fedorov had already approached quite closely to concepts developed later by Hartmann.

We cannot overlook how vigorously Fedorov participated in establishing the early results from x-ray structure studies; and although in all his theoretical studies he started from the basis that a crystal consists of molecules, he at once and without visible signs of regret abandoned this idea much more decisively than did chemists, as is clear from his purely crystallo-chemical studies.

We will not deal with a small series of papers concerned with crystallization from melts, but merely mention that in one of them (1904) he describes a process very similar to that of zone melting. In [11] he writes "By repeated partial melting I have enabled a new crystal to grow at the expense of a liquid, which moves along to one end, where it accumulates at the end of the process."

From his observations on melt crystallization he formulated his concept of crystal growth via thick layers, which he reported in [12], where he speaks of the complete impossibility of representing crystallization as a more or less continuous accumulation of individual particles laid down in a strictly parallel position. In a melt or highly supersaturated solution, the crystal is at once covered in layers visible under the microscope, each of which consists of tens or hundreds of thousands of elementary layers. He describes nonuniformities arising in such growth, and at the end of the 1930s these observations were revised and extended by Ansheles.

Criticism would hardly be in place in a jubilee survey, but two comments may be made in respect of the above papers. The basic topic in one series of papers was the solubility difference between crystal faces, which Fedorov considered as absolutely proven. At the present time, there is no doubt as to the identical solubility of the faces of the hypothetical equilibrium form, but there is no strict extrapolation of this concept to the faces of a real crystal.

Also, Fedorov's experiments usually lack quantitative results, and this might suggest a need to repeat or verify some observations with current experimental techniques. To this one raises the objection whether it is necessary; is it necessary to return to what was done 60 years ago? In my view, it is necessary because some topics that Fedorov considered need further extension, in spite of the enormous advances made recently.

This survey has omitted the historical background. We need to remember that Fedorov worked before Volmer, Kossel, and Stranski, when in essence there was no theory of crystal growth. Growth questions interested him mainly not for themselves but in relation to crystallo-chemical analysis. However, his papers indicate that he clearly understood the relation between structure and crystal growth, and the structure often existed only as a theoretical construction, while there was no theory of growth at all. Fedorov appears to us not only as a great theoretician but also as a talented experimenter, whose ingenious experiments can sometimes scarcely be bettered.

Literature Cited

1. E. S. Fedorov, Izv. Akad. Nauk, series 5, 15:519 (1901).
2. E. S. Fedorov, Zap. Min. Obshch., 18:281 (1883).
3. E. S. Fedorov, Zap. Min. Obshch., 27:464 (1891).
4. E. S. Fedorov, Zap. Gorn. Inst., No. 1, 81 (1907).
5. E. S. Fedorov, Zap. Gorn. Inst., No. 1, 160 (1908).
6. E. S. Fedorov, Zap. Gorn. Inst., No. 1, 397 (1908).
7. E. S. Fedorov, Zap. Gorn. Inst., No. 2, 255 (1909).
8. E. S. Fedorov, Izv. Akad. Nauk, series 5, 18:155 (1903).
9. E. S. Fedorov, Zap. Gorn. Inst., No. 6, 161 (1917).
10. E. S. Fedorov, Izv. Akad. Nauk, series 5, 6, 10(part 2, No. 17):1675 (1916).
11. E. S. Fedorov, Ezhegodnik po Geol. i Min. Rossii, 7(Section 1):151 (1904-5).
12. E. S. Fedorov, Priroda, No. 12, 1471 (1915).

PART I

CRYSTAL GROWTH UNDER HYDROTHERMAL CONDITIONS AND DIAMOND SYNTHESIS

SOME RESULTS OF RESEARCH ON HYDROCHEMICAL CRYSTAL SYSTEMS AND GROWTH

V. P. Butuzov and A. N. Lobachev

We have now well-developed techniques for growing single crystals from aqueous solutions and melts, but only a few types of compound are represented by crystals grown by these methods, and there is therefore considerable importance in developing methods of hydrothermal synthesis for single crystals, because only in that way can one produce crystals of silicates, carbonates, germanates, and certain types of sulfides and oxides.

Hydrothermal synthesis is a form of growing from solution; it extends the thermodynamic range of conditions that can be used and opens good prospects for producing single crystals, because certain substances are virtually insoluble in water and other solvents at ordinary pressures. In some cases, difficulty arises because of polymorphic transformations. The hydrothermal method enables one to overcome these difficulties by conducting the operations at temperatures far from the melting point, which facilitates formation of uniform crystals low in stress.

The media used in hydrothermal synthesis are solutions of salts, alkalis, and acids, with very large concentration ranges; in some cases, for example, the alkali concentration may rise to 90%. The method can also be used to simulate natural mineral production processes, so such researches extend our knowledge in fields of geochemistry and genetic mineralogy. However, the advantages of the method are accompanied by a number of difficult or unsolved problems, as follows:

1. The synthesis of growth is usually conducted in a multicomponent system; we usually lack evidence on the phase equilibria, solubilities, and other characteristics necessary to choose rationally the experimental conditions to be used.

2. The high pressure and temperatures employed mean that the apparatus is fairly complex and has a large working volume, which is needed if large single crystals are to be grown. Recent studies have seen steady extension of the work to higher temperatures and pressures.

3. At the temperatures and pressures used, the working solutions tend to react rapidly with the vessels, which raises the problem of protecting them adequately.

4. The crystallization is conducted without visual control, or most other forms of monitoring, i.e., one does not know the supersaturation, the growth rate, or the quality of the growing crystals except from experience with previous runs; this means in each particular case that one has to build up laboriously the relationships in order to obtain good results. Production of a large crystal is always a very lengthy process, and may last a considerable fraction of a year, and the conditions must be kept constant for this time.

5. The problem of seed crystals is extremely important; one needs fairly large high-grade seeds, which often control the quality of the final crystals. Any defects in the seed crystal are usually inherited in the finished product.

However, there are no insuperable obstacles in hydrothermal synthesis, and good results can be ultimately attained if all the necessary growth conditions can be provided.

The following are the basic trends and results in hydrothermal synthesis as derived from a survey of the published evidence.

There are many papers that deal directly or indirectly with genetic and experimental mineralogy, in particular numerous studies on mineral destruction in the presence of water and determination of the stability ranges as functions of temperature, water pressure, partial pressure of carbon dioxide or oxygen, and effects of isomorphous substitution of polymorphic transformation. All of these topics are closely related and have a direct bearing on problems in hydrothermal chemistry.

Very often, hydrothermal synthesis is conducted with the sole purpose of producing a compound as a crystal for further study and application.

In the last 25 years there have been many papers on the physics and technology of hydrothermal synthesis; for instance, there have been many studies on the synthesis of alkali aluminosilicates in the system $R_2O-Al_2O_3-SiO_2-H_2O$ (where R is Li, Na, or K), which have arisen from the need for zeolites in petroleum processing. Barrer and his colleagues [1] in Britain have done much in the development of industrial methods of hydrothermal synthesis for zeolites. The synthesis conditions and isomorphous substitution have been employed to produce a whole range of zeolites, which have been fairly widely applied in the purification and separation of gaseous and liquid products. In this country, this work is particularly associated with the names of Senderov and Khitarov [2], who performed the hydrothermal synthesis of the acid-resistant zeolite called mordenite.

Interest in asbestos synthesis has led to research on compounds of serpentine type; we may note particularly studies on magnesium compounds here, and also some on forms of absestos containing cobalt and nickel. Many studies in this region have been made by Fedoseev and his coworkers, who recently published a monograph [3], where a complete listing is to be found of the papers on artificial asbestos. The hydrothermal synthesis of blue asbestos confirms the wide scope for producing high-performance fibrous materials based on silicates.

At the present time, the All-Union Mineral Raw Material Synthesis Research Institute has collaborated with the Institute of Silicate Chemistry in developing the laboratory technology for making amphibole short-fiber asbestos (up to 0.5 mm), and experimental batches have been produced for special technical tests.

There are especially many papers on the synthesis of calcium hydrosilicates, hydroaluminates, and hydroaluminosilicates; these compounds are of interest in relation to setting and strength of cement. Particular difficulties are encountered in this region on account of identification of the products; the first systematic studies on hydrated calcium silicates were made by Flint, McMurdie, and Wells in Britain and the United States, which covered a wide range in the ratio of CaO to SiO_2. These researches enabled one to establish the mineral composition of solidified Portland cement. Later Taylor [4] made detailed studies on the stability of calcium hydrosilicates and the hydrolysis conditions for these. The synthesis of calcium hydroaluminosilicates led to the discovery of the hydrogarnets, which are of some interest in silicate crystallochemistry. Garnets have proved of considerable interest in physics, in particular $Y-Fe$ garnets; commercial hydrothermal synthesis of these has been carried out in the United States [5].

Semiconductor and infrared techniques have required methods of synthesizing the extensive class of compounds known as chalcogenides. Considerable number of compounds in this class have been made by Cambi and his coworkers [6] in Italy, who used double decompositions and other exchange reactions. Robinson [7] in the USA has made a number of simple and compound sulfides of lead and antimony by a direct reaction between the components. The number of papers on chalcogenides is small compared with that for silicates, and the studies have not been systematic, the reason being that sulfide media are extremely corrosive and often also toxic. Considerable advances have been made in this region as regards methods of crystallizing compounds in the groups $A_2^V B_3^{VI}$ and $A^V B^{VI} C^{VII}$ (where A = Sb, Bi; B = S, Se, Te; C = Cl, Br, I); the first group consists of semiconductors, while the second consists of very interesting semiconducting ferroelectrics [8].

Ikornikova [9, 10] at the Institute of Crystallography has done much valuable work on carbonate synthesis; the hydrothermal production of calcite has been extended to rhodochrosite, ottawite, and sphaerocobaltite. Carbonate synthesis is important not only in mineralogy, for conditions have been found where one can manufacture industrially materials of value in optics and other fields of physics.

Pogodin [11] at the All-Union Mineral Raw Materials Research Institute has grown crystals of optical-grade calcite weighing up to 200 g.

Various compounds are now made under hydrothermal conditions, extending from simple ones such as those of silver and copper to complex silicates such as tourmaline; the number of compounds increases every year, and now covers representatives of almost all classes of minerals: oxides and hydroxides, molybdates, arsenates, borates, etc.

A considerable branch of the work in hydrothermal chemistry, if not the principal one, is the production of large single crystals; recent advances in this field have failed to keep pace with the demands of technology and electronics, where one mostly requires large single crystals. Really striking advances in growth from seeds have been made in the production of quartz crystals.

Fig. 1. Synthetic quartz crystal.

We now have available a method of routine production of single crystals weighing over 10 kg (Fig. 1). Amongst other compounds, it is difficult to find any where the advances have been comparable; laboratory production has been undertaken with seeds for BeO (bromellite) [12], ZnO (zincite) [13], Al_2O_3 [14, 15] (corundum and ruby), Na_2ZnGeO_4 [16], which has been called phase D, and certain others.

Consider in more detail work carried out entirely or mainly by two organizations: the All-Union Mineral Raw Material Synthesis Institute and the Laboratory of Hydrothermal Synthesis at the Institute of Crystallography. These laboratories have made physicochemical studies on the properties of pure solvents for growing crystals, as well as on the properties of systems where the crystal and solvent interact.

In hydrothermal synthesis, one needs to know the PTFC relationships for the solutions; such diagrams are of considerable practical importance, because they contain information that indicates the degree of hydration of ions and thus has a bearing on the theory of solutions. In practice, PTFC diagrams are used to determine the pressure and specific volume in measurements on solubility and in surveys on crystal synthesis, i.e., when one is using an autoclave without a pressure gauge and where one needs to know the functional dependence of one parameter on the others.

Aqueous chloride solutions are widely used in hydrothermal synthesis, so the Laboratory of Hydrothermal Synthesis has made systematic studies on the PTFC relationships of chlorides for alkali metals and ammonium [17]. The results may be compared with those from the All-Union Institute in the case of certain analogous solutions [18], and this has shown that the results obtained by different methods agree within the error of experiment.

Hydrothermal synthesis employs a wide range of high-pressure apparatus; the thermal gradient method is a basic one for prolonged processes, and for this one has laboratory equipment working up to 3000 atm and 500-550°C with substantial working volumes. Systems and solubilities have been measured in an apparatus capable of maintaining temperatures to 0.1°C, which allows the solubility isotherm to be recorded in a single run [19] (Institute of Crystallography). General surveys at the Institute of Crystallography have been based on a hydrothermal gravimetric apparatus [20], which enables one to observe the transport of material from the dissolution zone to the growth zone, and to control the transport by adjusting Δt. At both of these institutes there are equipments for use up to 7000-10,000 atm at high temperatures; these use internal heaters, cooled seals, counterpressure, etc. At the All-Union Institute, hydrothermal syntheses are carried out on a commercial scale at pressures up to 1500 atm and 400°C with volumes of thousands of liters in a production plant, which can provide hundreds of kilograms of crystals in one run. At the Institute of Crystallography, there are autoclaves with windows for observing the growth [19]. Designs have been developed for protective linings for a synthesis when especial purity is required. At the All-Union Institute, various designs have been developed for protective linings of corrosion-resistant materials that allow one to perform very clean syntheses in volumes up to 200 liters.

All studies on hydrothermal synthesis may be divided into two basic groups: the first is to produce known or novel crystalline compounds by spontaneous crystallization, which yields in a short time crystals comparatively simple in composition that have physical properties of considerable scientific and practical interest. This method also enables one to establish the regions where these phases are formed and to produce small crystals for seed or raw-material purposes.

The object of the second group of studies is to produce large uniform crystals by prolonged growth on seeds on the basis of accumulated experience; a good example of this is the work on quartz.

Fig. 2. Zinc oxide single crystal grown from a seed parallel to (0001), × 2.

We now consider some details of the synthesis of some groups of compounds; the main materials here are from the groups of oxides, carbonates, germanates, silicates, and chalcogenides.

Of the oxides, the most interesting is ZnO [13], which have been grown as fairly large single crystals on seeds (Fig. 2). This material amplifies ultrasound and is therefore used in acoustic devices; it can also be used in electrophotography and as a catalyst in certain reactions.

Much interest attaches to crystals of $RBi_{12}O_{20}$ (where R = S; Ge, Ti, etc.) [21], which are crystals having little ultrasound absorption, which thus provide good conditions for transmitting the latter. Single crystals of sillenite (Fig. 3) can be grown in solutions of NaOH at 350-400°C and 600-800 atm [22]. The problem of high-grade seeds is a major one in the production of large single crystals of this material.

A major advance in the synthesis of refractory oxide is the hydrothermal production of single crystals of bromellite BeO (melting point 2547°C [12]), which can be used as a piezoelectric and an ultrasonic amplifier.

Fig. 3. Sillenite single crystals, ×10.

The Institute of Crystallography has made crystals from the group of trigonal carbonates as dimagnetic materials activated by paramagnetic ions, as well as magnetic carbonates with characteristic antiferromagnetic properties. It is not always necessary for hydrothermally grown crystals to be large in size, e.g., quartz; one often requires small crystals a few millimeters in size for physical purposes. A typical example of this is the carbonates of Co, Ni, and Mn, which have been made on a laboratory scale. Advances in science and technology have accelerated research for new crystals for use in instrumentation, e.g., silicates of the type $Na_xMe_ySi_pO_q$, where Me = Al, Zn, Cd, Mn. The results for crystals in the system $Na_2O - ZnO - SiO_2 - H_2O$ [23] show that there is considerable scope for synthesis at high temperatures and pressures for a number of new zinc silicate phases of previously unknown composition, which fluoresce in the yellow and green regions of the spectrum when activated by Mn^{2+}, and these have attracted attention especially because such compounds do not occur in nature.

The germanium is crystallochemically very similar to silicon, which indicates that one should be able to make germanates analogous to silicates and having similar properties.

The basic structural unit in a silicate is the SiO_4 tetrahedron, where the germanates have two such units, namely the tetrahedron and octahedron, i.e., it is possible to make germanium enter a not only tetrahedral coordination but also octahedral. The presence of both types of coordination simultaneously expands the scope for production of new groups of germanates.

A series of germanates and zinc-germanates of sodium and potassium have been made with the general formula $Na(K)Zn_yGe_pO_q$, with various relationships between the oxides in the initial mixture; these have been produced as euhedral crystals 1-10 mm in size. The crystallization and stability ranges of these phases have been established.

Most of the members of this series are phosphors, and crystals activated by Mn^{2+} fluoresce in response to UV, cathode rays, x-rays, and also friction. Crystals of phase D, composition Na_2ZnGeO_4, also show electroluminescence and the piezoelectric effect [24]. Phase D allows crystals to be grown on seeds (Fig. 4). There is no marked anisotropy in the growth rates (in a certain temperature range), which makes it possible to use spontaneously produced crystals as seeds. Preliminary results on the properties of the crystals show that they are promising for use in optical and piezoelectrical devices, and such crystals are at present under study at a number of institutions in this country.

Numerous silicates have been made hydrothermally in order to obtain new varieties for examination of their physical properties; recent studies have concerned the system $Na_2O - MeO - SiO_2 - H_2O$, with MeO = ZnO, CdO, MnO, Al_2O_3.

In these experiments, the principal variable has been the Na_2O concentration, with sometimes the MeO/SiO_2 ratio; these systems have yielded about 20 compounds, of which 15 have not previously been recorded. The crystals grow readily at 400-500°C and 1000 atm or above.

Fig. 4. Na_2ZnGeO_4 single crystal grown under hydrothermal conditions, × 2.

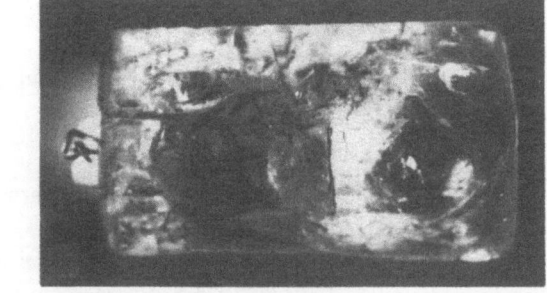

Fig. 5. Sodalite single crystal grown
from a seed parallel to a tetrahedron
face, × 1.5.

Manganese doping in the zinc system has given new phosphors with green emission, which is particularly pronounced for Na_2ZnSiO_4. The manganese system has given the phase of composition $Na_2Mn_2Si_2O_7$ with a bright red emission [25].

Also, most of the compounds have novel structures, which has provided for advances in the chemical crystallography of silicates with large cations.

Interesting results have been obtained as regards the crystallization conditions of sodium alumina silicates [26]; crystals in this group grow over a wide range in temperature and also in pressure, but the best conditions for nepheline are a T of 400°C with ΔT of 20-30°C, while for cancrinite they are T of 300-350°C and ΔT of 20-30°C, and for sodalite they are T of 250°C and ΔT of 10-20°C. The pressure range is 300-500 atm.

Nepheline, cancrinite, and sodalite are compounds lacking a center of symmetry in the crystal, so they show piezoelectric and electrooptic response; sodalite has cubic symmetry, so it is most promising for electrooptic purposes. Quite good crystals of the compound have now been grown (Fig. 5).

The silicates with aluminum in sixfold coordination are difficult to produce, because they require comparatively high pressures and temperatures, and only two representatives of the group have been examined: tourmaline and spodumen [27, 29]. It has been found that they will grow at elevated temperatures and pressures (above 700° and above 3000 atm) with (001) single-crystal seeds of iron-free tourmaline (elbaite) which have proved suitable for preliminary examination of the physical properties. Measurements have been made on the effects of temperature, pressure, boric acid concentration, and composition of the initial products as regards the crystallization of ferruginous and also iron-free tourmalines, together with the phase compositions in boron-fluoride, boron-chloride, and mixed boron-fluoride-chloride acid systems (weakly acid, pH 4-4.5). It has been found that elevated pressures and temperatures are desirable, while for the iron-free tourmalines with maximal Al_2O_3 content they are absolutely essential.

Major phases accompanying the tourmaline in these systems have been various fluoro-aluminates, which are of interest in the activated state as phosphors and as substances with very low refractive indices; other compounds also present are fayalite, hercynite, mullite, and various other substances.

In spite of extensive researches, it has so far proved impossible to make spodumene: in all cases where it has been claimed that spodumene has been made, some varieties of this have been produced with considerable replacement of Al^{3+} by other cations, and the synthesis conditions for these differ substantially from those for pure spodumene, as has been shown also in the case of substituted tourmalines.

We examined the crystallization of spodumene in the system $Li_2O-Al_2O_3-SiO_2-H_2O$, and we examined the effects of four variable parameters: component ratio in the initial mixture, temperature (350-700°), pressure (500-10,000 atm), and the pH (3-13). We found that five

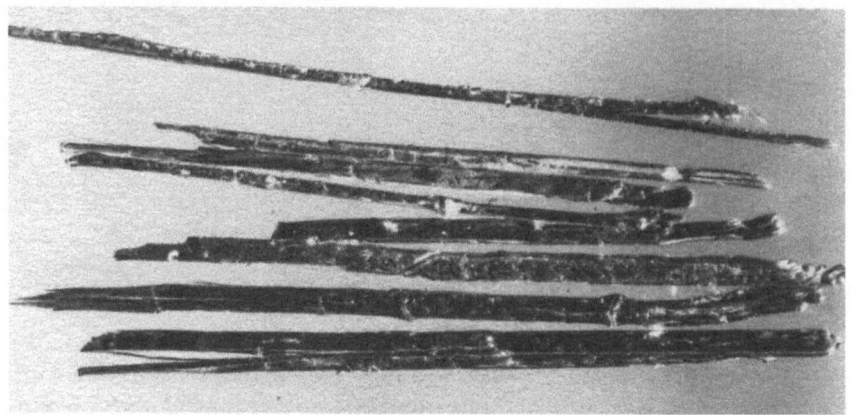

Fig. 6. SbSI single crystals, ×8.

compounds are formed in the system: α-eucryptite, lithium feldspar, petalite, lithium silicate, and spodumene.

Below 3000 atm, we found neither α-spodumene nor the β variety, and one takes the view that all previous claims for synthesis of spodumene under such conditions relate to varieties with isomorphous replacement of Al^{3+} by other cations. Spodumene single crystals of size 1–1.5 mm were produced by growth on seeds at pressures above 5000 atm.

Many of the compounds are familiar as natural products, but syntheses have also yielded a large group of completely new compounds, in particular SbSI (Fig. 6), SbSeBr (Fig. 7), Sb_2S_3 (Fig. 8), etc. These materials are semiconductors, ferroelectrics, piezoelectrics, and photoconductors, and their general formula is $A^V B^{VI} C^{VII}$, where A = Sb, B; B = S, Se, Te; C = Cl, Br, I [29]. The syntheses were performed under comparatively mild conditions of 150–300°C and 200–400 atm, with the solvents being aqueous solutions of hydrogen halides, usually in the presence of hydrogen sulfide.

The physicochemical aspect of the method consists in stabilizing each of the ions participating in the reaction at the appropriate valency, and this condition basically determines the working parameters. Each of the compounds from this group requires a definite pH and redox potential.

A notable feature of the compounds in this group is that they are produced via intermediate metastable substances of the type antimony-halogen or hydrogen sulfide, and the low T

Fig. 7. SbSeBr single crystals, ×8.

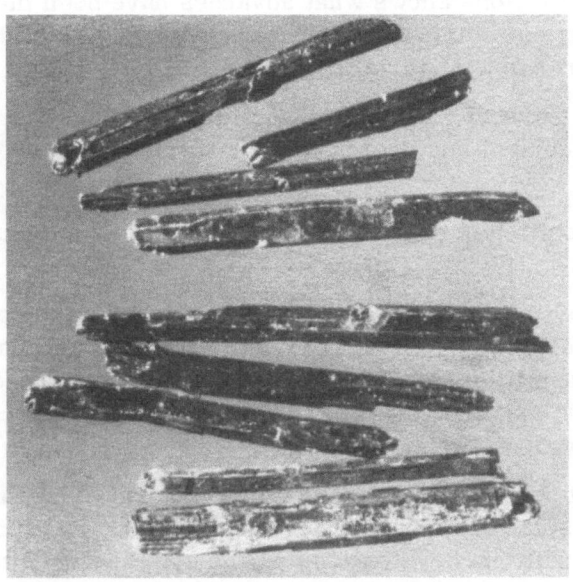

Fig. 8. Sb_2S_3 single crystals, ×8.

and P make it possible to examine these interesting transport reactions by visual observations on autoclaves with windows.

The kinetic trends have been reported for the formation of SbSI and BiSBr; the rates of formation are determined by the solvent concentration, temperature, and the ratio of the initial components. The activation energies calculated for these compounds show that the crystallization occurs in the kinetic region.

Detailed studies of the systems make it possible to establish which compounds in these systems are likely to be of practical use, and what direction should be taken in seeking new materials with valuable combinations of ferroelectric, piezoelectric, and semiconductor properties.

Fig. 9. Synthetic quartz crystals in routine production.

This survey of the work of only two institutions shows what advances have been made in hydrothermal crystal synthesis, and the progress in this area is most obvious as regards synthetic quartz as made by the All-Union Mineral Raw Material Synthesis Institute.

There are now commercially available several varieties of single-crystal quartz, and there are numerous Russian and other producers of and users of this material (Fig. 9).

Most such crystals are used in electronic devices, and the crystals cut for this purpose are as good as regards Q and temperature coefficient of frequency as ones cut from natural quartz [30].

Further, research on the growth of quartz crystals has made it possible to make ones with a guaranteed Q, which has made for considerable economies in device production and material utilization. Seeds of special orientation have been used to make available in quantity quartz for components that require unusual sizes and orientations, which were extremely scarce and expensive when natural crystals were employed.

However, much synthetic quartz cannot be used in optical devices because there are variations in refractive index. Microscopy, selective etching, and other methods of examining these quartz crystals made under various conditions have shown that the optical nonuniformities are due to impurities and structural defects. Measurements have been made on the conditions of incorporation of impurity and structure and defect formation, and techniques have been devised for eliminating these [31].

A team at the All-Union Institute has made optically uniform single crystals, which have been compared as regards optical and electronic properties with natural and other synthetic ones. This synthetic optical quartz has high uniformity in refractive index and good ultraviolet transparency (Fig. 10); it is a unique optical medium for making lenses, prisms, cells, polarizers, compensators, windows, and other components for spectrographic interference, polarization, and modulation devices. Also, synthetic optical quartz can be used in crystals having Q of the order of $50 \cdot 10^6$, which are used in quartz frequency standards.

Some varieties of synthetic optical quartz have high radiation stability and essentially represent a new material never previously available in the form of natural quartz.

In certain cases, this synthetic quartz has been used to make special quartz glasses; for this purpose one requires relatively cheap crystals of high chemical purity [32].

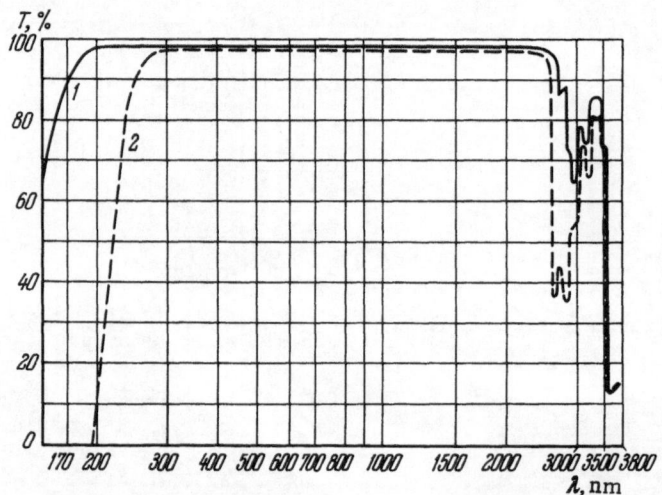

Fig. 10. Transmission curves: 1) synthetic quartz,
2) natural quartz.

The method has also yielded colored varieties of quartz (blue, green, yellow, brown, smoky, citrine, and amethyst), which have been used in ornaments and jewelry.

Literature Cited

1. F. R. S. Barrer, Trans. Brit. Ceram. Soc., 56:155 (1957).
2. É. Senderov, "Crystallization of sodium zeolites in relation to origin and preparation," Thesis, Vernadskii Institute of Geochemistry and Analytical Chemistry, Moscow (1967).
3. A. D. Fedoseev, L. F. Grigor'eva, and G. A. Makarova, Fibrous Materials [in Russian], Nauka, Moscow (1966).
4. H. F. Taylor, Fourth International Congress on the Chemistry of Cement.
5. R. A. Laudise and E. D. Kold, J. Amer. Ceram. Soc., 45:51 (1962).
6. Cambi, Mario, Atti Accad. naz. Lincei, Rend. Cl. Sci., fis., mat., natur., 40:241 (1966).
7. S. C. Robinson, Econ. Geol., 4:293 (1948).
8. V. I. Popolitov, "Hydrothermal synthesis of $A_2^V B_3^{VI}$ and $A^V B^{VI} C^{VII}$ semiconductor compounds," Thesis, Institute of Crystallography, Academy of Sciences of the USSR, Moscow (1969).
9. N. Yu. Ikornikova, in: Hydrothermal Synthesis of Crystals, Consultants Bureau, New York (1971), p. 80.
10. N. Yu. Ikornikova, Kristallografiya, 5:745 (1961).
11. Yu. V. Pogodin and V. M. Sergeev, in: Growth of Crystals, Vol. 7, Consultants Bureau, New York (1969), p. 163.
12. V. G. Hill and R. J. Harker, J. Electrochem. Soc., 115:294 (1968).
13. I. P. Kuz'mina, A. N. Lobachev, and N. S. Triodina, in: Crystallization Processes under Hydrothermal Conditions, Consultants Bureau, New York (1973), p. 27.
14. R. A. Laudise and A. A. Ballman, J. Amer. Chem. Soc., 80:11 (1958).
15. V. A. Kuznetsov and A. A. Shternberg, Kristallografiya, 12:336 (1967).
16. I. P. Kuz'mina, A. N. Lobachev, and N. S. Triodina, in: Crystallization Processes under Hydrothermal Conditions, Consultants Bureau, New York (1973), p. 211.
17. N. Yu. Ikornikova and V. M. Egorov, in: Hydrothermal Synthesis of Crystals, Consultants Bureau, New York (1971), p. 34.
18. L. A. Samoilovich, Handbook on the Relation between Pressure, Temperature, and Density for Salt Solutions [in Russian], Geologiya, Moscow (1969).
19. N. Yu. Ikornikova, A. N. Lobachev, A. R. Vasenin, V. M. Egorov, and A. V. Antoshin, in: Crystallization Processes under Hydrothermal Conditions, Consultants Bureau, New York (1973), p. 241.
20. A. A. Shternberg, in: Crystallization Processes under Hydrothermal Conditions, Consultants Bureau, New York (1973), p. 225.
21. P. V. Lenzo, E. G. Spencer, and A. A. Ballman, Appl. Opt., 5:1688 (1966).
22. B. N. Litvin, Yu. V. Shaldin, and I. E. Pitovranova, Kristallografiya, 13:1106 (1968).
23. L. M. Belyaev, B. N. Litvin, I. M. Dianova, and O. K. Mel'nikov, Kristallografiya, 11 (2):334 (1966).
24. I. P. Kuz'mina, O. K. Mel'nikov, and B. N. Litvin, in: Hydrothermal Synthesis of Crystals, Consultants Bureau, New York (1971), p. 99.
25. B. N. Litvin, I. M. Dianova, and A. A. Kazan, Kristallografiya, 9:571 (1964).
26. B. N. Litvin and O. K. Mel'nikov, Kristallografiya, 14:101 (1969).
27. I. E. Voskresenskaya and M. A. Barsukova, in: Hydrothermal Synthesis of Crystals, Consultants Bureau, New York (1971), p. 126.
28. L. V. Kuznetsov, A. A. Shternberg, and G. K. Ivanova, in: Crystallization Processes under Hydrothermal Conditions, Consultants Bureau, New York (1973), p. 173.
29. V. I. Popolitov and B. N. Litvin, in: Crystallization Processes under Hydrothermal Conditions, Consultants Bureau, New York (1973), pp. 57, 73.

30. V. P. Butuzov, P. Ya. Tsigarov, Yu. K. Aleksandrov, and N. A. Petrov, Trudy TSNII Svyazi MO SSSR, No. 11-12 (1962).
31. L. N. Tsinober, V. E. Khadzhi, L. A. Gordienko, and M. I. Samoilovich, in: Growth of Crystals, Vol. 6A, Consultants Bureau, New York (1968), p. 25.
32. L. A. Gordienko, V. G. Lushnikov, M. I. Samoilovich, L. N. Tsinober, V. E. Khadzhi, and E. M. Tsygarov, in: Growth of Crystals, Vol. 7, Consultants Bureau, New York (1969), p. 297.

ROLE OF ALKALI METALS AND GERMANIUM IN PRODUCING GERMANATES UNDER HYDROTHERMAL CONDITIONS

L. N. Dem'yanets, V. I. Ilyukhin, I. P. Kuz'mina,
A. N. Lobachev, and N. V. Belov

The natural abundance of silicon and the commonness of Si minerals are reflected in the numerous studies in experimental mineralogy designed to establish the natural conditions of silicate formation and the physical chemistry of simple and complex silicate systems. A major result from these studies has been the production of new silicon compounds, with identification of their crystallization and stability ranges. A very similar analog of silicon is germanium, but this cannot compete with the latter as regards the number of natural minerals, mainly on account of its low natural abundance, but under laboratory conditions it should show at least as many germanates as there are silicates.

Germanium falls in group IV in the periodic table, i.e., it is an analog of Si, Ti, and also diagonally Al, so it can act in compounds either as a cation (when it enters octahedra) and as an anion (in tetrahedra). This duality in the purely crystallochemical function of germanium greatly increases the scope for synthesis of stable compounds, the more so since there is no reason why Ge should not enter both types of position in a single structure simultaneously.

There have been major advances in the hydrothermal synthesis of quartz and silicates from alkali solutions, so one expects similar results for analogous germanate systems; the analogy with silicates indicates that most of the results are to be obtained at elevated temperatures and pressures, and the simplest binary germanate system is here GeO_2-H_2O, while the simplest ternary ones are $A_2O-GeO_2-H_2O$, and the simplest quaternary ones are $A_2O-Me_xO_y-GeO_2-H_2O$ (A = Li, K, Na). These systems will be the subject of discussion here.

The $GeO_2 - H_2O$ System. Germanium oxide GeO_2 exists in three modifications: α-GeO_2, which has a structure of α-quartz, β-GeO_2 (β-quartz), and tetragonal GeO_2 with the structure of rutile, which will be denoted by $GeO_2(t)$. The $\alpha \rightarrow \beta$ transition occurs at about 1049°C under dry conditions, while $GeO_2(t)$ is stable below 1035°C under these conditions [1].

The first study of the GeO_2-H_2O system under hydrothermal conditions is in [2], where it was found that α-GeO_2 becomes $GeO_2(t)$ in the presence of water at 185°C.

We have examined the hydrothermal crystallization in this system over the range 150-400°C for pressures from the saturation vapor pressure of water up to about 1000 atm. We found with high reproducibility that there is an irreversible transition α-$GeO_2 \rightarrow GeO_2(t)$ at 185 ± 10°C; below that temperature one gets crystallization of α-GeO_2 with a structure of quartz type (Fig. 1), while above this one gets only crystals of $GeO_2(t)$ (Fig. 2).

Fig. 1. Structure of α-GeO$_2$ in terms of polyhedra.

Fig. 2. Structure of the rutile-type form of GeO$_2$(t).

The System A$_2$O $-$ GeO$_2$ $-$ H$_2$O. We examined the hydrothermal crystallization in simple ternary systems* at 200-500°C and pressures up to 1500 atm, with a filling coefficient of 0.7-0.8. Addition of a third component to GeO$_2$$-H_2$O, especially in alkali at a concentration of 0 to 80 wt.%, expands the scope for synthesis and leads ultimately to the production of alkali germanates (see the T$-$C diagrams in Fig. 3).†

The T$-$C diagrams for the Li, Na, and K systems show shifts in the boundaries for crystallization of GeO$_2$(t) toward higher concentrations of alkali as the radius of the alkali element increases: LiOH 1, NaOH 3, and KOH 10 wt.% [5].

The set of alkali germanates under hydrothermal conditions at these T and P is much smaller than the set for dry anhydrous systems [6-9], where one finds the compound Li$_4$GeO$_4$, Li$_2$GeO$_3$ as the Ge content increases in the Li system, while in the Na system one finds Na$_4$GeO$_3$, Na$_4$GeO$_4$ (the latter is metastable, splitting up into Na$_4$Ge$_9$O$_{20}$ and Na$_2$GeO$_3$), and in the K system K$_2$GeO$_3$, K$_2$Ge$_2$O$_5$, K$_2$Ge$_4$O$_9$, K$_4$Ge$_9$O$_{20}$ (metastable).

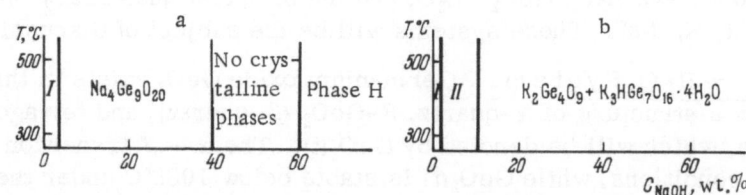

Fig. 3. Crystallization ranges in the systems (a) Na$_2$O$-$
GeO$_2$$-H_2$O and (b) K$_2O-GeO_2$$-H_2$O; I) GeO$_2$(t); II) GeO$_2$(t) +
K$_2$Ge$_4$O$_9$.

---

*When this work was started, it was known [3, 4] that GeO$_2$ + Na$_2$O will react with water to give the compounds Na$_4$Ge$_9$O$_{20}$ and Na$_3$HGe$_7$O$_{16}\cdot$4H$_2$O under hydrothermal conditions.
† The synthetic solid phases were identified by chemical, spectrographic, and x-ray methods on single crystals.

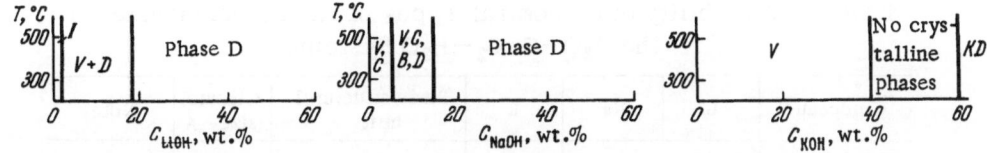

Fig. 4. Crystallization ranges in the systems $Na_2O(K_2O, Li_2O)-GeO_2-ZnO-H_2O$; V) Zn_2GeO_4; B) $A_2Zn_2Ge_2O_7$, C) $A_2Zn_3[GeO_4]_2$, D) $A_2ZnGeO_4(A - Li, Na, K)$; I) $V + GeO_2(t)$.

Under hydrothermal conditions, one gets the following compounds without a stage of production of simple alkali germanate as one increases the alkali concentration in the solution and hence also the GeO_2 concentration, because the alkali acts as a solvent. This is evidently due partly to the transition of α-GeO_2 to $GeO_2(t)$ as the temperature is raised, then in the Na system we have

$$\alpha - GeO_2 \rightarrow GeO_2(t) \rightarrow Na_4Ge_9O_{20} \rightarrow Phase\ H*$$

and in the K system

$$\alpha - GeO_2 \rightarrow GeO_2(t) \nearrow \begin{array}{l} K_3HGe_7O_{16} \cdot 4H_2O \\ \\ K_2Ge_4O_9 \end{array}$$

As in the dry systems, the Na enagermanate is stable, as is the K tetragermanate.

The K system has the zeolytic germanate of composition $K_3HGe_7O_{16} \cdot 4H_2O$ as the stable solid phase, whereas this compound does not occur in the Na system (the Na zeolytic germanate is formed in the $Na_2O-GeO_2-H_2O$ system as a metastable intermediate product).

The System $A_2O - MeO - GeO_2 - H_2O$. Of systems of this type, the ones most fully examined are those containing Zn [10] (Fig. 4); the fullest set of germanates is observed in the Na system.

The number of stable germanium compounds is much reduced if the radius of the alkali element deviates from this, i.e., if Na is replaced by Li or K.

Replacement of Zn^{2+} by another divalent cation (Cd, Co, Fe, Mn, etc.) results in a shift in the boundaries of the crystallization range and sometimes to production of new crystalline phases.

All of these systems have been examined over the range 200-500° at alkali concentrations up to 80% and filling factors of 0.7-0.85; the feature common to them is the production of germanates of composition A_2MeGeO_4 with a structure of phase D, and such compounds have been found for the Zn, Co, and Fe systems.

In the $T-C$ diagram, these phases have wide crystallization ranges: 15-80 wt.% NaOH for the Zn phase, 5-40 wt.% for Cd, and 25% and above for Co.

Hydrothermal Stability and Chemical

Crystallography of Germanate Phases

The basic GeO_2-H_2O system has germanium as both anion and cation, with the first predominant at low T (below 185°C). At higher temperatures, one sees from Figs. 1 and 2 that the

*There are no analogs in the dry system; the composition is under examination.

TABLE 1. Stability of Structural Types of Alkali Germanates in
the $A_2O-GeO_2-H_2O$ Systems

Compound	Li	Na	K	Basic structural unit	Channel size, Å	Notes
$A_3HGe_7O_{16} \cdot 4H_2O$	Stable	Meta-stable	Stable	$[Ge_4O_{16}[+[GeO_4]$	3.2—4	In channels $Li+H_2O$
$A_4Ge_9O_{20}$	Stable	Stable	Not stable	$[Ge_4O_{16}]+[GeO_3]_\infty$	2.5—2	—
$A_2Ge_4O_9$	Not stable	Less stable	Stable	$[Ge_3O_9]+[GeO_6]$	—	In cavities $Na+H_2O$

α-GeO_2 is converted from its ordinary acidic function to $GeO_2(t)$, where Ge resembles Mg in olivine, Ti in rutile or brookite and anatase, or Al in diaspor, where it enters into octahedra as an active part of the structure.*

The dual behavior of germanium is characteristically seen in products from the A_2O- GeO_2-H_2O systems particularly; the compounds $A_3HGe_7O_{16} \cdot 4H_2O$; $A_2GE_4O_9$; $A_4Ge_9O_{20}$ [9, 11, 12] have Ge located in polyhedra with coordination numbers of 6 and 4, i.e., some of the germanium acts as a cation and the rest as an anion (Table 1).

The basic structural units in the first structure is the $[Ge_4O_{16}]$ group, which is a tetrahedron composed of four germanium octahedra linked by their edges; the upper pair through 90° (symmetry $\bar{4}$, Fig. 5). Each such set of four is related to translationally identical ones along all three directions via germanium atoms contained in tetrahedra (GeO_4 groups). There are large channels 3.2-4 Å in diameter, which contain the K or Li and the H_2O.

The alkali cations in a structure of this type act as do the large cations in zeolites (sodalite). They serve as seeds for the Ge units, similar to the sodalite structures. The resulting cellular framework is so strong that it becomes possible to replace the cations without disturbing the structure generally, which is confirmed by the isostructural nature of the zeolytic alkali germanates from Li^+ to Ag^+ and Tl^+.

A more detailed study of the stability of the alkali germanates enables one to define more closely the role of the A cations, as in beryllium, where the steric factor is the principal one. The diameters of the trunk columns are 3.2-4 Å, which allows large cations such as Cs and K

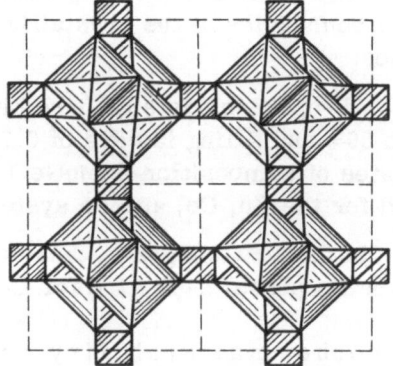

Fig. 5. Structure of zeolitic $A_3HGe_7O_{16} \cdot 4H_2O$ in projection on (001) showing groups of four Ge octahedra and empty channels.

---

*At T > 185°C one gets only $GeO_2(t)$ with Ge in sixfold coordination, which indicates that the solution under hydrothermal conditions contains not $Ge(OH)_{4tetr}$ but $Ge(OH)_{6oct}$, or even complexes consisting of 2, 3, 4, or more octahedra.

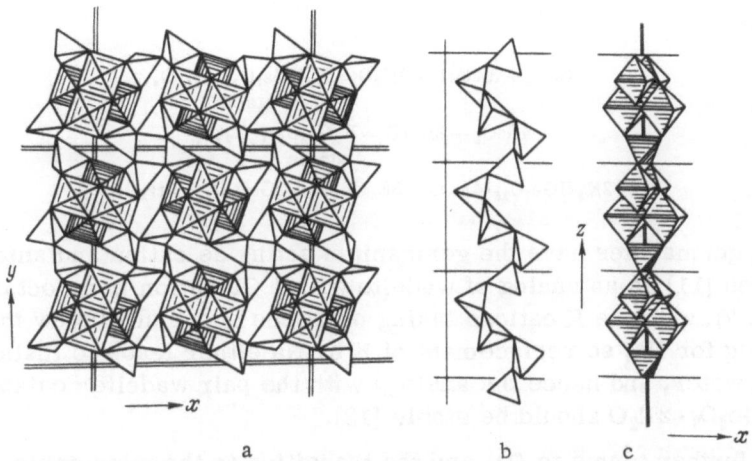

Fig. 6. Structure of $A_4Ge_9O_{20}$, Na enagermanate. a) Groups of four octahedra (Fig. 5) and linking of Ge tetrahedra; b) screw chain of Ge tetrahedra; c) column of four octahedra plus Ge linking tetrahedra.

to enter without causing distortions; the H_2O particles in this case are randomly distributed, and another possibility is Li + H_2O. The cavity is too big for the Na ion, whereas the Na + H_2O pair exceeds the size of the channel, which is why hydrothermal syntheses yield the stable compounds $A_3HGe_7O_{16} \cdot 4H_2O$, where A is Li, K, or Cs, whereas $Na_3HGe_7O_{16} \cdot 4H_2O$ is only metastable. This last is a zeolytic germanate that readily demonstrates the transition to the stable compound $Na_4Ge_9O_{20}$, whose structure is shown in Fig. 6.

Removal of the water involves some changes in the structure; if we are correct in considering that the germanium cations predominate (Ge in octahedral holes), then the basic structural unit in the altered structure is still $[Ge_4O_{16}]$, which should persist. Also, the links of these groups along the Z axis are single Ge tetrahedra, which should also persist. In the other two directions, the Ge link tetrahedra are joined into a spiral chain (Fig. 7).

Then here again the basic part is played by Ge cations, while the Ge anions have a subordinate position; the alkali cations still lie in the empty channels, but now these are bent and much narrower (2 - 2.5 Å). The channel diameter is too small for K but is sufficient for Na, so the Na germanate $Na_4Ge_9O_{20}$ should be stable under hydrothermal conditions, whereas the K one is unstable. This means that the hydrothermal reactions most probably occur as follows at T > 185°C:

in the Na system

$$7GeO_2 + 3NaOH + 3H_2O = Na_3HGe_7O_{16} \cdot 4H_2O,$$
$$\text{metastable}$$
$$4Na_3HGe_7O_{16} \cdot 4H_2O = 3Na_4Ge_9O_{20} + GeO_2(t) + 18H_2O;$$

(1)

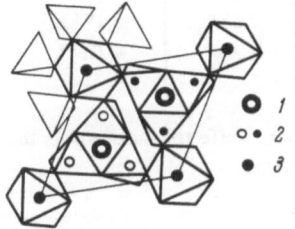

Fig. 7. Structure of $A_2Ge_4O_9$, Ge octahera and tetrahedra shown: 1) Element A; 2) Ge in tetrahedral positions at various levels in unit cell; 3) Ge in octahedral positions.

in the K system

$$7GeO_2 + 3KOH + 3H_2O = K_3HGe_7O_{16} \cdot 4H_2O,$$
$$\text{stable}$$
$$4GeO_2 + 2KOH = K_2Ge_4O_9 + H_2O, \tag{2}$$
$$\text{stable}$$
$$2K_3HGe_7O_{16} \cdot 4H_2O = 3K_2Ge_4O_9 + 2GeO_2(t) + 9H_2O.$$

The $A_2Ge_4O_9$ germanates have the germanium acting as cation and anion; the $K_2Ge_4O_9$ phase was described [11] as an analog of wadeite with 1 Ge cation in an octahedron and 3 anions in tetrahedra (Fig. 7), with the K cations taking positions above and below the threefold ring. The cavity is too big for Na, so replacement of K by Na either leads to instability or requires the introduction of water, and hence the analogy with the pair wadeite—catapleiite means that the compound $Na_2Ge_4O_9 \cdot 2H_2O$ should be stable [13].

One can add further terms to (1), and the transition to the metastable zeolytic germanate can be represented as in (3) with two different final products:

$$4Na_3HGe_7O_{16} \cdot 4H_2O \rightarrow 3Na_4Ge_9O_{20} + GeO_2(t) + 18H_2O;$$
$$2Na_3HGe_7O_{16} \cdot 4H_2O \rightarrow 3Na_2Ge_4O_9 \cdot 2H_2O + 2GeO_2(t) + 3H_2O. \tag{3}$$

These structures also illustrate the role of the alkali cations in producing crystalline phases in the systems $A_2O - GeO_2 - H_2O$, in particular the compound $A_3HGe_7O_{16} \cdot 4H_2O$; $A_2Ge_4O_9$; $A_4Ge_9O_{20}$. The stability of the first two types increases with the cation radius, whereas the stability of $A_4Ge_9O_{20}$ increases as the cation becomes smaller (Table 1).

Then germanium behaves in two ways in the presence of alkalis in the $A_2O - GeO_2 - H_2O$ systems: as a cation in waves analogous to that of titanium in ramsayite and sphene, whereas the Ge itself determines the core of the structure, and as an anion producer similar to silicon. The alkali metals (particularly in the first case) fill holes or channels of zeolite type.

The closest analog of germanium, namely silicon, is located only in tetrahedra, and therefore it is natural that the analogous silicate systems lack certain compounds; in the tetrahedral basis, some silicates of spinel or scheelite type do not occur, whereas there exist

TABLE 2. Coordination of Ge in Hydrothermal Products
(T > 185°C)

System	$GeO_2 - H_2$	$A_2O - GeO_2 - H_2O$			$A_2O - MeO -$ $-GeO_2 - H_2O$
Structure type	GeO_2 (t)	$A_3HGe_7O_{16} \cdot$ $\cdot 4H_2O$	$A_4Ge_9P_{20}$	$A_2Ge_4O_9$	A_2MeGeO_4
A_2O/GeO_2 in crystalline product	0	1,5:7	2:9	1:4	1:1
Ge coordination	$Ge^{VI} +$ $+ {}_0Ge^{IV}$	${}_4Ge^{VI} +$ $+ {}_3Ge^{IV}$	${}_4Ge^{VI} +$ $+ {}_5Ge^{IV}$	${}_1Ge^{VI} +$ $+ {}_3Ge^{IV}$	${}_0Ge^{VI} +$ $+ Ge^{IV}$

Note. The superscript (${}_4Ge^{IV}$) is the coordination number, while the subscript is the number of atoms in the unit cell with this coordination number.

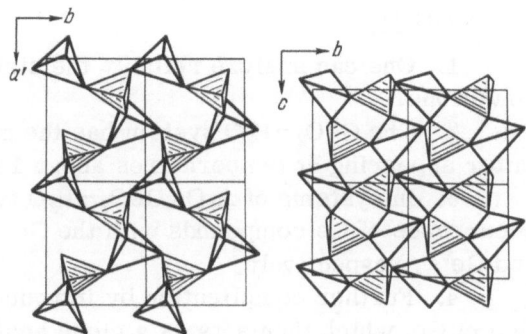

Fig. 8. Structure of A_2ZnGeO_4 showing the Zn–Ge tetrahedron link. The Zn coordination polyhedron is hatched.

germanates with such structures, by virtue of the ionic dimorphism of Ge^{IV} and Ge^{VI}: Fe_2GeO_4, Co_2GeO_4 of spinel type,* and $CeGeO_4$, $ZrGeO_4$, $HfGeO_4$ of scheelite type [9].

If to $A_2O-GeO_2-H_2O$ hydrothermal system we add a fourth component with a prominent cation function, such as Zn, Co, or Cd, the cation function of the germanium tends to be suppressed and the latter behaves as an anion, i.e., its acidity becomes more prominent, and in that respect germanium does not differ from silicon (Table 2). Not all types of Si compounds occur in the germanium systems, which is again due to the steric factor: $r_{Ge} = 0.53$ Å, whereas $r_{Si} = 0.39$ Å, and so certain structural types with complex anion radicals cannot be formed under the conditions found for silicon.

If $Na_2O_2 : Ge_2O = 1:1$ or more [or if $(A + Me^{2+}) : Ge = 3:1$], we get somewhat unexpected features in the behavior of the alkali cations; for instance, solutions rich in alkali result in the coordination number of Na becoming 5, 4, or occasionally 6, or rather we get the coordinations divided into two subspheres: I with Na–O distances of 2.30-2.50 Å and II with distances of 2.86-3.20 Å. This differentiation markedly increases the edge length of the Na polyhedra, and some of them become comparable with the lengths of the edges of the elongated [Me + Si(Ge)]·O_7 group.†

We have already noted that all the $A_2O-MeO-GeO_2-H_2O$ systems have compounds with the D structure [14, 15], and in this type the Zn (Co, Fe, etc.) participates on the same basis as the germanium in constructing the three-dimensional basic unit, which is a mixed diortho group[(Me, Ge)O_7], with the Na polyhedra joined into corrugated chains extending along $[10\bar{1}]$ in the (101) planes, in which the polyhedra are linked by the edges, while the chains (the initial one and the one reflected in glide plane) are linked via the vertices of the $Na_{(1)}$ and $Na_{(2)}$ polyhedra; to each common vertex there converge $(2Na_{(1)} + Na_{(2)})$ polyhedra. The characteristic feature is the adaptation of the framework of Zn + Ge tetrahedra to the Na unit (Fig. 8). Compounds with this structure have been found also in silicate systems, and replacement of Si by Ge hardly alters the Zn–O–Si (Ge) angle: the mean angle for Si is 125°, while that for Ge is 124°. The framework elasticity is related to the sodium, and this is confirmed by the possibility of replacing the Zn by Co, Fe, and various other valent elements. On the other hand, replacement of Na by other alkali cations does not always result in a D structure; for instance, in a chain of K polyhedra there are no elements comparable with the (Zn + Ge)O_7 group and the D type K phase is not formed in the Zn–Ge system.

A detailed crystallochemical analysis has been performed also for other phases made in the quaternary silicate and germanate systems (phases B, C, etc.).

---

*The scope for transformation of olivine Mg_2SiO_4 into a spinel modification is a major aspect of research into the upper mantle.

† This reduction in the coordination number is most prominent in solid Na_2O, which has obvious coordination number 4 (antifluorite type).

Conclusions

1. One can analyze reliably the stability of solid germanate phases under hydrothermal conditions.

2. The $GeO_2 - H_2O$ system has the germanium acting in anion and cation capacities, the latter appearing at temperatures above 185°C.

3. In systems of $A_2O - GeO_2 - H_2O$ type, the dual behavior of germanium is seen as the occurrence of Ge compounds with the Ge in octahedral and tetrahedral positions, namely Ge^{IV} and Ge^{VI}, respectively.

4. Further complication by introduction of divalent cations suppresses the cation function of Ge, which then acts as a close analog for silicon, the principal role in the structure being taken by the large alkali cations in phase D etc., which is especially so for Na, whose polyhedra are surrounded by adapted anion groups of Ge tetrahedra.

Literature Cited

1. I. V. Tananaev and I. Yu. Shpirt, Germanium Chemistry [in Russian], Khimiya, Moscow (1967).
2. R. Schwarz and E. Huf, Z. anorg. Chem., 203:188 (1932).
3. E. R. Shaw, J. F. Corwin, and J. W. Edwards, J. Amer. Chem. Soc., 80(7):1536 (1958).
4. J. White, E. Shaw, and J. F. Corwin, Analyt. Chem., 31:315 (1959).
5. I. P. Kuz'mina, "Crystallization of sodium and potassium germanates and zincogermanates under hydrothermal conditions," Thesis, Institute of Crystallography, Moscow (1968).
6. P. P. Budnikov and S. T. Tresvyatskii, Dokl. Akad. Nauk SSSR, 99:761 (1954).
7. S. T. Tresvyatskii, Dok. Akad. Nauk Ukr. SSR, 3:295 (1958).
8. R. Schwarz and F. Meinrich, Z. anorg. Chem., 205:43 (1932).
9. A. Wittman, Fortsch. Mineral. 43(2):120 (1955).
10. I. P. Kuz'mina, O. K. Mel'nikov, and B. N. Litvin, in: Hydrothermal Synthesis of Crystals, Consultants Bureau, New York (1971), p. 99.
11. N. Nowotny and A. Wittman, Monatsh. Chem., 84:701 (1953); 85:558 (1954); 87:654 (1956).
12. N. Ingri and G. Lundgren, Acta Chem. Scand., 17:617 (1963).
13. B. K. Brunovski, Acta Physicochem. USSR, 5:863 (1936).
14. V. V. Ilyukhin, A. V. Nikitin, and N. V. Belov, Dokl. Akad. Nauk SSSR, 171:1325 (1966).
15. É. A. Kuz'min, V. V. Ilyukhin, and N. V. Belov, Kristallografiya, 13:976 (1968).

SYNTHESIS OF SINGLE-CRYSTAL OPTICAL QUARTZ

V. P. Butuzov, L. A. Gordienko, V. E. Khadzhi, L. I. Tsinober, and V. S. Doladugina

Much synthetic quartz cannot be used for optical purposes mainly because the refractive index varies within a single crystal; there are two basic groups of optical nonuniformity in such material:

I. Optical nonuniformities related to growth forms:

a) zonal nonuniformity in the distribution of structural and other impurities [1], which arises from change in the parameters of the growth process;

b) sector nonuniformity, which arises because of differences in the uptake factors for impurities in the faces of the various simple forms; the discontinuities in the refractive index are seen on passing from one growth pyramid to another;

c) nonuniformity related to uneven capture of impurities by different parts of the relief on the growing surface (preferential uptake at the boundaries of accessory growth faces on the basal plane). A particular case of this form of nonuniformity is the secondary sector form, which is produced by transformation of the ordinary accessory relief into multiheaded growth surfaces. This degeneration is characteristic of a rapidly growing basal plane, which is stable at relatively low synthesis temperatures.

II. Nonuniformities related to dislocations.

Shadow pictures of such nonuniformities take the form of thin wavy lines (Fig. 1) that run throughout the thickness of the deposited layer and that are oriented almost normal to the growing faces in all growth pyramids [2]. As a rule, these linear nonuniformities are either inherited from ones in the seed or generated at various defects such as gas −liquid and solid inclusions, or else cracks, etc. Selective etching [3, 4] and x-ray topography [5] shows that these line defects are of dislocation nature; the $\langle c \rangle$ and $\langle r \rangle$ pyramids have these dislocations: mainly edge ones whose Burgers vectors are parallel to $[11\bar{2}0]$.

Research on the conditions producing these defects has provided various technological means of producing large optically uniform quartz crystals (Fig. 2); for instance, reduced supersaturation enables one to eliminate completely the nonstructural impurities commonly incorporated into such crystals. If the initial components are purified as far as possible, one gets crystals with a minimal content also of structural impurities. Uniform distribution of impurities within the growth pyramids is provided by careful stabilization of the growth parameters. Efficient means have also been found for removing growth dislocations [6]. The crystals grown under such conditions have transmission in the ultraviolet, visible, and infrared regions superior to that of ordinary synthetic or natural quartz, and the material also

31

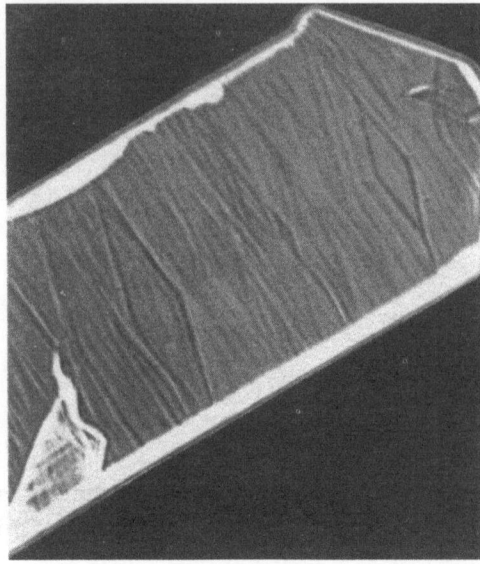

Fig. 1. Streaks revealed by the shadows from
a point source, actual size.

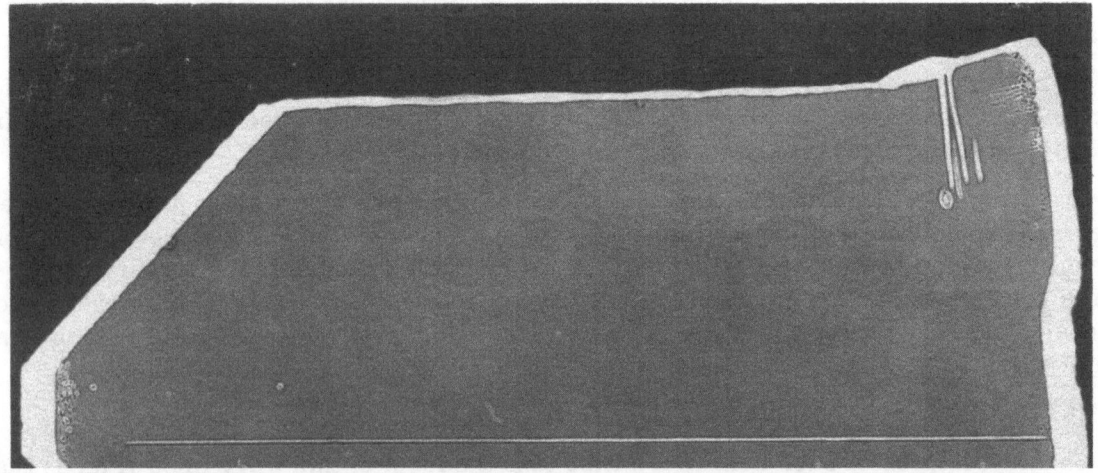

Fig. 2. Optically homogeneous synthetic quartz, point source apparatus,
actual size.

Fig. 3. Relief on the basal plane of a synthetic
quartz crystal containing growth dislocations,
×4.

Fig. 4. Relief on the basal plane of dislocation-
free synthetic quartz, ×4.

has high radiation stability; it is the unique optical material suitable for making lenses, prisms, cells, polarizers, compensators, windows, and other such parts of spectroscopic, interference, polarization, and modulation devices. The Q of crystals cut from synthetic optical material is considerably above that normally found for synthetic quartz.

A distinctive feature of this synthetic optical quartz is the relief of the basal plane; an ordinary synthetic crystal containing dislocations has a basal surface composed of conical growth accessories with sharp points and concentrically layered structures (Fig. 3), whereas dislocation-free quartz has a basal surface consisting of flattened accessories of cellular structure without sharp vertices (Fig. 4).

Literature Cited

1.   L. I. Tsinober, V. E. Khadzhi, L. A. Gordienko, and M. I. Samoilovich, in: Growth of Crystals, Vol. 6A, Consultants Bureau, New York (1968), p. 25.
2.   G. M. Safronov and V. E. Khadzhi, Trudy VNII Sinteza Min. Syr'ya, Vol. 2, No. 1 (1958).
3.   E. V. Tsinzerling and Z. A. Mironova, Kristallografiya, 8:117 (1963).
4.   V. E. Khadzhi, Min. Sborn. L'vov. Gos. Univ., 20:418 (1966).
5.   A. R. Lang and V. F. Miuskov, Growth of Crystals, Vol. 7, Consultants Bureau, New York (1969), p. 112.
6.   L. A. Gordienko, V. F. Miuskov, V. E. Khadzhi, and L. I. Tsinober, Kristallografiya, 14:539 (1969).

CRYSTAL MORPHOLOGY, KINETICS, AND GROWTH MECHANISM

A. A. Shternberg

Crystal nucleation and growth may result in perfect single crystals or skeletal formations, the decisive factors being composition and state of the medium, the temperature and the kinetics of the process. There is also great variety in the surface micromorphology, which becomes the more varied the more in detail one examines it. Also, if we consider crystals of different substances grown under comparable conditions, we often find a surprising similarity in the surface morphology (Fig. 1).

The form reproducibility and analogous origin indicate that there is a single process for production of crystals of all substances over a very wide range of conditions; however, existing theories of crystal growth cannot explain much of the evidence on kinetics and morphology, so here we survey the available evidence and select mechanisms suitable for such description.

It is generally recognized that the main faces of a crystal grow in two stages: production of steps and propagation of layers from these over the surface. If the growth rate of a step and the rate of tangential spread of the layers are identically dependent on the supersaturation or composition, any change in these independent variables would only affect the face growth rate, not the morphology; but it is familiar that any change in supersaturation or solution composition alters the face morphology and crystal habit as a whole.

This enables us to define the problem. We need to define two mechanisms common to all crystals: one that explains nucleation and growth of the ridges, and a second that explains the tangential growth of the layers.

If the conditions are such that the normal growth rate is very much predominant, one gets only projections at active points on the surface, with resultant skeletal forms of crystal of maximum specific surface. If tangential growth dominates one gets crystals with flat faces. Both types of crystal are fairly common, so there are equal probabilities for conditions that produce the two growth mechanisms, so we can consider them separately.

Normal Growth

If the supersaturation is very low, one gets tangential growth of the layers, and the crystal faces are flat or nearly ideal planes, while normal growth ceases. However, if such conditions persist for a long while, one finds that the faces of some of the crystals continue to grow slowly. For instance, if one is growing large numbers of crystals of Rochelle salt, one finds that point depressions in the seed do not usually grow at low supersaturations, but some of them slowly develop into columns elongated along the C axis and some acquire the shape of unsymmetrical plates perpendicular to the C axis or else grow as triangular patches in the BC

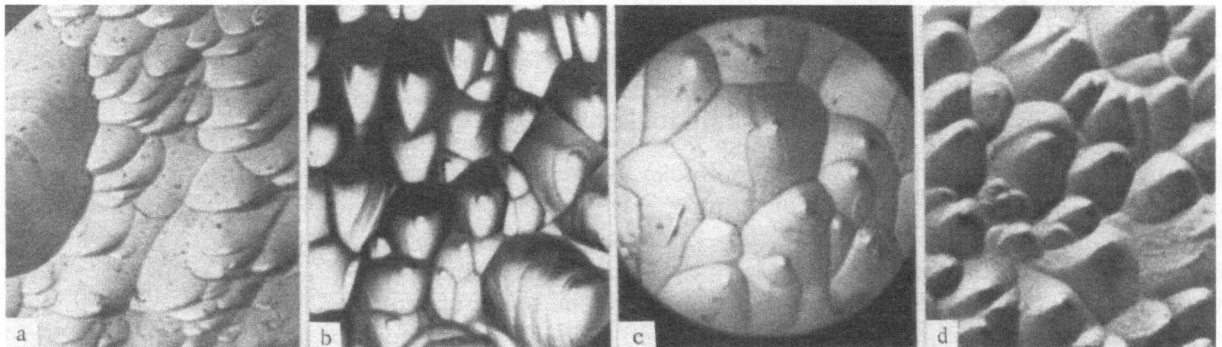

Fig. 1. Faces of crystals of (a) corundum, (b) quartz, (c) sodalite, and (d) calcite
grown under comparable conditions. Various magnifications.

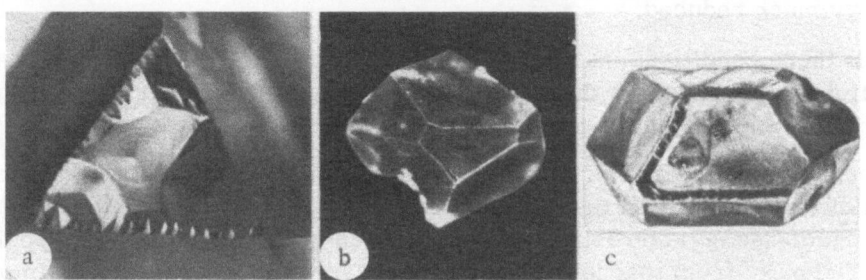

Fig. 2. Highly developed vicinal faces at edges and corners
of crystals of (a) quartz, (b) calcite, and (c) sphalerite at
small supersaturations.

plane. If the supercooling is increased by a few hundredths of a degree, all the crystals begin to grow at the same rate and acquire in time the same isometric habit. A similar picture
can always be observed also with quartz crystals, whose rhombohedron faces do not grow when
the supercooling is less than 5°, and one gets ideal planes on account of normal growth of
visually wavy (0001) surfaces. On the other hand, some rhombohedron faces that are turned
upwards and are unprotected from deposition of solid particles do grow slowly, and on such
faces one always observes flat projections or vicinal faces, which are clearly associated with
structure defects.

Steps with edges (Fig. 2) appear when the supersaturation is increased; they arise at the
edges of rapidly growing faces and give the impression [1] that the slowly growing faces are
overgrown by the fast-growing ones. These edge vicinals arise not only at the actual edges of
the crystals but also along open cracks, at holes in the seeds, and in fact at all convex relief forms.

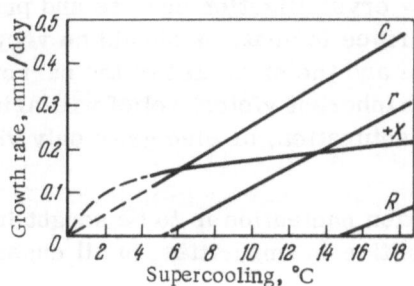

Fig. 3. Growth rates of quartz crystal faces
as functions of supersaturation.

One gets reproducible growth of all faces with a given index when supersaturation critical for that set of faces is reached [2, 3]. The face growth rate increases linearly with the supersaturation above the critical value (Fig. 3) and is independent of defects; crystals without dislocations grow at the same rate as defective ones [4].

When the critical supersaturation is reached, all the faces in a set with a given index become covered with microscopic hummocks, which are not related to edges or defects, so one can call such relief the inherent vicinal relief of the face. As normal growth proceeds, the faces of the vicinals, which overlap, become larger; the slope of their sides is, however, unaltered, and this slope increases with the supersaturation. If it is more than 30-40°, the vicinals do not coalesce at the base and one gets the onset of multiheaded growth. The flanks of the completely distinct vicinals are covered not with basic faces of the crystals but with spherical express faces, which will be considered later on. Unfavorable conditions such as low temperatures or growth-hindering impurities retard the tangential growth of the layers but do not reduce the speed with which the vertices of the vicinals rise, so the flanks of the vicinals acquire steeper slopes. As a result, the region of supersaturations giving uniform crystalline products is very much reduced.

Under all these conditions, the normal growth rate is determined by the rate of rise of the vicinal relief, which indicates that all the evidence on this topic is best discussed in terms of the dependence of the projection growth rate as a function of supersaturation.

This morphology is characteristic of normal growth, and to explain it we consider the slow-electron diffraction data on the surface layer structure [5]; the important point is that even a pure surface has structures that differ for each face, in which the disposition of the atoms differs radically from that in the underlying layers. The presence of impurities (even the solvent) will facilitate the formation of surface structures; the superposition of a new structure on a basic one should generate a net of surface defects, which act as points for preferential attachment of atoms (potential growth centers), which result in regional nucleation of vicinal relief. The displacement of the atoms in the surface layers is greatest at the corners and edges of the crystals; this is because one gets here more active potential centers and the vicinal nucleation occurs at large supersaturation, the resulting vicinal projections being the vertices, which determines the viability of the resulting vicinal relief.

The surface layers become buried as the growth proceeds; the structures are incorporated into the basic structure of the crystal, while the new surface layers continually fill their places with sites of preferred addition of particles. The frequency of attachment of particles should be proportional to the concentration in the solution; proper crystalline particles, which provide solution saturation, do not enter into the process, and we are interested only in the excess dissolved particles that produce the supersaturation. The concentration of these is proportional to the supersaturation, which is the basis for the linear dependence of the projection growth rate on the supersaturation.

The need to transform the surface structures into the base structure by self-diffusion predetermines the lower temperature limit, which can be fairly high, below which the growth of a single crystal is difficult. One gets nucleation of new crystallization centers and production of spherulites. The presence and structure of the surface formations should be very much dependent on the physical relation between their structure and the structure of the surrounding medium; if the latter two are similar it is less likely that inherent vicinal relief will arise, and the crystals will not grow over a wide range in supersaturation, or else grow only via dislocation mechanisms [6].

This theory indicates that the mechanism of projection nucleation is to be sought in the upper atomic layers of the crystal and therefore is insensitive to impurities; in all cases it produces only projections above the surface and provides for their growth no matter what the mode of behavior of the face as a whole. The volume between the projections is filled by the tangential layer growth; if these layers have time to level out the depressions between the growing projections, one gets isometric crystals, but if the tangential growth is slow or is stopped, one gets skeletal ones.

Tangential Growth

Projections at edges and on faces result in negative forms of relief on the surface; geometrical considerations mean that negative surfaces, spheres, and reentrant angles occur over the area of the face, and therefore one finds higher activity in faces that grow at the highest rates [7], i.e., express faces (Fig. 4). Therefore, tangential growth is in essence the normal growth of expressed faces.

The growth rate on the express face is maximal, so such a face does not participate in the final habit of the crystal and therefore such faces have so far not attracted attention; they arise on each projection, and transform these into vicinal hummocks, with provision for rapid extension to the edge of a face, while they level out each face to an ideal plane, while the crystal as a whole takes up a fairly perfect form.

The surfaces of express faces are spherical; they fuse one with another, and they often resemble semifacetted growth forms.

Express faces can be seen in their natural form on the steep flanks of vicinals on corundum or quartz; impurities such as germanium retard the growth of the express faces that could participate in producing the relief on pinacoidal surfaces of quartz, and so the vicinals take on the clearer forms of three-faced pyramids, and it is clearly seen that their faces involve two forms of spherical express face. The sphericity is regular, because they are perpendicular to the directions of most rapid growth, and therefore deviation of part of a face from the basic position retards the growth of the deflected part and increases the sphericity. The faces are thus free from any process that would level them off.

The reentrant angles between express faces also do not stimulate the growth rate; conversely, the hindered supply of material to them means that they grow poorly, and if their angles of encounter are small, they form slots and inclusions of any material that may be present.

Although the express faces that produce the tangential growth result in production of almost all the crystalline material, very little is known about them; at the present time, use is made of only one express face, $+X\,(\bar{2}110)$, for growing quartz crystals (Fig. 5). This face has an unusual form of growth rate as a function of supersaturation (Fig. 3).

There is no evidence available for low supersaturations, but it is clear that the growth rate of the express face increases rapidly with the supersaturations when conditions are such that the main faces do not yet grow, and the rate for the express faces will be high when the normal growth mechanism becomes active, and that rate is almost independent of supersaturation.

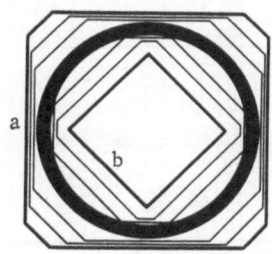

Fig. 4. Regeneration of a seed cut as a ring: a) Face developing on outside with minimal growth rate; b) face on inside with maximal growth rate.

Fig. 5. Surface of a fast +X face in quartz.

It is common to all crystals that the steepness of the vicinals increases with the supersaturation, so one expects an analogous relationship for the entire class of express faces.

The unusual dependence of growth rate on supersaturation for express faces can be well explained by concepts developed by Petrov and others [8-10], namely that the surface of the crystal is covered with a layer of chemisorbed impurities or solvent; the lifetime of the sorbed particles on the surface of a face is limited, so vacant sites periodically arise, and the addition of inherent crystal particles to such vacant sites is the more likely the higher the concentration in the solution. On the other hand, the particles take up practically all vacancies when supersaturation is fairly high, and the growth rate is then limited by the number of vacancies arising, i.e., by the rate of disorption of the impurity, which is independent of the supersaturation.

This mechanism is applicable owing to the growth of express faces, not basic ones; it predetermines the high sensitivity to the composition of the solution, namely the presence of

Fig. 6. Graphical representation of the relation of morphology to supersaturation and solution contamination (explanation in text).

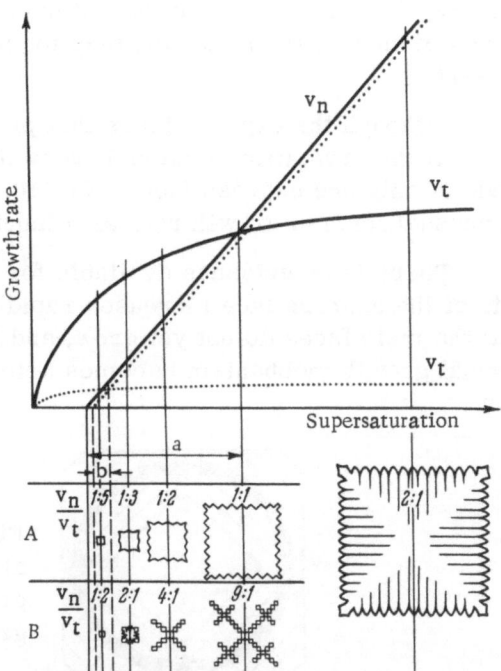

impurities. We therefore see that the single process of crystal growth is the joint result of two completely different mechanisms: normal growth at projections, whose speed increases linearly with the supersaturation and is largely independent of impurities, and tangential growth of ridges, which occurs as a result of express faces present at the edges of ridges, whose growth rate is largely independent of the supersaturation but is very sensitive to impurity contents.

Geometrically, this relationship is expressed as a variation in the ratio of the growth rates of the ridges or projections and express faces; the variations in this ratio predetermine all the variety of forms of crystals, as Fig. 6 shows, which represents the normal growth rate v_n and the tangential growth rate v_t as a function of supersaturation in a pure solution (full lines) and a contaminated one (broken lines). The ratio v_n/v_t increases with the supersaturation, and hence the steepness of the vicinal flanks increases; if $v_n/v_t < 1$, we may expect to obtain visually uniform crystals without solution trapping. As the ratio increases, one gets inclusions and then dendritic growth; the curves show that the width of the supersaturation region in which $v_n/v_t < 1$ is much wider for pure solutions (case a) than it is for contaminated ones (case b). Underneath the curves are shown crystals that grow in identical time periods at various supersaturations and hence at various v_n/v_t for pure solutions (A) and for contaminated ones (B).

If the conditions are such that the absolute values of the normal and tangential growth rates are large compared with those that can be provided by diffusion or heat transfer, the latter processes will make a contribution to the production of the faces and the crystal as a whole.

I am indebted to N. N. Sheftal', whose criticism facilitated formulation and exposition of the above concept.

Literature Cited

1. V. N. Voitsekhovskii, Thesis, Leningrad Mining Institute (1966).
2. G. G. Lemmlein and L. I. Tsinober, Trudy VNIISIMS, Vol. 4 (1962).
3. O. K. Mel'nikov and B. N. Litvin, Kristallografiya, Vol. 10, No. 2 (1965).
4. L. A. Gordienko, V. F. Miuskov, V. E. Khadzhi, and L. I. Tsinober, Kristallografiya, 14(3):539 (1969).
5. V. F. Dvoryankin and A. Yu. Mutyagin, Kristallografiya, 12(6):1112 (1967).
6. A. P. Kasatkin, Kristallografiya, 9(2):302 (1964).
7. V. T. Ushakovskii, K. F. Kashkurov, and A. V. Simonov, Kristallografiya, 13(3):559 (1968).
8. T. G. Petrov, Kristallografiya, Vol. 9, No. 4 (1964).
9. V. A. Kuznetsov, in: Hydrothermal Synthesis of Crystals, Consultants Bureau, New York (1971), p. 52.
10. A. A. Chernov and V. A. Kuznetsov, Kristallografiya, Vol. 14, No. 5 (1969).

FEATURES OF AMETHYST CRYSTALS SYNTHESIZED
IN THE $K_2O-CO_2-SiO_2-H_2O$ SYSTEM

E. M. Tsyganov, V. E. Khadzhi, A. A. Shaposhnikov,
M. I. Samoilovich, and L. I. Tsinober

Synthetic amethysts may be made hydrothermally in the system $K_2O-CO_2-SiO_2-H_2O$ in the presence of iron [1, 2]; when the $\langle R \rangle$ and $\langle r \rangle$ pyramids of such crystals are exposed to ionizing radiation,* they acquire an amethyst color, and the strength of the color in the $\langle R \rangle$ pyramid is usually much greater than that in the $\langle r \rangle$ pyramid. An analogous color distribution is characteristic of natural amethyst crystals. We have defined closely the formation conditions for amethyst crystals in this system, and have compared some physical properties of synthetic and natural amethyst. The amethyst color purity is better when the synthesis temperature is low; the characteristic color is related to the presence of iron, but aluminum is also taken up as a structural impurity during the synthesis, and this is [3, 4] related to hole centers producing smoky color. The absorption spectra for the two types of center are superimposed in the irradiated crystal, which naturally results in a poorer amethyst color. It has been shown [5] that the trapping factor for the structural impurity is directly related to the growth temperature, whereas the uptake factor for the iron is largely independent of temperature in the range covered. Therefore, one reduces the crystallization temperature to reduce the proportion of impurity and thus to improve the amethyst color. The same result can be obtained by more careful removal of traces of aluminum; then a pure and strongly colored amethyst material may be grown at higher temperatures.

Amethyst is made in a system containing potassium, and many natural amethysts have elevated amounts of potassium, so it is of interest to consider the role of this element in producing the amethyst color centers. The physical properties of amethysts give no evidence indicating that potassium enters into the color centers; also, it is difficult to expect the large potassium ion (r = 1.33 Å) to enter the structural channels, whose radius is about 1 Å, so one supposes that the potassium merely provides conditions favorable to the entry of the iron. In fact there are no stable iron silicates of potassium at the P and T used in the synthesis, whereas stable ferruginous silicates of the type acmite and riebeckite are formed in the soda system, the latter at elevated temperatures. Consequently, the iron in a potassium system remains free and can be taken up by the quartz as a structural impurity; under these conditions, a small proportion of potassium may be taken up as a nonstructural impurity, which is related to the elevated content found in amethysts.

---

*The following letter symbols are used for the faces of quartz: R — the principal positive rhombohedron; r — the principal negative rhombohedron. The growth pyramids are denoted by the letter symbols for the corresponding faces enclosed in angle brackets.

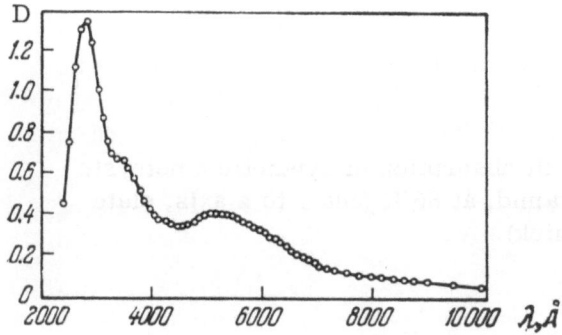

Fig. 1. Optical absorption of synthetic amethyst, ⟨r⟩ pyramid.

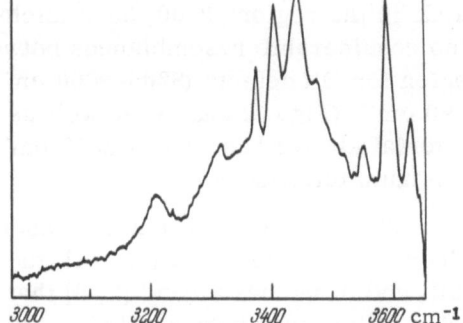

Fig. 2. IR absorption of synthetic amethyst, ⟨r⟩ pyramid, at 85°K (cut ⊥ to z axis, plate 2 mm thick).

All natural and synthetic amethyst crystals so far examined* have a characteristic ESR spectrum associated with Fe^{3+} ions, which isomorphously may replace Si^{4+}. The constants have been given for the spin Hamiltonian that describes this spectrum [6-8]; the ESR line strength for Fe^{3+} in the initial (unirradiated) crystals is directly proportional to the strength of the amethyst colors produced by ionizing radiation. Usually, the Fe^{3+} ESR spectrum is not seen after irradiation in synthetic amethysts, so the synthetic variety has all the isomorphous Fe^{3+} converted to another state by gamma irradiation, which substantially alters the ESR spectrum.

There are the following possible reasons for the change in the ESR spectrum of Fe^{3+} on irradiation.

1. The valency of the Fe^{3+} is altered, i.e., it goes over to the state Fe^{2+} [8] or Fe^{4+} [7, 9].

2. A paramagnetic O^- center is formed in a tetrahedron in which the silicon is replaced by iron; the ESR spectrum of amethysts can be observed at room temperature, while it is unlikely that Fe^{2+} and Fe^{4+} exist under these conditions [7], so we suppose that the second model for the center is the more likely one. The paramagnetic O^- with spin $S = {}^1/_2$ in the immediate environment of Fe^{3+} will undoubtedly affect the ESR spectrum; the dipole−dipole interaction of the two close paramagnetic centers can alter the magnitude and symmetry of the intracrystalline field, and it can also make the ESR spectrum of the O^- centers unobservable. In this case, one of the oxygen atoms in the Fe^{3+} tetrahedron is no longer equivalent to the others, so the symmetry of the environment is reduced, which increases the number of equivalent paramagnetic complexes in the unit cell.

_____

*The ESR spectrum of amethysts has a quartet structure with a series of lines.

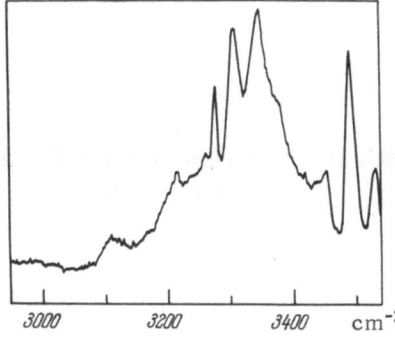

Fig. 3. IR absorption of synthetic amethyst,
⟨R⟩ pyramid, at 85°K (cut ⊥ to z axis, plate
2 mm thick).

3. The UV visible spectra of synthetic and natural amethysts are identical and have
bands in the regions 2800, 3500 (dichroic crystals, $D_e > D_a$), and 5400 Å (Fig. 1). There are
also considerable resemblances between the absorption spectra of amethysts and the stretching
region for OH defects (3200-3600 cm^{-1}); both varieties have bands at 3400, 3440, 3550, and
3590 cm^{-1} (Figs. 2 and 3), as well as a broad diffuse band near 3430 cm^{-1}. The synthetic
material always has a weak 3370 cm^{-1} band, which is related to OH groups lying in respective
aluminum tetrahedra.

The amethyst color has an anomalous pleochroism, which has long been discussed in the
mineralogical literature [10, 11]; the physical essence of the effect has been elucidated via
ESR, and it has been found [7, 8] that the anomalous pleochroism is a consequence of the popu-
lation differences between three structurally equivalent tetrahedra containing Fe^{3+} ions, which
are related to the amethyst color center. This distorts the optical symmetry of the crystals
and results in most natural amethysts being biaxial. An analogous effect has been observed
for the aluminum centers of the smoky color [12, 13].

The extent of the anomaly in the pleochroism is [14] readily characterized via the intensity ratios for the Fe^{3+} ESR lines from the three equivalent but differently populated
groups of tetrahedra: it is substantially dependent on the crystallization temperature. The
most diachroic of the crystals at our disposal in the case of natural amethysts has a ratio of
$1:3:10$; for synthetic amethysts the range is usually $1:1:2$ to $1:2:4$. Figure 4 shows optical

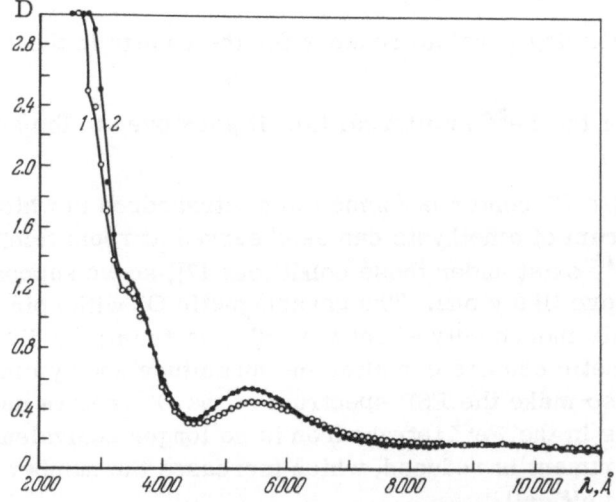

Fig. 4. Optical absorption of synthetic amethyst
crystals with pronounced pleochroism: 1) R,
2) r.

absorption spectra for synthetic amethysts with the most pronounced anomalous pleochroism; the degree of anomaly increases as temperature is reduced, as one would expect, and the very strong anomalous pleochroism in most natural amethysts merely indicates a low temperature of formation.

4. The amethyst color becomes saturated at comparatively low radiation doses (about 10^5 R, as against $5 \cdot 10^6$ R for the smoky color); the reason is that the number of centers for the amethyst color is only about 10^{18} cm^{-3} even in the most deeply colored crystals, which enables us to explain the existence of natural amethysts that become additionally colored in smoky or amethyst-smoky tones on further irradiation.

It has been found for synthetic amethysts that smoky color centers in these crystals reduce the thermal stability of the amethyst color, which again shows that the amethyst centers resemble the Al ones in having a hole nature, i.e., are related to loss of an electron from [FeO$_4$]. In fact, a high concentration of Al centers leads on heating to release of many electrons from the corresponding traps, so the amethyst color is bleached out at a comparatively low temperature, near the temperature of annealing for the smoky color (about 350°C). On the other hand, natural amethysts include ones with very much greater color stability, the bleaching temperature being about 450°C, which is due to the presence of electron traps of a different type having a higher thermal activation energy.

These studies show that synthetic amethyst crystals grown in the system $K_2O-CO_2-SiO_2-H_2O$ are closely analogous to natural amethysts as regards various physical properties.

Literature Cited

1. L. I. Tsinober and L. G. Chentsova, Kristallografiya, 4:633 (1959).
2. L. I. Tsinober, V. E. Khadzhi, L. A. Gordienko, and M. I. Samoilovich, in: Growth of Crystals, Vol.6A, Consultants Bureau, New York (1968), p. 25.
3. I. H. E. Gliffiths, I. Owen, and I. M. Ward, Nature, 173:439 (1954).
4. M. C. M. O'Brien, Proc. Roy. Soc. London, A239:404 (1955).
5. V. E. Khadzhi and M. V. Lelekova, in: Growth of Crystals, Vol. 8, Consultants Bureau, New York (1969), p. 43.
6. D. D. Hutton, Phys. Letters, 12:310 (1964).
7. T. J. Barry, P. McNamara, and W. I. Moore, J. Chem. Phys., 42:2599 (1965).
8. L. G. Chentsova, L. I. Tsinober, and M. I. Samoilovich, Kristallografiya, 11(4):236 (1966).
9. M. Schlesinger and A. V. Cohen, J. Chem. Phys., 44:3146 (1966).
10. I. Lietz and W. Murchberg, Neues Jahrb. Mineral. Monatsch., 1:17 (1958).
11. C. V. Raman, Current Sci., 23:379 (1954).
12. A. V. Shubnikov, Kristallografiya, 6:319 (1961).
13. L. I. Tsinober, M. I. Samoilovich, L. A. Gordienko, and L. G. Chentsova, Kristallografiya, 12:65 (1967).
14. L. I. Tsinober, M. I. Samoilovich, V. E. Khadzhi, and M. V. Lelekova, Dokl. Akad. Nauk SSSR, 176:676 (1967).

GROWTH CONDITIONS AND THE MORPHOLOGY OF VICINAL FACES OF THE PRINCIPAL RHOMBOHEDRA OF QUARTZ CRYSTALS

V. S. Balitskii, V. V. Bukanov, and T. A. Karyakina

A given face on different quartz crystals often will have differing vicinals; this is particularly characteristic of the faces of the $r\{10\bar{1}1\}$ and $z\{01\bar{1}1\}$ basic rhombohedra, in which the vicinals can be represented by two types of three-face pyramids [1, 2] and growth hummocks (Figs. 1 and 2). Vicinals indicate closely changes in growth condition [3], although it must be borne in mind that the details of their structure reflect the state of the mineral-forming region only during the last stages of growth; nevertheless, the vicinal morphology in relation to growth conditions can provide additional information in elucidating natural crystal origins.

We have examined the vicinal structures on the r and z faces of quartz crystals from a variety of deposits, as well as artificial ones; the r and z faces of crystals from rock-crystal pegmatites usually have vicinals as growth humps or three-faced pyramids of type I. Fairly often, these faces do not have macroscopic sculpturing, and only in rare cases do they have type II vicinals [2]. Quartz crystals from hydrothermal rock-crystal deposits may have vicinals as humps and as pyramids [1, 4]. For instance, crystals from the rock-crystal veins in the Pamir and Aldan most characteristically have vicinals in the form of humps, while crystals from the Polar Ural and amethysts from hydrothermal veins have type I vicinals. Similar vicinals and growth humps occur on quartz crystals from mercury, polymetallic, gold-ore, tin-ore, and tungsten deposits. Type II vicinals are most usual for amethyst crystals from deposits of agates related to basic lavas.

The vicinal structures for quartz from rock-crystal veins in various rocks do not show any definite relationship between the vicinal type and the rock composition; moreover, even within a single vein the crystals in adjacent parts sometimes have different types of vicinals. The various vicinals in some cases can be observed even on a single face, or else when one generation of quartz replaces another. In these cases, one sees under the crust of the late generation residual vicinals of type I, whereas the external surface is coated with type II vicinals (Fig. 1). Less often one finds the reverse sequence. There are gradual transitions between the pyramids and the growth humps.

Measurements on gas−liquid inclusions for quartz from pegmatite and hydrothermal deposits indicate that the last zones in the quartz crystallized at 100–180°C, no matter what the type of deposit; but it has proved impossible to determine the temperature at which the formation of the vicinals terminated. Comparisons of a more general character have failed to reveal the relationship between the crystal growth temperature and the type of vicinal.

Fig. 1. A z face of a quartz crystal with type I
vicinal faces under a layer of quartz incrusta-
tion of a later generation with type II vicinal
faces, ×3.

Observations on natural quartz crystals have thus failed to reveal any unambiguous relation between the type of vicinal and the geological conditions of crystal formation; we have
used synthetic crystals to elucidate vicinal formation, which were grown by the usual hydrothermal method with a temperature difference in refractory autoclaves of capacity 150 to
600 cm^3. The media were corrosive, so linings of PTFE were used. The seeds were single-
crystal plates cut strictly parallel to the r or z faces and the c {0001} pinacoid [5]. Set conditions were maintained in the oven with chromel—alumel thermocouples and ÉPV-2-11a potentiometers. The temperatures in the autoclaves were measured with chromel—copel thermocouples and KP-59 potentiometers, the error of measurement being ±1.5°. The necessary
pressure was provided by filling with the appropriate amount of solution [6]. The conditions

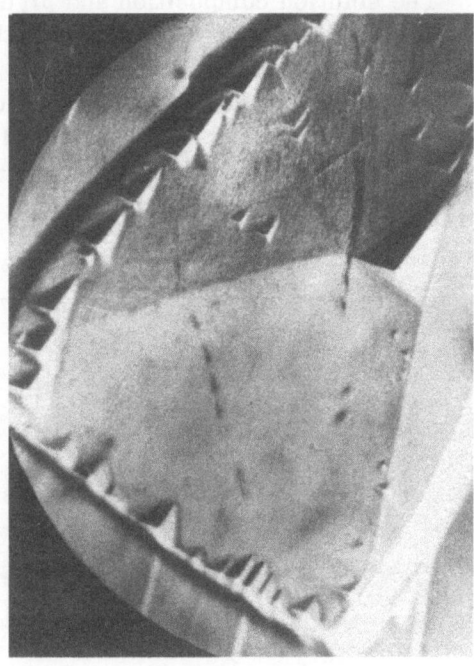

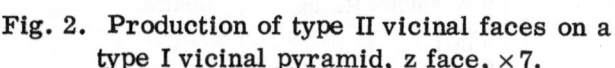
Fig. 2. Production of type II vicinal faces on a
type I vicinal pyramid, z face, ×7.

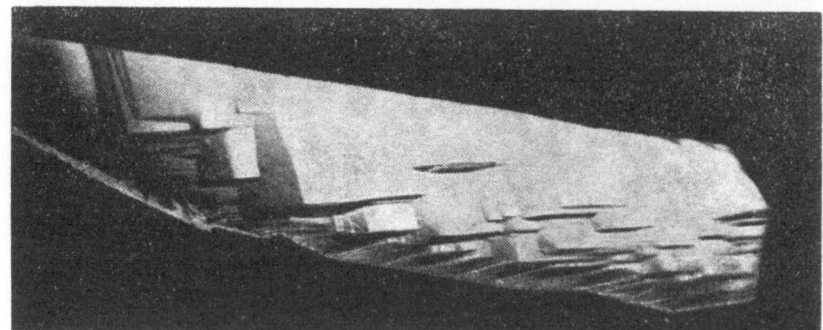

Fig. 3. Type II vicinals on an r face of a quartz crystal grown in neutral fluoride solution, ×4.

were chosen such as to allow one to trace the variation in the vicinals in relation to temperature, solution composition, pH, total silica content, and supersaturation. We grew crystals at temperatures between 250 and 320°C and pressures from 200 to 700 atm in aqueous solutions of sodium and potassium sulfides, and also in weakly alkaline (pH 8.5) chloride-bicarbonate-sodium (and potassium) solutions, which were close in composition to the gas—liquid inclusions found in natural quartz. These experiments showed that the r and z faces became coated with typical growth humps similar to those formed in strongly alkaline solutions [7]. If we used weakly alkaline (pH 8-9) solutions of sodium and potassium fluorides at 320° and 500 atm, the r and z faces also had vicinals in the form of growth humps; at 250°C and 100-200 atm we got vicinal sculptures transitional from humps to type I pyramids on r faces and type II on z faces. If the quartz was grown in weakly acid or neutral fluoride solutions at 250-320°C and 100-500 atm, the r faces had typical type II vicinals, while the z faces had growth humps and forms transitional to type I vicinals (Fig. 3). The form of the vicinals was unchanged on reducing the crystallization temperature to 90°C.

    As the crystals had different vicinals, and the growth was carried out under identical temperatures and pressures, we assumed that these parameters are not by themselves decisive.

    The solution composition and pH determine the content and transport form of the silica [8], as well as the supersaturation and the face growth rates; Table 1 gives the conditions and

TABLE 1. Growth Conditions for Quartz Crystals at 280°C
and a Filling Factor of 0.80

Solution	ΔT, °C	Silver, g/liter	Normal growth rate, mm/day			Vicinal type on faces		
			z	r	c	z	r	c
Neutral fluoride pH = 7.0	40	35	0.04	0.02	0.22	Humps	Type II	Stepped
	20	33	0.025	0.01	0.11	»	»	»
	10	32	0.005	0.003	0.06	»	»	»
	5	32	0	0	0.04	»	»	»
7% Na_2CO_3 pH = 10	40	15	0.03	0.015	0 06	Humps	Humps	Humps
	20	14	0.02	0.005	0.02	»	»	»
	10	13.6	0	0	0.01	»	»	»
	5	13.3	0	0	0		No vicinals	
7% $NaHCO_3$, pH = 8.5	40—20	4	0	0	0.06	Humps	Humps	Humps
	10—5	4	0	0	0		No vicinals	
4% NaF, pH = 8	40	1.2	0	0	0.03	Humps	Ridged	Humps
4% KF, pH = 8	40	1.3	0	0	0.02	»	»	»

results, which show that the various types of vicinals arise in ways having no direct relation to any of the factors, except perhaps the silica transport form. The transport forms differ substantially [9, 10] as between alkaline, neutral, and fluoride solutions. It may be that the composition and structure differences between the silicon-bearing complexes determine the type of vicinal that is produced.

Conclusions

1. The types of vicinal sculpture seen on r and z faces of natural quartz crystal are not related to the genetic type of deposit, the composition of the country rocks, or the temperature conditions for crystal growth.

2. There is no clear regularity in the sequence of vicinal sculptures in the final growth stage of natural quartz crystals, although the statistical evidence indicates that it is usual for growth humps to be replaced by type I vicinals and the latter by type II vicinals.

3. Experiment indicates that the occurrence of a given type of vicinal in the temperature and pressure ranges used is independent of the crystallization temperature, the silica content of the solution, and the supersaturation. The transitions between vicinal sculptures on faces of natural quartz crystals may be considered as consequences of composition and pH change in the mineralizing solutions.

Literature Cited

1. G. Kalb, Z. Kristallogr., 74:65 (1930).
2. G. M. Virovlyanskii, Zap. Vses. Min. Obshch., ser. 2, 67(3):446 (1938).
3. V. D. Kuznetsov, Crystals and Crystallization [in Russian], Gostekhteorizdat, Moscow (1953).
4. G. G. Lemmlein, Trudy Lomonosov. Inst. Akad. Nauk SSSR, No. 6, 13 (1935).
5. J. Dana, E. S. Dana, and C. Frondel, System of Mineralogy: Vol. 3, Silicon Minerals [Russian translation], Mir, Moscow (1966).
6. V. P. Butuzov, A. N. Kovalevskii, and L. A. Samoilovich, Trudy VNII Sint. Min. Syr'ya, Vol. 6 (1962).
7. G. G. Lemmlein and L. I. Tsinober, Trudy VNII Sint. Min. Syr'ya, 6:13 (1962).
8. V. S. Balitskii, in: Problems of Metamorphogenic Ore Formation [in Russian], Naukova Dumka, Kiev (1969).
9. R. K. Ailer, Colloid Chemistry of Silica and Silicates [in Russian], Stroiizdat, Moscow (1959).
10. I. G. Ryss, Chemistry of Inorganic Fluorine Compounds [in Russian], Goskhimizdat, Moscow (1956).

CRYSTALLIZATION KINETICS OF ZINCITE UNDER HYDROTHERMAL CONDITIONS

M. M. Lukina and V. E. Khadzhi

The literature carries only restricted information on the crystallization kinetics of zincite under hydrothermal conditions; we find [1, 2] only the face growth rates for monohedra as functions of temperature difference ΔT in solutions of KOH and NaOH. Preferential growth along the polar L_6 axis leads ultimately to loss of the positive monohedron face. Prolonged growth causes the crystal to be covered with the slowly growing faces only, and the overall growth rate becomes low, which is particularly noticeable when small-area seed plates are used. Introduction of lithium into the solution alters the growth rates of the (0001) and ($10\bar{1}0$) faces; this has been reported as advantageous also in the growth of perfect zincite crystals [2, 3].

We have examined the growth rate of zincite single crystals as functions of crystallization temperature, supersaturation, solvent concentration, and amount of LiOH added to the solution.

Methods

The crystals were grown in 2-9 m solutions of KOH containing 0.1-3.0 m LiOH for autoclaves of volume 0.4-1 liter, which were lined with silver and PTFE. The tests were done with constant and increasing ΔT, which varied over the range 10-25°C in accordance with the size of the reaction volume. A constant filling factor of 0.86 was used, and the runs lasted 10-30 days.

The initial mixture consisted of tablets of oxide zinc fired at 1100°C, which were crushed to 1-3 mm in size. The seeds were plates cut from hydrothermal crystals perpendicular to the L_6 axis. The growth rate was determined from the thickness of the deposited layer on one side of the seed and expressed in millimeters per day.

The results were processed statistically, the formula corresponding to systems of values of v_i, T_i, and v_iT_i, where v_i is the zincite growth rate, T_i is the temperature in degrees Celsius, and i = 1, 2, 3, ..., n, with $n \geq 10$ [4]. The coefficients in this formula were determined by least squares, and we determined also the deviation σ of the observed points on the empirical curve. The height of the bars in the graph varies from 0.5 to 3σ and reflects the spread in the experimental points due to the hydrodynamic factors in the process.

Temperature Dependence of the Growth Rate

We examined the monohedron face growth rates at 270-380°C in 5 m KOH plus 1 m LiOH with $\Delta T = 15$°C; the growth rate increased exponentially (Fig. 1). Figure 2 shows $\ln v = f(1/T)$,

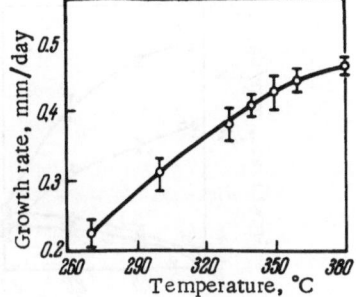

Fig. 1. Growth rate on (0001) as a function of crystallization temperature.

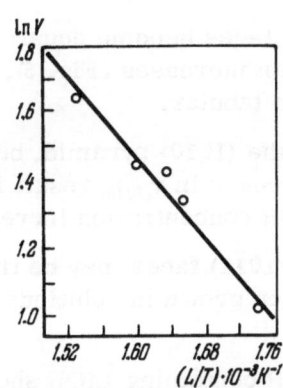

Fig. 2. Relation of ln V to 1/T (°K) for (0001).

which implies an activation energy [5] of 19.75 kcal/mole. Similar values have been reported for the following: (0001) faces of quartz crystals 19.9 in 0.5 m NaOH [6], $(11\bar{2}0)$ of sapphire 17.5 in 10% Na_2CO_3 [7], and $(10\bar{1}1)$ of calcite 16.6 under hydrothermal conditions.

Growth Rate as a Function of Supersaturation

The solubility of zinc oxide increases linearly with temperature between 200 and 500°C for alkali solutions [8], so the supersaturation in the growth zone was proportional to ΔT, so one could examine the effects of supersaturation. The growth rate of the (0001) faces in 5 m KOH containing LiOH was linear at 300°C (Fig. 3), and the slope of the straight lines gradually decreased as the LiOH concentration increased.

Effects of LiOH Concentration on the Absolute

and Relative Growth Rates of (0001)

and $(10\bar{1}0)$ Faces

The monohedron growth rate is appreciably reduced as the LiOH concentration increases in 5 m KOH, whereas the growth rate of the $(10\bar{1}0)$ prism increases (Fig. 4). At a certain

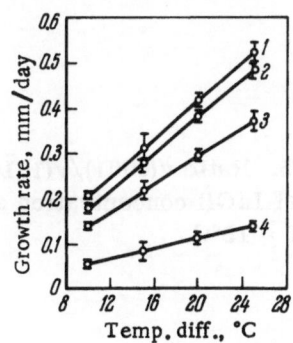

Fig. 3. Growth rate on (0001) as a function of supercooling ΔT for KOH containing LiOH concentrations (m) of (1) 0, (2) 1, (3) 2, and (4) 3.

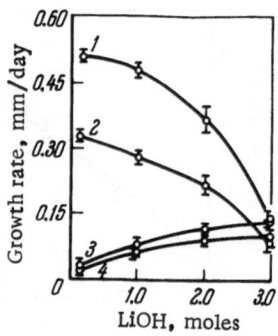

Fig. 4. Absolute growth rates of (0001) and (10$\bar{1}$0) in KOH containing LiOH at 300°C: 1, 2) v(0001) $\Delta T = 25$°C; $\Delta T = 15$°C; 3, 4) v(10$\bar{1}$0), $\Delta T = 25$°C; $\Delta T = 15$°C.

LiOH concentration, the growth rates of the (0001) and (10$\bar{1}$0) faces become equal; the ratio $v_{(0001)}/v_{(10\bar{1}0)}$ decreases parabolically as the LiOH concentration increases (Fig. 5), which reflects the change in the habit of the crystals from prismatic to tabular.

We did not make direct measurements of the growth of the (10$\bar{1}$0) pyramid, but geometrical considerations show that the reduction in $v_{(0001)}$ and the increase in $v_{(10\bar{1}0)}$ result in increase in $v_{(10\bar{1}1)}$, so the ratio $v_{(0001)}/v_{(10\bar{1}0)}$ also falls as the LiOH concentration increases.

The shift in the relative growth rates of the (0001) and (10$\bar{1}$1) faces may be the main reason for elimination of the degeneracy of the (0001) face when grown in solutions containing lithium.

These results for the solubility of zinc oxide in 5 m KOH containing LiOH showed that the solubility increases linearly from 3.7 to 5.6 wt.% as the LiOH concentration goes from 1 to 3 m; consequently, supersaturation in the growth zone should not fall as the LiOH concentration increases, but the growth rate of the main growing face falls, probably because of specific interaction of the lithium ions with this face.

Lithium as 2-2.5 m; LiOH allows one to grow seeds of small area (5 × 5 mm) and even almost point seeds (spontaneous crystals), which is impracticable with quartz.

We found that LiOH at ΔT of 25-35°C gave an increase in the absolute growth rate of (10$\bar{1}$0) faces, whereas the growth rate of the (0001) monohedron remained within the limits for potash media containing 2-2.5 m LiOH.

Effects of KOH Concentration on

Monohedron Face Growth Rates

The KOH concentration was varied from 2 to 7 m with otherwise fixed conditions (300°C, $\Delta T = 25$°C, 1.0 m LiOH), which revealed the following tendencies in the monohedron growth rates. Between 2 and 5 m KOH, there was an increase in the growth rate of (0001) faces from

Fig. 5. Ratio v(0001)/v(10$\bar{1}$0) in KOH as a function of LiOH concentration at: 1) $\Delta T = 25$°C; 2) $\Delta T = 15$°C.

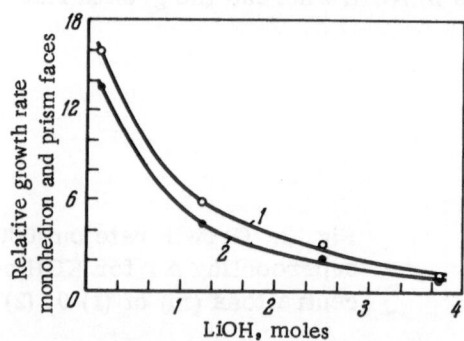

0.32 to 0.53 mm/day; in 7 m KOH, the rate had fallen to 0.28 mm/day, although the supersaturation should not have been reduced on account of the linear increase in the solubility of zinc oxide as the KOH concentration increases [8]. The growth rates of the polar monohedra remained practically equal. There was also a change in the ratio of the growth rates of the positive monohedron and the pyramid, and as a consequence the area of the $(10\bar{1}1)$ face was appreciably reduced, while the (0001) face was reduced to a smaller extent, which is an extremely important factor in the continuous growth of zincite single crystals.

Literature Cited

1.   R. A. Laudise, E. D. Kold, and A. J. Caporaso, J. Amer. Ceram. Soc., 47:9 (1964).
2.   I. P. Kuz'mina, Kristallografiya, 13:854 (1968).
3.   E. D. Kold and R. A. Laudise, J. Amer. Ceram. Soc., 49:302 (1 966).
4.   B. P. Demidovich, I. A. Maron, and É. Z. Shuvalova, Numerical Analysis Methods [in Russian], Fizmatgiz, Moscow (1963).
5.   J. W. Mallin, Crystallization [Russian translation], Metallurgiya, Moscow (1965).
6.   E. D. Kold and R. A. Laudise, J. Amer. Ceram. Soc., 48:342 (1965).
7.   V. A. Kuznetsov, Kristallografiya, 10:663 (1965).
8.   R. A. Laudise and E. D. Kold, Amer. Mineralogist, Vol. 48, No. 5/6 (1963).

ROLE OF THE STABILITY OF ALUMINATE SOLUTIONS IN HYDROTHERMAL PRODUCTION OF CORUNDUM

V. N. Rumyantsev, I. G. Ganeev, and I. S. Rez

Acid—alkali differentiation between the growth and solution zone is [1, 2] an important factor affecting the stability of aluminate solutions and producing the scope for hydrothermal crystallization of corundum.

Detailed descriptions have been given [3] of the crystallization of corundum, while the acid—alkali differentiation was examined with an apparatus modified from one previously described [1], with the tube trap replaced by a high-pressure valve allowing one to shut down at the end of the experiment the communication between the lower and upper autoclaves. Tables 1 and 2 give the results.

The role of the differentiation is clear when one uses a solution of sodium tetraborate; in this case the solubility and temperature coefficient of the latter exceed those for aqueous solutions of sodium carbonate [4, 5], but the corundum does not crystallize. The reason is that the acid-alkali differentiation does not make itself felt in a solution of $Na_2B_4O_7$ on account of the absence of a volatile in the salt or hydrolysis products, in conjunction with the high buffer capacity of the solution.

There is another characteristic feature of the crystallization of corundum from aqueous alkali solutions: strongly alkaline solutions (aqueous NaOH) do not cause corundum to grow on a seed even at temperature differences up to 75°; the lack of crystallization is due not only to the absence of acid—alkali differentiation (which we have confirmed by experiments) but also to the high alkalinity, which [6] increases the stability of aluminate solutions.*

The growth rate is markedly dependent on pH when one uses aqueous sodium carbonate or mixed solutions of moderate alkalinity; the growth rate increases as the pH decreases, and the tendency persists [8, 9] on going to solutions of even lower alkalinity, such as sodium or potassium bicarbonate. The reason appears to be that aluminate solutions become less stable as the alkalinity decreases.

Corundum dissolves in aqueous NaOH and also in salts that hydrolyze to give caustic soda such as Na_2CO_3, $NaHCO_3$, and $Na_2B_4O_7$, and under hydrothermal conditions the oxide reacts with the hydroxide [10], so the mechanism may be put as

$$Al_2O_3 + 2NaOH + 3H_2O \rightarrow 2Na[Al(OH)_4].$$

---

*For comparison we note that the temperature coefficient of the solubility for solutions in sodium hydroxide equals or even exceeds that of quartz and NaOH in the range 430-600°C, and the latter is sufficient for hydrothermal growth of quartz [5, 7].

TABLE 1. Crystallization of Corundum in Aqueous Solutions
at 1500 atm*

Solution zone temp., °C	$\triangle\text{T}_{,\text{°C}}$	Component	Concentration, g/liter	Initial pH	$(10\bar{1}1)$ growth rate, mm/day
440	40	NaOH	100	13,50	0
	50	NaOH	200—300	Over 14	0
520	50	NaOH	100	13.50	0
	60—75	NaOH	50	13.25	0
	20	Na_2CO_3	80—100	10,70—10,90	0.04
	20	Solution is		9,95	0.42
	20	mixture of components†		9,76	0.50
	50	$Na_2B_4O_7$	100	9.14	0

*$Na_2B_4O_7$ runs at 1000 and 1500 atm.
†Sodium carbonate with added sodium or ammonium bicarbonate.

The resulting aluminate ions have a regular tetrahedral structure and may be considered as typical complexes inclined to polymerization, which is all the more likely the lower the concentration and activity of the OH ions. The $[Al(OH)_4]$ ions are stable [11] in the absence of CO_2, which explains why corundum does not crystallize from aqueous solutions of sodium hydroxide and sodium tetraborate, whereas solutions of carbonate or bicarbonate produce acid—alkali differentiation and the reaction becomes reversible in the growth zone, while the polymerization gives rise to aluminate ions that interact with the surface of the seeds, which ultimately leeds to crystallization of α-Al_2O_3.

The concentration also has a marked effect on the stability of aluminate solutions; we consider that the elevated stability of aluminate solutions at high concentrations may explain why corundum does not crystallize when one uses strong (30%) solutions of sodium or potassium carbonate, when the corresponding aluminate is formed as the stable crystalline phase [9].

Conclusions

1. Aluminate solutions show high stability and high supersaturation at elevated temperatures and pressures.

2. A temperature difference is needed in the hydrothermal growth of corundum, but it is not a sufficient condition for growth; the material does not crystallize if one does not provide conditions corresponding to low stability for aluminate solutions, such as low alkalinity and a

TABLE 2. Acid—Base Differentiation in Aqueous Solutions
at Elevated P and T, Lower Autoclave at 520°

Temp. (°C) of upper autoclave	$\triangle\text{T}_{,\text{°C}}$	Pressure, atm	Concentration, g/liter	pH			$\triangle$pH	Run, hr
				before	after			
					lower	upper		
500	20	1500	—80Na_2CO_3	10 90	10.90	10.78	0.12	11
480	40	1500	—80Na_2CO_3	10.90	11.00	10.72	0.28	33
500	20	1500	—80Na_2CO_3+ +20NaHCO_3	9.92	10.05	9.80	0.25	8
470	50	1000	—100$Na_2B_4O_7$	9.14	9.26	9.22	0.04	11

difference between the growth and solution zones. These conditions are realized in aqueous solutions of carbonates, bicarbonates, and mixtures of these.

3. The effects of pH on the crystallization rate indicate that the stability of the aluminate solution should be considered as a major kinetic factor.

We are indebted to I. V. Churikov for assistance and to A. A. Shternberg for valuable comments.

Literature Cited

1. I. G. Ganeev and V. N. Rumyantsev, Abstracts for the Seventh International Crystallography Congress and Symposium on Crystal Growth [in Russian], Nauka, Moscow (1966).
2. V. A. Shorygin, V. N. Rumyantsev, and I. G. Ganeev, "The physics of rock processes," Abstracts for the Conference of Technical Colleges and Research Institutes of the USSR, Section Rock Chemistry and Biochemistry [in Russian], Mining Institute, Moscow (1969).
3. I. G. Ganeev, B. K. Kazurov, É. N. Karaul'nik, V. N. Rumyantsev, and I. S. Rez, in: Crystallization Mechanisms and Kinetics [in Russian], Nauka i Tekhnika, Minsk (1969), p. 523.
4. S. Levinson, G. Douglas, and L. R. Johnson, Amer. Min., 50(3-4):403 (1965).
5. R. L. Barns, R. A. Laudise, and R. M. Shields, J. Phys. Chem., 67:835 (1963).
6. S. I. Kuznetsov and V. A. Derevyankin, The Physical Chemistry of Alumina Production by Bayer's Method [in Russian], Metallurgiya, Moscow (1964).
7. R. A. Laudise, J. Amer. Chem. Soc., 81:562 (1959).
8. V. A. Shorygin, Kristallografiya, 8(5):808 (1963).
9. V. A. Kuznetsov, "Crystallization of corundum under hydrothermal conditions," Thesis, Institute of Crystallography, Academy of Sciences of the USSR, Moscow (1967).
10. V. N. Rumyantsev and I. G. Ganeev, Abstracts for the Eighth All-Union Conference on Experimental and Technical Mineralogy and Petrography [in Russian], Novosibirsk (1968).
11. K. F. Jahr and I. Pernoll, Ber. Bunsenges. Phys. Chem., 69(3):221 (1965).

GROWTH OF OPTICAL CALCITE FROM SEEDS WITH VARIOUS ORIENTATIONS

Yu. V. Pogodin and V. V. Dronov

Methods of making small calcite crystals were suggested in the past century [1-3], and Vater [3] made the most complete study of the morphology of artificial calcite and the variations between the crystallographic forms in response to growth conditions. The attention of researchers on calcite has been concentrated on reliability in laboratory apparatus and methods [4-15].

Recently, the All-Union Mineral Raw Material Synthesis Institute has designed lined autoclaves and determined the physicochemical conditions for hydrothermal synthesis, which has provided for continuous production of single-crystal calcite on seeds, while eliminating spontaneous crystallization [10, 11].

Tests on growing calcite crystals on platy seeds with various orientations have been performed in autoclaves lined with platinum or PTFE, the outside material being stainless steel and the temperature-difference method being employed. The object was to determine the growth rates along various crystallographic directions, and also to elucidate the effects of growth conditions on the uniformity of the grown layer and the habit of the crystals.

The seeds were cut parallel to the most prominent faces of crystals of Iceland spar from the Siberian platform: $(10\bar{1}1)$, $(02\bar{2}1)$, $(01\bar{1}1)$, $(01\bar{1}2)$, (0001), $(10\bar{1}0)$, $(11\bar{2}0)$, $(8.8.\bar{16}.3)$, $(4.8.\bar{12}.5)$, $(35\bar{8}4)$, and these as far as possible were monopyramidal.

Seeds on $(10\bar{1}1)$ and $(02\bar{1}1)$ rhombohedra gave single-crystal growth over a wide range of supersaturations (Figs. 1 and 2). The other eight sections usually gave multiheaded growth under analogous physicochemical conditions (Figs. 3 and 4). One found uniform growth on such seeds only at low supersaturations (Fig. 2) or after the pyramids from the multiheaded growth on the $\{10\bar{1}1\}$ and $\{02\bar{2}1\}$ rhombohedron faces had vanished (Figs. 5 and 6).

In all cases, no matter what the seed orientation, the crystals were covered by $\{10\bar{1}1\}$ and $\{02\bar{2}1\}$ rhombohedron faces, and sometimes pinacoid ones; in prolonged growth, the $\{10\bar{1}1\}$ faces were gradually displaced by the slow-growing $\{02\bar{2}0\}$ faces, as is evident from the structure of the growth pyramids and the habit of the crystals from cycles of various lengths (Fig. 7).

Visually uniform synthetic calcite differs from the natural material in being almost completely colorless and having high light transmission in the ultraviolet region [10, 11]. The main macroscopic defects in individual crystals were cracks, streaks, and zones with gas–liquid inclusions at the interface with the seed and at the junctions between $\langle 10\bar{1}1\rangle$ and $\langle 02\bar{2}\rangle$ growth pyramids, which have now been largely eliminated.

55

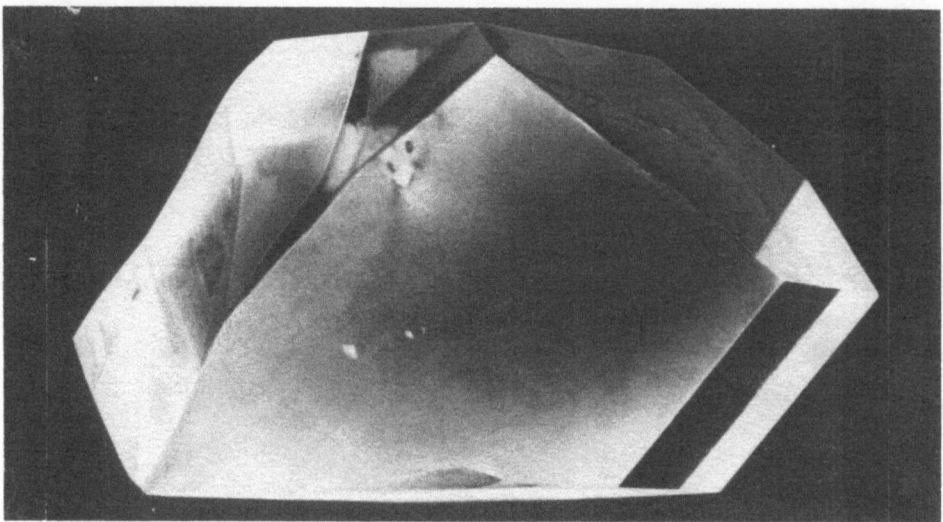

Fig. 1. Single crystal of optical calcite grown from seed oriented
on $(10\bar{1}1)$, natural size.

The faces of the $\{10\bar{1}1\}$ rhombohedron and pinacoid produced during the growth were usual-ly flat, smooth, or with minute accessory steps; the $\{02\bar{2}1\}$ faces had stepped growth and vici-nal development, with the edges represented by forms close to the $\{16\bar{7}4\}$ scalenohedron [11].

In multiheaded crystallization, there was splitting of the growth pyramids from the seed places into a multiplicity of crystals retaining the orientation of the substrate and covered with faces of the $\{10\bar{1}1\}$ and $\{02\bar{2}1\}$ rhombohedra, which are typical of synthetic calcite, these some-times being accompanied by pinacoid faces. The quality of the layer then deteriorates consider-ably on account of large-scale formation of gas-liquid inclusions and microcracks, but often the individual crystals are visually uniform. When seeds on $(01\bar{1}\bar{2})$ were used, the subindivi-

Fig. 2. Single crystals of optical calcite grown from seeds oriented on
$(8.8.\overline{16}.3)$, $(35\bar{8}4)$, $(4.8.\overline{12}.5)$, and $(02\bar{2}1)$, natural size.

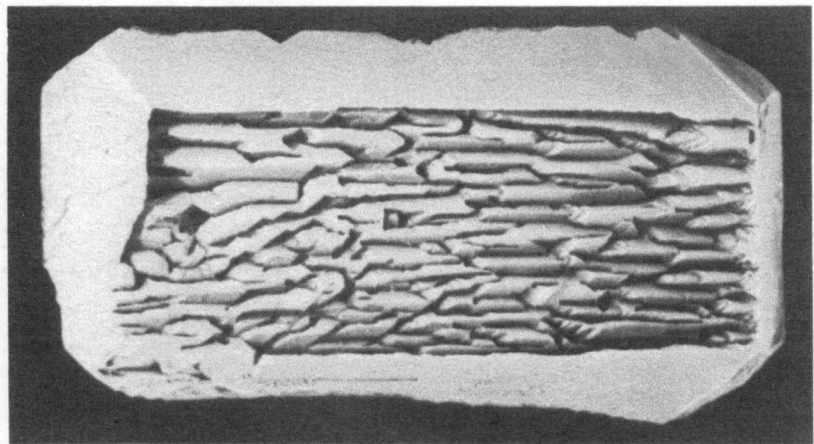

Fig. 3. Growth figures of calcite on a platy seed oriented
on (01$\bar{1}$2), natural size.

duals were usually elongated and oriented in planes parallel to the edges of the cleavage rhombohedron (Fig. 3), while for the other sections no such regularity could be observed.

In certain runs with seeds cut on the cleavage rhombohedron, we observed accelerated
growth on the polar edges, along which there was marked reduction in the size of the subindividuals even as far as formation of single-crystal layers and zones, together with appearance of
parts of the faces of the principal rhombohedron. This effect was more prominent on other
sections; for most of these, the tops of the subindividuals were covered with thin single-crystal
layers of thickness 0.1-2 mm, which represented the basis for production of (10$\bar{1}$1) and (02$\bar{2}$1)
faces. There was a fast tangential growth of the layers, and as a result one obtained unusual
growth forms above the subindividuals (Fig. 6).

The main cause of multiheaded growth on seeds under these conditions was elevated
supersaturation well above the optimal value; we often found multiheaded growth when the

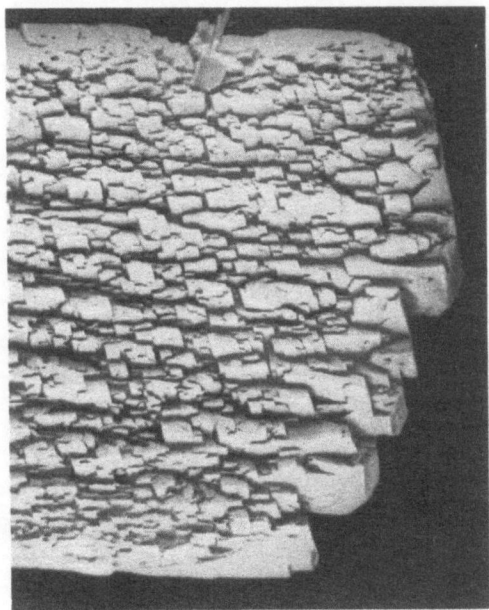

Fig. 4. Growth figures of calcite on a platy
seed oriented on (11$\bar{2}$0), natural size.

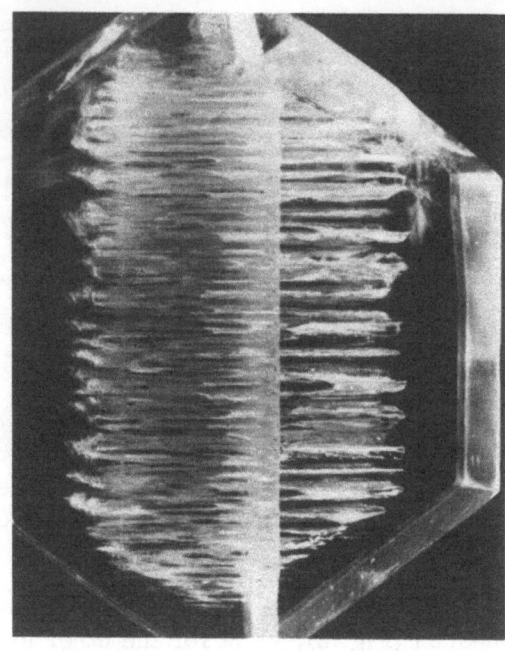

Fig. 5. Single crystal of optical calcite grown
from seed oriented on (01$\bar{1}$1).

temperature difference between the growth chamber and the solution one was 1.5-2 times the optimal one (Fig. 8).

We may now consider the evolution of the habit forms of natural calcite.

It has been observed [16-22] that the sequence of the main habit forms of calcite in time is $\{0001\}$ pinacoid → $\{10\bar{1}1\}$ principal rhombohedron → $\{10\bar{1}0\}$ prism, followed by various scalenohedra with the $\{01\bar{1}2\}$ rhombohedron → $\{02\bar{2}1\}$ acute rhombohedron.

The tests with seeds of various orientations gave the mean growth rates for sections parallel to the principal habit forms (Table 1).

We concluded that the sequence of production of new simple forms (Table 1) is such that the average growth rate decreases during the overall growth.

Fig. 6. Calcite crystal grown from platy seed made on
(0001), natural size. The subindividuals from multipoint
growth overlap as plates on (10$\bar{1}$1) and (02$\bar{2}$1).

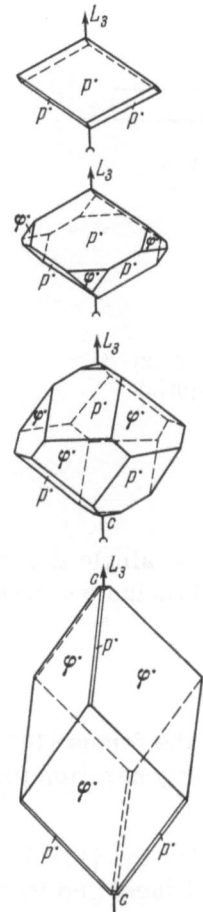

Fig. 7. Habit forms of artificial calcite in relation to growth time, platy seed on $(10\bar{1}1)$; P* face $\{10\bar{1}1\}$, φ* face $\{02\bar{2}1\}$, C face $\{0001\}$.

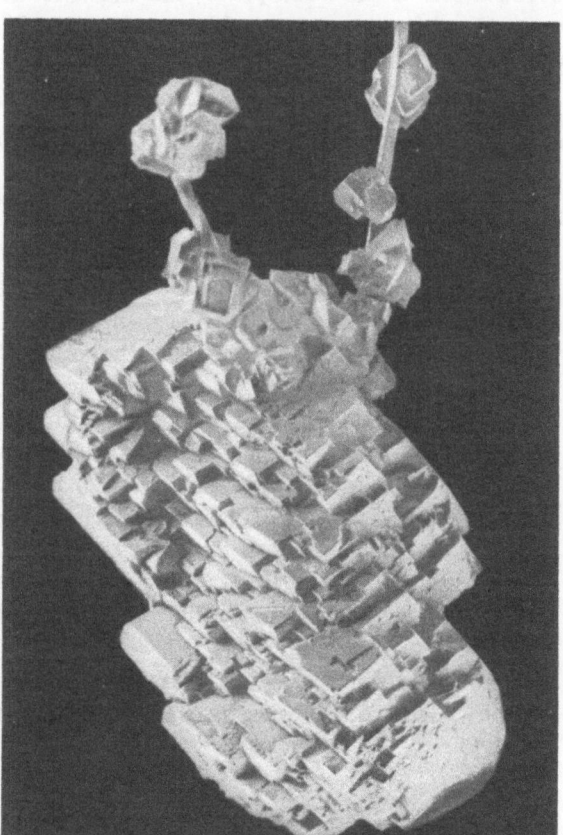

Fig. 8. Multipoint growth of calcite from a platy seed oriented on $(10\bar{1}1)$, natural size.

TABLE 1. Growth Rate and Shape Change for Calcite Crystals

Synthetic calcite		Natural calcite	
seed orientation	mean growth rate on two sides of seed, mm/day	crystallization stage	sequence of habit forms
(0001)	0.60	I	{0001}
(11$\bar{2}$0)	0.55	—	—
(01$\bar{1}$1)	0.37	—	—
(10$\bar{1}$1)	0.35	II	{101$\bar{1}$}
(01$\bar{1}$2)	0.30	—	{10$\bar{1}$0} + {01$\bar{1}$2}
(358$\bar{4}$)	0.25	III	{hkil} + {01$\bar{1}$2}
(8.8.$\overline{16}$.3)	0.20	—	—
(02$\bar{2}$1)	0.15	IV	{02$\bar{2}$1}
(4.8.$\overline{12}$.5)	0.10	—	—

The reasons for this can be established by examining clumps and single crystals of calcite from various deposits in conjunction with tests on gas —liquid inclusions and simulation of the processes by hydrothermal synthesis.

Conclusions

1. Crystals of artificial optical calcite grow via combination of the forms {10$\bar{1}$1}, {02$\bar{2}$1}, {000$\bar{1}$} without relation to the orientation of the seed plate; the final regeneration form is the {02$\bar{2}$1} acute negative rhombohedron.

2. Multiheaded growth is replaced by single-crystal growth after the {10$\bar{1}$1} and {02$\bar{2}$1} rhombohedra have been displaced by the growth pyramid of the initial faces, no matter what the orientation of the seed plate; the multiheaded growth is replaced by single-crystal growth by formation of some convex forms as a result of tangential growth of the {02$\bar{2}$1} faces at high speeds.

3. The sequence of mean growth rates for seeds of various orientations is in agreement with the way in which the main habit forms of natural calcite occur.

Literature Cited

1. G. Rose, Pogg. Ann., 42:353 (1837).
2. L. Bourgeois, Reproduction artificielle des minéraux, Paris (1884).
3. H. Vater, Z. Kristallogr., 21:442 (1893).
4. T. Warynski and S. Kouropatwinska, J. Chim. Phys., 14:328 (1916).
5. N. Yu. Ikornikova and V. P. Butuzova, Dokl. AN SSSR, 111:105 (1956).
6. F. V. Syromyatnikov, Zap. Vses. Min. Obshch., series 2, 90:697 (1961).
7. J. Kašpar, in: Growth of Crystals, Vol. 2, Consultants Bureau, New York (1959), p. 57.
8. V. G. Lushnikov, Trudy VNIIsinteza mineral'nogo syr'ya, 8:173 (1964).
9. C. Barta and J. Žemlička, Abstracts for the Seventh International Crystallography Congress and Symposium on Crystal Growth [in Russian], Nauka, Moscow (1966), p. 261.
10. Yu. V. Pogodin and V. M. Sergeev, Abstracts for the Seventh International Crystallography Congress and Symposium on Crystal Growth [in Russian], Nauka, Moscow (1966), p. 266.
11. Yu. V. Pogodin and V. M. Sergeev, Abstracts for the Seventh International Crystallography Congress and Symposium on Crystal Growth [in Russian], Nauka, Moscow (1966), p. 276.

12. Yu. V. Pogodin, N. I. Andrusenko, and V. B. Naumov, Abstracts for the Third All-Union
     Conference on Mineral Thermobarometry and the Geochemistry of Plutonic Processes
     [in Russian], Moscow (1968).
13. P. M. Grusensky, J. Phys. Chem. Solids, Suppl. 1, 365 (1967).
14. K. Yasushi and K. Nobuko, Geochem. J., 1:1 (1966).
15. J. F. Nester and J. B. Schroeder, Amer. Mineralogist, 52:276 (1967).
16. G. Kalb, Zbl. Mineral., part A, 10:337 (1928).
17. V. F. Vasilevskii, Trudy Uzb. Geol. Upr., Vol. 14 (1939).
18. I. Sunagawa, Rept. Geol. Surv. Japan, 155 (1953).
19. N. Z. Evzikova, Zap. Vses. Min. Obshch., ser. 2, No. 87 (1958).
20. D. P. Grigor'ev, Ontogeny of Minerals [in Russian], Izd. L'vov. Univ., L'vov (1961).
21. R. M. Aliev, in: Origin of Mineral Crystals and Clumps [in Russian], Nauka, Moscow
     (1966).
22. E. Naidenova, God. Sof. Univ. i Geol.-Geogr. Fak., 1964-5, 59(1):211 (1966).

SYNTHESIS OF PYROLUSITE CRYSTALS

S. P. Fedosova

Manganese dioxide occurs in nature as true crystalline or powdery sooty masses, single crystals being very rare [1]; it exists in several polymorphic forms, which differ in structural order and which are usually not stoichiometric [2]. All the modifications, apart from the most stable one β-MnO_2 (pyrolusite), lose water at 105°C and above and are completely dehydrated at 500°C. Pyrolusite exists up to 500°C, but then it begins to lose oxygen and at 800°C it becomes β-MnO_2 [6]. Pyrolusite is a ferroelectric whose Curie point is +49°C [3].

Here we present some preliminary results on the hydrothermal crystallization of pyrolusite. It is necessary to use a solvent that is highly oxidizing in order to stabilize the Mn (IV); we used 20% HNO_3, whose oxidizing action is due to autocatalysis by decomposition products, the exact oxidation potential being dependent on the concentration and temperature [4]. The solution is highly corrosive, so one needs to protect the autoclave walls carefully. We found that titanium (grade BTT-1) was least attacked, and this was the material used. The initial product was high-purity Mn_2O_3. Crystals were made in 5-20% nitric acid at 450°C, and these were black and had a metallic luster and perfect cleavage; the dimensions were about 1-2 mm. We observed twins similar to those of rutile (Fig. 1). The faces were covered with striations.

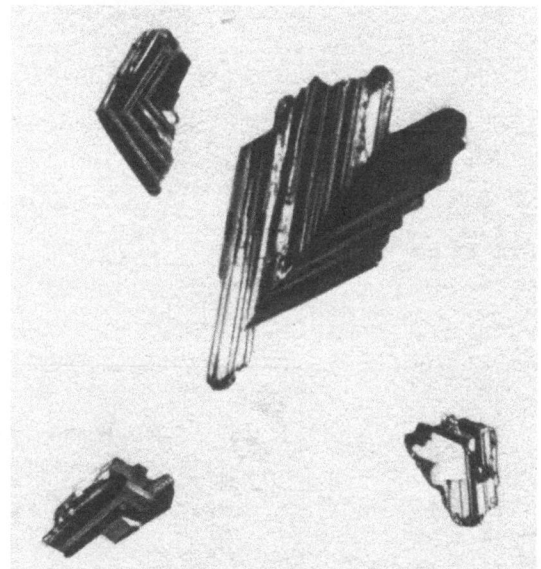

Fig. 1. Hydrothermal pyrolusite crystals.

TABLE 1. x-Ray Phase Analysis of Synthetic Pyrolusite*

I		II		III	
I	d/n, Å	I	d/n, Å	I	d/n, Å
				1	3.48
				10	3.14
10	3.07	9	3.05		
5	2.40	7	2.38	5	2.41
2	2.197	2	2.18	1	2.21
1	2.10	6	2.09	3	2.13
1	1.96	3	1.94	2	1.98
				5	1.81
				5	1.63
3	1.55	5	1.55	3	1.56
2	1.43	5	1.43	2	1.43
1	1.39	3	1.39	2	1.40
3	1.30	7	1.30	2	1.31
		1	1.19	5	1.20
		1	1.12	5	1.12
		2	1.10	5	1.10
		1	1.06	2	1.05

*I) our results, II) published values, III) from ASTM card index.

The crystals were identified by x-ray phase analysis with a URS-50 IM apparatus with Cu K_α radiation. The interplanar distances of the artificial crystals agreed with those of the natural material (Table 1).

In this way we have made artificial single crystals of pyrolusite of size 1-2 mm at 450°C in 5-20% nitric acid.

I am indebted to B. N. Litvin for direction in this work.

Literature Cited

1. A. G. Betekhtin, Mineralogy [in Russian], Moscow (1950).
2. B. Ettel and V. Ieprek-Siska, Chemicke listy, 57:785 (1963).
3. V. G. Bhide and R. V. Daml, Physica 26:33 (1960).
4. E. A. Kanevskii and A. P. Fimenov, Radiokhimiya, 6:732 (1964).
5. E. Ya. Rose, Oxygen Compounds of Manganese [in Russian], Izd. AN SSSR, Moscow (1952).
6. R. Huder and A. Schmier, Electrochim. Acta, 3:127 (1960).

CRYSTALLIZATION KINETICS OF ANTIMONITE IN AQUEOUS SOLUTIONS OF NH_4Cl, NH_4Br, and Na_2S

V. I. Popolitov and B. N. Litvin

We have examined the growth rates of (001) and (010) faces of Sb_2S_3 single crystals as functions of temperature under hydrothermal conditions, and also in relation to compositions and concentration of the solvent (NH_4Cl, NH_4Br, Na_2S) at constant temperature difference of 20°C. The autoclaves were of batch type and lined with titanium. The temperature was 350-450°C, and the pressure was 600-1500 atm, with runs lasting from 24 to 120 hours. The initial mixture consisted of elemental sulfur, antimony, and reagent-grade Sb_2S_3 of especial purity. The concentrations of the aqueous solutions of NH_4Cl, NH_4Br, Na_2S ranged from 2 to 12 wt.%. The method of experimentation and growth rate measurements have been described [1]; the following are the general functional relationships for the (001) and (010) faces in these solutions:

$$v_{001} = f(C, \ T)_{\tau \ = \ const, \ \Delta T \ = \ const,}$$
$$v_{010} = f(C, \ T)_{\tau \ = \ const, \ \Delta T \ = \ const.} \tag{1}$$

We used the reduced variable in (1) to construct a series of generalized relationships for the temperature−concentration dependence of the growth rates of these flashes of Sb_2S_3; Figs. 1 and 2 show for 350-450°C these relationships approximated by a linear equation of the form

$$\log v = \log K + n \log C,$$

where v is the growth rate in millimeters per day, K is kinetic coefficient, n is the order of crystallization (n ≈ 2), and C is the solvent concentration in moles per liter.

Figures 1 and 2 show that the growth rates of these faces increase linearly with the solvent concentrations, and more strongly for NH_4Cl than for NH_4Br and Na_2S, i.e., the ratio of the growth rates of these Sb_2S_3 faces for these solutions may be represented by

$$v_{Sb_2S_3(NH_4Cl)} > v_{Sb_2S_3(NH_4Br)} > v_{Sb_2S_3(Na_2S)},$$

and this applies throughout the range of concentrations used for both faces.

Figure 3 shows log v for these faces as a function of 1/T for these aqueous solutions, the values being closely fitted by an erroneous equation. The slopes of the straight lines give the activation energies for the (001) face as 12, 13, and 15 kcal/mole, respectively, while those for (010) are 2.5, 2.8, and 3.1 in the same units.

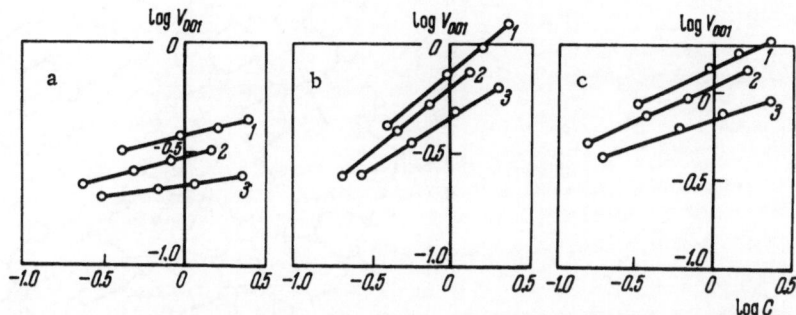

Fig. 1. Growth rate on (001) as a function of solvent concentration at (a) 350, (b) 400, and (c) 450°C; 1) NH_4Cl, 2) NH_4Br, 3) Na_2S.

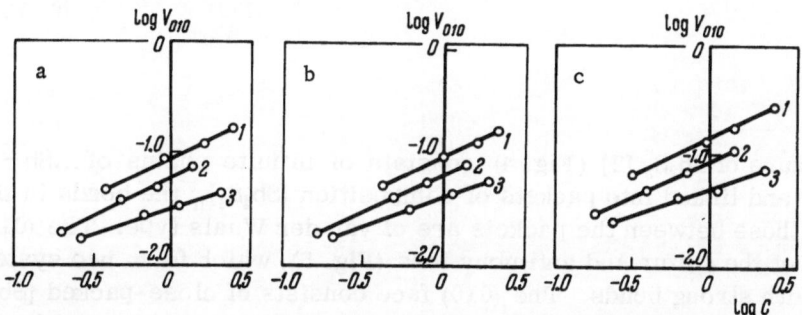

Fig. 2. Growth rate on (010) as a function of solvent concentration at (a) 350, (b) 400, and (c) 450°C; 1) NH_4Cl, 2) NH_4Br, 3) Na_2S.

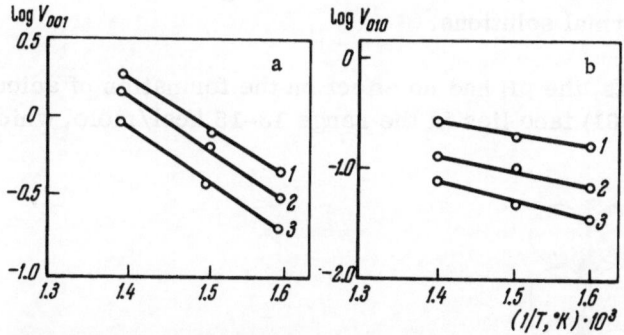

Fig. 3. Growth rate for (a) the (001) face and (b) the (010) face of Sb_2S_3 as a function of 1/T: 1) NH_4Cl, 2) NH_4Br, 3) Na_2S.

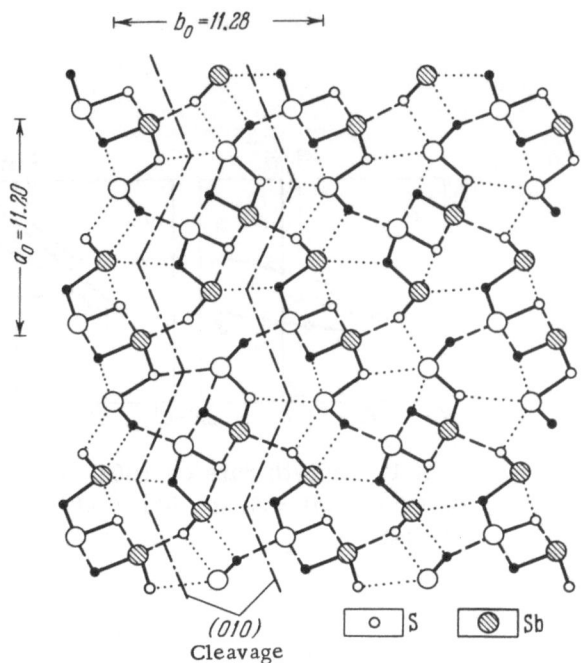

Fig. 4. Structure of Sb_2S_3.

| ○ S | ⊘ Sb |

(010)
Cleavage

Discussion

The structure of Sb_2S_3 [2] (Fig. 4) consists of infinite chains of ...Sb−S−Sb... lying along the c axis and linked into packets of composition $[Sb_4S_6]_n$; the bonds in the chains are covalent, while those between the packets are of van der Waals type. The (010) plane has the closest packing of the sulfur and antimony ions (Fig. 5), which form two systems of chains ...Sb−S−Sb... with strong bonds. The (010) face consists of close-packed $[Sb_4S_6]_n$ packets, so it can grow only by van der Waals interaction between packets, which can be considered as the basis for formation of two-dimensional nuclei. The (001) face is a K face and so grows rapidly on account of the ease and independence of attachment of particles by the normal-growth mechanism. The preferential growth of Sb_2S_3 single crystals along the c axis is therefore a result of the different growth mechanisms of the (001) and (010) faces. An additional explanation is needed for the formation of (001) faces over wide ranges in temperature and concentration for these aqueous hydrothermal solutions.

In our experiments, the pH had no effect on the formation of acicular crystals; the activation energy for the (001) face lies in the range 13-15 kcal/mole, which is direct evidence

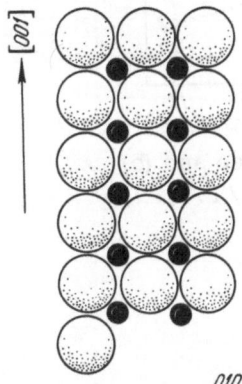

Fig. 5. Disposition of antimony and sulfur ions in the plane of a (010) face.

that the growth is controlled by a chemical reaction at the interface with the solution. In fact, the different activation energies for the solutions of NH_4Cl, NH_4Br, Na_2S for Sb_2S_3 indicate that the growth of (001) faces is related to the nature and rate of deposition of the material in the solution. Complexes exist in the reaction zone [1], so the growth of a (001) face may be considered as a heterogeneous reaction; in our case such reactions will correspond to decomposition of the complexes

$$\left[Sb\!<^S_S\right]^- + \left[S\!-\!Sb\!<^S_S\right]^{2-} + 2\left[^S_S\!>\!Sb\!<^S_S\right]^{5-} + 5H^+ \rightarrow \left[Sb\!<^S_S\!>\!Sb\!<^S_S\!>\!Sb\!<^S_S\!>\!Sb...\right] + 2S^2 + 5HS^-.$$

$$4\left[^\Gamma_\Gamma\!>\!Sb\!<^\Gamma_\Gamma\right]^- + 6S^{2-} \rightarrow \left[...Sb\!<^S_S\!>\!Sb\!<^S_S\!>\!Sb\!<^S_S\!>\!Sb\right] + 16\Gamma^-$$

$$(\Gamma = Cl,\ Br).$$

The growth rate of (001) varies with the type of solvent [1–6]; the straight lines for NH_4Cl and NH_4Br are almost parallel, i.e., the order of the reaction does not vary with the type of solvent. The displacement of the straight lines can be due only to difference in the stability of the chloride and bromide complexes, i.e., to difference in decomposition rates. The straight line for Na_2S solutions has a somewhat different slope, and this is naturally so, because here the reaction involves different complexes. We may therefore suppose that the growth rate in the direction of the (001) face will be limited for the various solvents by the complex decomposition rate. The activation energies calculated for growth of (010) faces of Sb_2S_3 are almost all equal to about 3 kcal/mole, which is evidence that here the growth is limited by diffusion.

Conclusions

1. It is found that the logarithm of the growth rate for (001) and (010) faces of Sb_2S_3 single crystals is represented by a straight-line relation to the logarithm of the concentration of NH_4Cl, NH_4Br, or Na_2S at 350, 400, and 450°C;

$$\log v = \log K + n \log C.$$

2. The growth rates along (001) and (010) directions obey the following inequalities, other things being equal:

$$v_{Sb_2S_3(NH_4Cl)} > v_{Sb_2S_3(NH_4Br)} > v_{Sb_2S_3(Na_2S)}.$$

3. The calculated activation energies for growth of (001) and (010) faces show that the growth rate of the first is limited by the complex decomposition rate, while that for the second is limited by diffusion.

Literature Cited

1. V. I. Popolitov, Kristallografiya, 14:545 (1969).
2. W. Hafmann, Z. Kristallogr., 86:225 (1933).

GROWTH KINETICS OF DIAMOND FROM SOLUTION IN A MOLTEN METAL

Yu. A. Litvin and V. P. Butuzov

A major problem in solid-state physics is to grow reasonably large and uniform diamond crystals; however, the growth of a single-crystal diamond has hardly been examined.

One can understand the transport and recrystallization of carbon when this reacts with a molten metal on the basis of a one-dimensional hydrostatic model (Fig. 1). Consider the cases where the carbon source is a thermodynamically stable phase, diamond (examples 1-4), and where it is an unstable one, graphite (examples 5-10), since the parameters of the processes are related to the P-T area for diamond. The results are shown in Figures 1 and 2 and also in Table 1.

The supersaturation ΔN for diamond is a function of the form

$$\Delta N^{S,T,P} = A\Delta N^S \pm Bf(\Delta T) \pm Cf(\Delta P), \tag{1}$$

where the constants A, B, and C are related to the nature of the interacting substances. The superscripts are to be understood as implying a functional dependence. The first term characterizes the contribution to the supersaturation from the solubility difference in the carbon sources: metastable graphite and diamond (superscript S for the latter) [1, 2]. This is zero if diamond is the carbon source. The other terms are related to the gradients in the thermodynamic parameters (superscripts T and P). Table 1 and Figure 1 give the basic definitions and general form for (1) subject to the condition P = constant. The supersaturation corresponding to the metastable limit is denoted by ΔN_{mS}.

Figure 2 shows the phase diagram for a metal with carbon, in which allowance is made for metastable solid phases, and one sees here the nominal paths of the processes for carbon concentration change in the solution (for $\Delta T = 0$ these are solid lines, while for $\Delta T > 0$ they are broken lines). For clarity we show also the vectors (concentration gradients) dN/dT (short arrows). The P-T diagram is shown in the top left corner; this combines the equilibrium curve for graphite and diamond and the melting curves for the solvent metal and eutectic. We also show the relative position of the metastability boundary for diamond [3]. The lower P-T point for diamond crystallization on a seed is to be taken as the intersection of the melting line for the eutectic (metastable) with the diamond—graphite equilibrium curve; in the case of spontaneous crystallization, this point will be the intersection of the melting line with the metastability limit. It is readily seen that the diagram is a section of the more general P-T-N supersaturation diagram [2].

Then the reason for crystallization may be the supersaturation arising either on account of solubility differences between metastable graphite and diamond or on account of gradients

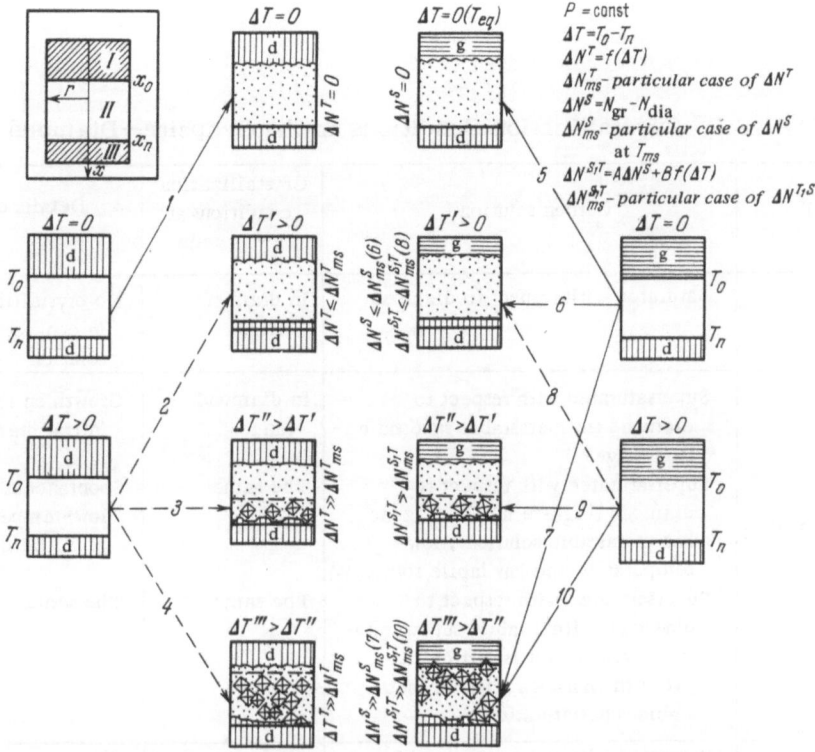

Fig. 1. One-dimensional model for interaction of diamond and graphite with a solvent (see also Table 1): I) carbon source (d, diamond; g, graphite); II) solvent; III) diamond seed (diffusion of dissolved carbon solely in the direction x, r = constant); x_0, dissolution surface; x_n, growth surface.

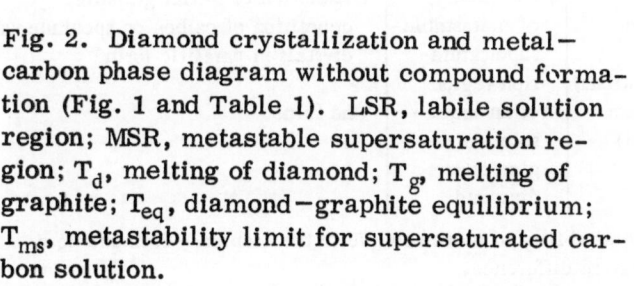

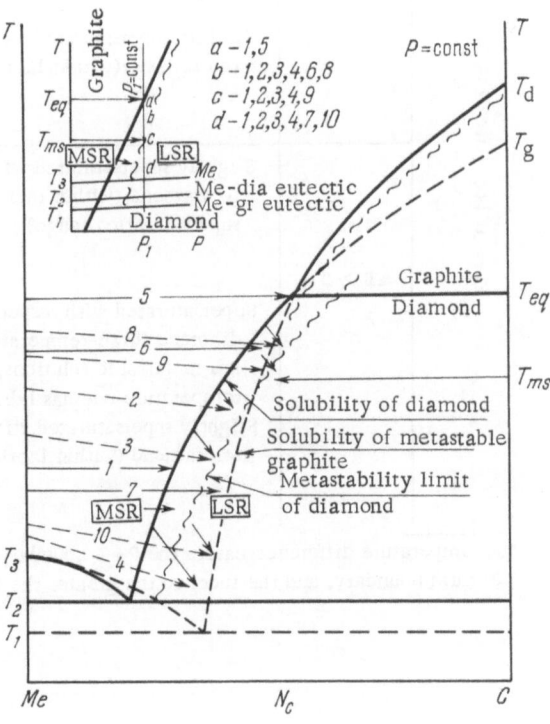

Fig. 2. Diamond crystallization and metal-carbon phase diagram without compound formation (Fig. 1 and Table 1). LSR, labile solution region; MSR, metastable supersaturation region; T_d, melting of diamond; T_g, melting of graphite; T_{eq}, diamond-graphite equilibrium; T_{ms}, metastability limit for supersaturated carbon solution.

TABLE 1. Crystallization Conditions in the Graphite—Diamond System*

Carbon source	ΔT	Carbon solution	Crystallization conditions at P—T point	Details of crystallization and results
Diamond	$\Delta T = 0$	Saturated with respect to diamond	In diamond range	No crystallization: diamond in contact with saturated solution
	$\Delta T > 0$	Supersaturated with respect to diamond (in metastable supersaturation range)	In diamond range	Growth on seed and solution of initial diamond
		Supersaturated with respect to diamond (high-temperature zone has metastable solutions, low-temperature one has labile solutions)	The same	Spontaneous crystallization in the low-temperature zone, dissolution of initial diamond
		Supersaturated with respect to diamond (within labile supersaturations, although it is difficult to avoid the narrow region of metastable supersaturations)	The same	The same
Graphite (metastable)	$\Delta T = 0$	Saturated with respect to graphite and diamond	On graphite—diamond equilibrium curve	No crystallization: graphite and diamond in contact with a saturated carbon solution
		Slightly supersaturated with respect to diamond (within metastable supersaturation range)	Within metastable supersaturation region for diamond	Growth on seed, dissolution of initial graphite
		Slightly supersaturated with respect to diamond (within labile range)	Within labile-solution region for diamond	Spontaneous crystallization on seed, dissolution of initial graphite near seed
	$\Delta T > 0$	Slightly supersaturated with respect to diamond (within metastable supersaturation range)	Within metastable supersaturation region for diamond	Growth of seed and dissolution of initial graphite
		Supersaturated with respect to diamond (high-temperature zone has metastable solutions, low-temperature one has labile solutions)	At boundary of metastable-supersaturation region	Dissolution of initial graphite, deposition of carbon on spontaneous crystals in parasitic form
		Slightly supersaturated with respect to diamond (within labile range)	Within labile-solution region for diamond	The same

*A temperature difference causes the P—T boundary for metastability to move relative to the graphite—diamond equilibrium boundary, and the more so the greater the temperature difference.

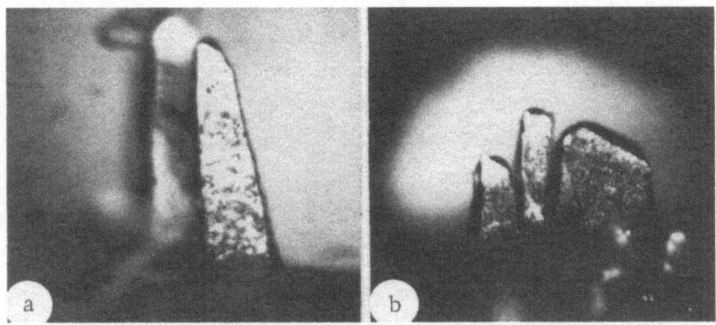

Fig. 3. Diamond needles, a) × 150, b) 100.

in the thermodynamic parameters (this is the only possibility if diamond is the carbon source), or else the simultaneous presence of these two factors. The dissolved carbon diffuses in atomic form, and the supersaturation with respect to diamond determines the kinetic features of the crystallization. If diamond is the carbon source, the supersaturation is a function of the supercooling; if the source is graphite, the supersaturation is a function of the state of the system (isothermal form) plus the supercooling function arising from the temperature gradient. In order to produce nuclei it is necessary for the process to be operated in the region of labile solutions; supersaturation with respect to diamond is not accompanied by spontaneous crystallization in the region of metastable supersaturations.

The crystallization of graphite is in general similar in the region where it is thermodynamically stable (it is symmetrical relative to the graphite–diamond equilibrium curve).

Tests show that single diamond crystals will grow on local areas of a seed crystal (Fig. 3), on individual faces (Fig. 4), and also on certain sets of faces even extending as far as all-around growth (Fig. 5). The form of experiment was as in example 6 in Fig. 1 in determining the growth rate of diamond crystals on [100]: the flat interface between the graphite and metal was set parallel to (100) in the seed (artificial diamond). The temperature difference between the solution and growth surfaces (0.5 mm) did not exceed 15°C; the measurements were made simultaneously at the surface of the graphite and at the lower face of the seed. The temperature in the chamber was 1150°C, and a thermocouple of Pt against Pt–Rh with 10% Rh was used, with corrections for the pressure [4]. The external force on the chamber remained unchanged during the experiment and was deduced from individual calibrations for pressure in each run by reference to standard transitions in Bi and Tl, and also to the Kennedy – Lamory [5]

Fig. 4. Diamond crystals made by growth on
[100] (seed below). a) In 15 min, ×20; b) 25 min,
×20.

Fig. 5. Diamond crystals obtained by all-round
growth. a) Cube, ×20; b) cube-octahedron, ×35.

scale (the initial pressure was 41.5 kbar at room temperature). The pressure was not monitored during the heating and constant-temperature periods.

These results gave the linear size x as a function of time τ (Fig. 6), which can be represented closely by

$$x = -a\tau^2 + b\tau,$$

which is correct for the range $0 \leq \tau \leq 33$ (in min), where a = 0.0003 and b = 0.020. Then the
curve for the normal growth rate plotted as $dx/d\tau$ against τ is linear (Fig. 7) and corresponds to

$$\frac{dx}{d\tau} = -2a\tau + b.$$

At τ = 33 min, one expects the curve for the change in the linear dimensions of the crystal to
reach the limiting value for these working conditions; the curves for x and $dx/d\tau$ against τ

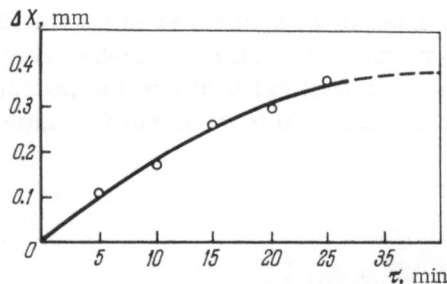

Fig. 6. Growth of a diamond single crystal on
[100] in a nickel−manganese alloy.

Fig. 7. Normal growth rate of diamond on [100].

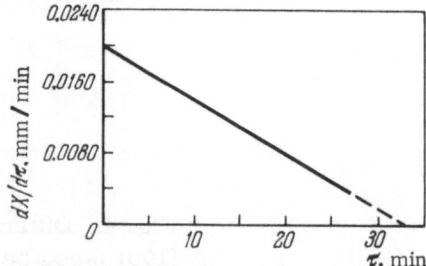

are determined by the continuous unmonitored fall in the supersaturation with respect to diamond on account of the fall in the pressure and the approach to the graphite—diamond equilibrium curve, which denotes reduced supersaturation with respect to diamond. This may be due to local fall in the initial pressure on account of volume reduction in the graphite—diamond phase transition; then the limiting value of x will be the initial point for the linear dependence of the form x = constant, because this transformation should cease when the system attains the graphite—diamond equilibrium curve. If some other factor is operative that is not represented in the phase diagram, the limiting value of x should be the peak on an asymmetric maximum; the resulting sign change in $dx/d\tau$ should denote passage through this turning point, i.e., the onset of dissolution as the system goes from the region where diamond is stable to the region where graphite is stable. One cannot explain the pressure reduction as a redistribution or equalization effect; the available evidence indicates that the high-pressure chamber has a positive pressure gradient from the center to the edge, which means that stress redistribution cannot reduce the pressure.

Literature Cited

1. Yu. A. Litvin, Izv. Akad. Nauk SSSR, ser. neorg. mat., Vol. 4, No. 2 (1968).
2. Yu. A. Litvin, Zap. Vses. Min. Obshch., 98:114 (1969).
3. Yu. A. Litvin, G. N. Bezrukov, and V. P. Butuzov, in: Crystallization Mechanisms and Kinetics [in Russian], Minsk (1969), p. 477.
4. R. E. Hanneman and H. M. Strong, J. Appl. Phys., 37:612 (1966).
5. J. Kennedy and P. Lamory, in: High-Pressure Physics, edited by K. Svenson [Russian translation], Mir, Moscow (1963), p. 313.

KINETIC FEATURES OF DIAMOND CRYSTALLIZATION

G. N. Bezrukov, V. P. Butuzov, K. F. Vorzheikin,
D. F. Korolev, and V. A. Laptev

There are only two papers on the kinetics of crystallization of diamond of value in understanding the growth mechanism; one gives the time dependence of the diffusion of carbon to the seed crystal via molten nickel [1], while the other gives the yield and quality of the crystals in relation to synthesis time [2]. Here we consider our relationships derived from chemical kinetics as regards the graphite—diamond transformation.

The tests were carried out with a high-pressure apparatus by the usual method [3]; the pressure was evaluated from phase transitions in bismuth, thallium, and barium without temperature correction. The correction at 50 kbar is $7 \cdot 10^{-3}$ kbar/°K [1]. The temperature was measured in each run with a platinum—platinum/rhodium thermocouple. The error of measurement in pressure was ±0.5 kbar, while that in temperature was ±20°C.

We performed two series of experiments at identical pressures of 45 kbar but at 1100 and 1200°C [3]; the catalyst was a mixture of nickel with manganese placed between layers of graphite, which provides the best reproducibility in the crystals, although it reduces the yield.

The container material was lithographic stone; various errors arise from penetration of container material into the charge, and there are errors in determining the temperature and pressure, and there are also other sources of error, with the result that there was considerable variation in the yield of diamond crystals; the variation in proportion from run to run was 15-25% for a process duration of 60 sec, but it rose to 50% for runs of 120 sec and more. The mean crystal yield was therefore deduced from the three maximal values from six or seven runs. The results given here therefore are of semiquantitative character.

Table 1 gives the results, which we used in drawing the kinetic curves; we characterize the process via the degree of transformation α, which was defined as the proportion of graphite that had become diamond: $\alpha = 2.25x/3.51w$, where x is the yield of diamond crystals in milligrams for a given process duration and w is the initial weight of gaphite, while the factors 2.25 and 3.51 represent the densities of graphite and diamond. The weight of the initial graphite was determined from the width of the reaction zone in the metal. The ratio of metal to reacting graphite was 1.4:1.

The yield of diamond crystals at 1200° was less than that at 1100°, while α was lower by almost an order of magnitude; the reduced yield at 1200° is due to approach to the thermodynamic equilibrium line at constant pressure, i.e., to approach to the stability range of graphite.

It appears that many polymorphic transitions occur as first-order or monomolecular reactions; this is true of the graphite—diamond transition. For this reason, we tried fitting Mampel's equation (1) and also a first-order equation (2), the two corresponding to different

TABLE 1. Yields of Diamond Crystals

Con-ditions	x, mg	α	t, sec	$K_1 \cdot 10^{-4}$ sec^{-1}	$K_2 \cdot 10^{-4}$ sec^{-1}	Con-ditions	x, mg	α	t, sec	$K_1 \cdot 10^{-4}$ sec^{-1}	$K_2 \cdot 10^{-4}$ sec^{-1}
1100° C, 45 kbar	6	0.01	30	1.3	3.6	1200° C. 45 kbar	1	0,0021	30	0.4	0.3
	7	0.015	60	0.8	2.6		4	0,0085	60	0.5	1.5
	24	0.051	120	1.4	4.4		12	0.025	120	0.7	2.2
	26	0.055	180	1.0	3.2		16	0.034	180	0.6	2.0
	27	0.057	300	0.6	2.0		17	0.036	330	0.4	1.3

mechanisms for nucleation of the new phase [3, 4]:

$$1 - (1-\alpha)^{1/3} = K_1 t, \tag{1}$$

$$-\ln(1-\alpha) = K_2 t. \tag{2}$$

Equation (1) describes a topochemical reaction in which there is preferential growth of nuclei; it is assumed that the nucleation rate follows a monomolecular law, while the interphase boundary propagates with the constant linear velocity.

The value of K_1 calculated from (1) at 1100°C and 180 sec runs varies from 0.8 to $1.3 \cdot 10^{-4}$ sec^{-1}, which represents satisfactory agreement, and therefore this equation can be used to describe the process. Similarly, K_2 varies from 2.0 to $4.4 \cdot 10^{-4}$ sec^{-1}; variations in K_2 are less those in K_1, so it is better to use the first-order equation; however, the calculations are only semiquantitative, so no final conclusions can be drawn on this topic. This explains why one cannot choose between the kinetic equations, and therefore why it is difficult to consider the processes unambiguously, especially the nucleation and growth stages, on the basis of merely a formal kinetic law. This ambiguity is characteristic of many topochemical reactions [3, 4], but there is a certain meaning in the equal applicability of the two equations: both involve the assumption that the nucleation rate follows a monomolecular law.

Figure 1 shows the kinetic curves; curve 1 is for 1100°C and has a short induction period, the yield of crystals increasing appreciably on extending the run time from 30 to 60 sec. The lower inflection lies on the gently sloping part of the curve. The crystal size in the later period varies from 0.03 to 0.1 mm, and it rises to 0.3-0.5 mm when the process lasts 180 sec; at 120 sec there is a clearer inflection, and thereafter the yield of crystals rises slowly. Curve II characterizes the result for 1200°C, and shows no appreciable induction period, while the degree of transformation of this part is lower by almost an order of magnitude, which is due to

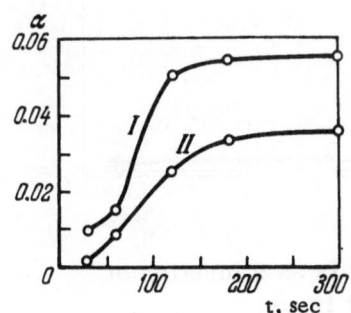

Fig. 1. Time dependence of α (degree of conversion) for the graphite—diamond system. I) 45 kbar and 1100°C, II) 45 kbar and 1200°C.

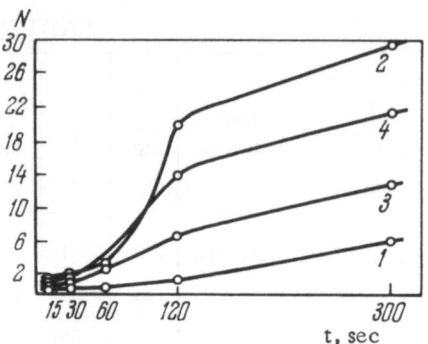

Fig. 2. Numbers of crystals in habit types as functions of synthesis time: 1) Cube with vestigial octahedron faces, 2) cube-octahedron, 3) octahedron with vestigial cube faces, 4) octahedron.

1200°C approaching the line of thermodynamic equilibrium, i.e., the stability region for graphite. Larger crystals are formed during growth on the nuclei (the part of the curve 60-120 sec), which occurs by fusion of small ones, which explains the steep slope of the curve, which indicates a high transformation rate.

We examined how N, the number of crystals of various habit types arising in a given volume, as a function of process duration (Fig. 2), which varied from 10 to 300 sec at 45 kbar and 1200°C; N for the predominant habit type (usually cube-octahedron) showed a considerable rise in the range 12-50 sec; the subsequent behavior is roughly the same for the various habit types (Fig. 3). The linear growth rate increases sharply during the first 20 sec and then falls sharply, and subsequently the crystals grow with a relatively constant rate of about 0.15 mm/min. The figure for 180 sec or more is even less, and there is virtually no growth after 20 min.

Diamond crystals arise in the presence of manganese and nickel via the carbides, which are produced as intermediate metastable compounds in solution in the metal; this is clear from the scope for making metastable carbides of iron, manganese, and nickel in this synthesis [6], together with the production of diamonds by decomposition of these carbides, and also the presence of these materials as oriented inclusions in the crystals [7]. Significant results

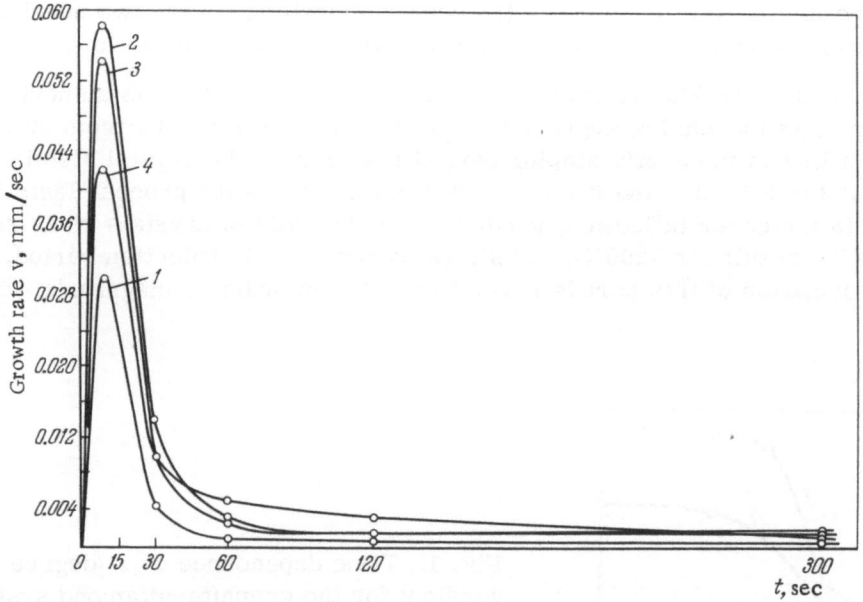

Fig. 3. Linear growth rate of diamond crystals as a function of synthesis time for various habit types.

follow in this respect from diamond crystals made from alloys of titanium or vanadium with copper, silver, or gold [8]; pure titanium, vanadium, or certain other transition metals (W, Mo, Zr, Hf) do not act as independent catalysts up to pressures of 95 kbar and 2500°C, but as alloys with group I metals they catalyze the conversion of graphite to diamond at 60–80 kbar and 1500–2000°C. This can be explained on the assumption that the synthesis occurs via carbides; titanium and vanadium carbides are very stable, and at atmospheric pressure they decompose at about 3200 and 2800°C respectively. The group I metals reduce the thermal stability of the carbides and so give rise to diamonds.

Conclusions

1. One can describe the conversion of graphite to diamond in the presence of nickel or manganese via a first-order equation or via Mampel's equation.

2. The transformation rate varies around 10^{-4} sec^{-1}, and the linear growth rate of the crystals rises sharply during the first 20 sec, then falls to 0.15 mm/min, and thereafter remains relatively constant.

Literature Cited

1. H. M. Strong and R. E. Hanneman, J. Chem. Phys., 46:3668 (1967).
2. V. N. Bakul' et al., Abstracts for the 8th All-Union Conference on Experimental and Technical Mineralogy and Petrography [in Russian], Novosibirsk, p. 209.
3. "Improvements on diamond growth," General Electric Co., Brit. Pat. No. 1,069,801 (January 4, 1963).
4. G. M. Zhalrova and V. A. Gordeeva, in: Kinetics and Catalysis [in Russian], Izd. AN SSSR, Moscow (1960).
5. P. Jacobs and F. Tompkins, in: Solid-State Chemistry [Russian translations], IL (1961), p. 245.
6. A. A. Giardini and J. E. Tydings, Amer. Mineralogist, 47:1393 (1962).
7. S. I. Futergendler and G. V. Khatelishvili, Kristallografiya, 13:120 (1968).
8. S. Takasu et al., Science and Technology of Industrial Diamonds, Oxford (1967).

EQUILIBRIUM IN THE OLIVINE–DIAMOND SYSTEM

V. S. Petrov

Magnesian olivine (forsterite) containing about 10% of the fayalite molecule is the main mineral component of a kimberlite magma. We have examined the thermodynamic stability of diamond in relation to natural olivine at very high P and T., i.e., when diamond is stable and can be made in a reducing medium in the usual way.

We used a platinum heater with natural olivine taken from a diamond-bearing deposit and also synthetic fayalite made in this laboratory. The second component was natural diamond.

The fayalite was made with a reducing atmosphere in a TPK-40/300 oven with an iron crucible at 1170-1200°C; the mixture consisted of quartz sand, iron oxide, and metallic iron in stoichiometric amounts. The reaction is

$$2Fe_2O_3 + 2Fe = 6FeO, \qquad 6FeO + 3SiO_2 = 3Fe_2SiO_4.$$

The product was examined petrographically. We used synthetic fayalite because its melting point is well below that of natural olivine, and hence one can readily produce a liquid in the pressure chamber, whereas natural olivine did not melt under the conditions we used.

In all we performed 24 runs at 55-60 kbar and 1400-1700°C with times from a few minutes to 2 hr. The results were as follows.

Synthetic Fayalite and Small Synthetic Diamonds (pale-yellow euhedral octahedra). In each run we took 250-400 diamonds, which were carefully mixed with the fayalite and placed in the middle of the pressure chamber. There was no visible change in the diamonds up to 1400°C, but between 1400 and 1500°C there was some attack on the edges and corrosion of the faces, while at 1500-1550°C all the diamonds were consumed in an hour.

Natural Olivine and Natural Diamond. These runs were done near the melting point of the platinum heater (1700-1750°C); the olivine did not melt, but the natural diamonds were always attacked at the edges, while the colorless and transparent ones became smoky.

Natural Olivine and Synthetic Diamond. We used pale-yellow diamonds of the largest size (0.3-0.4 mm); at 60 kbar and 1700-1750°C they were attacked in the same way as natural ones. In certain instances the diamonds penetrated into the wall of the platinum heater. These crystals were unaltered on the platinum side, but they became darker in contact with the olivine, and the edges became rounded. It seems clear that our tests were conducted within the thermodynamic stability range of olivine, since in the reducing platinum wall they were unaffected whereas they were attacked by the oxidizing olivine.

It is considered that the earth's upper mantle (depth 150-200 km) has pressures corresponding to graphite–diamond equilibrium and the melting point of the material exceeds 1590°C [1]. If this is so, diamonds cannot crystallize at these depths in ultrabasic kimberlite magmas.

Conclusions

1. Diamonds in natural olivine and synthetic fayalite are oxidized above 1400°C under conditions where they are thermodynamically stable when they are in reducing conditions.

2. Diamonds cannot exist in an ultrabasic kimberlite magma at the depths where they have been assumed to crystallize (150-200 km); at the temperatures found there, free carbon (diamond) is oxidized by olivine (the main mineral in the magma) to CO_2, as is evident from the juvenile CO_2 in the magma.

3. Experiment does not agree with generally accepted concepts of the natural origin of diamond at depths where the pressure represents graphite—diamond equilibrium.

Natural diamonds crystallize near the earth's surface, in the carbonate rocks of the sediment cover, where diamond-bearing kimberlite bodies are usually found.

Literature Cited

1.   S. I. Subbotin, G. L. Naumchik, and I. Sh. Rakhimova, The Earth's Mantle and Tectogenesis [in Russian], Naukova Dumka, Kiev (1968).

PART II

CRYSTAL GROWTH FROM SOLUTION

ANOMALOUS GROWTH RATES OF POTASSIUM CHLORIDE CRYSTALS FROM AQUEOUS SOLUTION

Yu. O. Punin and T. G. Petrov

One usually judges the effects of solution properties on crystal growth kinetics from the variation in the kinetic coefficient, e.g., from the adsorption of solvent on the crystal faces; but sometimes some features of the thermodynamics of the solution lead to a marked but neglected change in the crystallization driving force, which substantially affects the growth kinetics, as we shall show in relation to the growth of KCl crystals on aqueous solutions.

We have previously shown that small amounts of amines present in reagent-grade KCl have a marked effect on the growth on solution [1], so we used KCl of chemically pure grade that had been purified by zone melting in a graphite crucible under nitrogen with the zone passing at 0.2 cm/min and a total of 14 or 15 zone passes.

We used the purified material to examine the growth kinetics for KCl crystals in aqueous solutions with face convection; we used the microsystem described in [2], and in all we used 31 solutions of various concentrations with saturation temperatures from 7 to 86°C. At each saturation temperature we measured the growth rates of three seeds at seven supercoolings between 0.2 and 2.0°. The supercooling was determined to 0.75°, while the growth rate was determined within limits of 0.035 to 0.35 μm/min at the various supercoolings; consequently, the error in determining the mean growth rate from three seeds was 0.1-0.5 μm/min. At each saturation temperature we obtained three curves relating the growth rate to the supercooling; the discrepancies between the curves did not exceed the calculated errors, which indicates a small spread in the growth rates of the various seeds. The curves varied greatly in form for the various saturation temperatures: from nearly linear ones to curves with inflections. As we were unable to derive a general relation between the growth rate and the supercooling, we gave the curves no mathematical processing.

The $v = f(\Delta T)$ curves were used to construct graphs relating the growth rate to temperature at constant supercoolings from 0.2 to 2.0° by intervals of 0.2° (Fig. 1). The points indicate the experimental spread in the values. The curves reveal five clear anomalies, four of which take the form of sharp peaks followed by minima in the range 6-27°C, while the fifth consists of a broad rounded maximum and a following minimum in the range 50-70°C. The magnitudes of these anomalies are much larger than the experimental errors.

Anomalies of this kind occur not only for solutions of carefully purified KCl; Fig. 2 shows curves for technical KCl solutions containing very large amounts of inorganic and organic impurities; although there are few points on these curves, it is clear that anomalies of the above type are present.

Sipyagin [3] has observed similar anomalies for $NaClO_3$ and $KClO_3$ growing from aqueous solutions, and he supposed that these are due to some structural change in the solution; un-

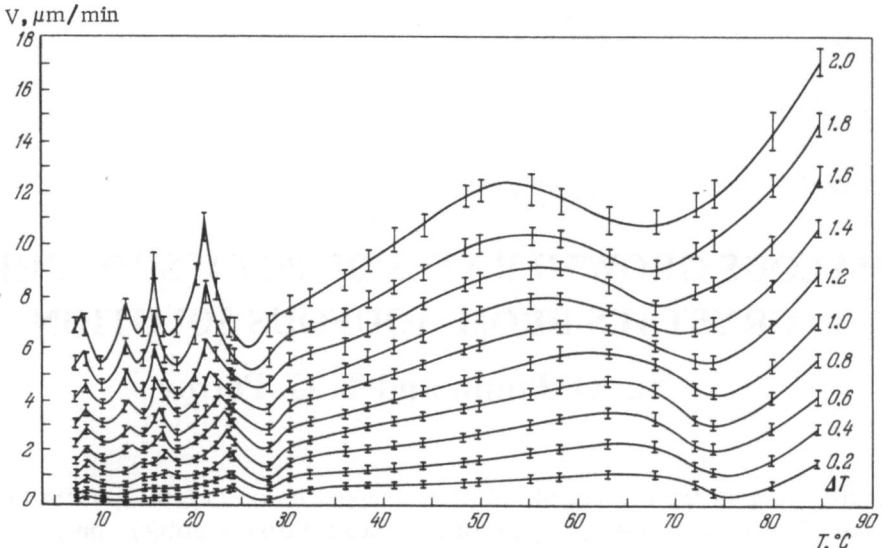

Fig. 1. Growth rate of KCl in pure solution as a function of temperature.

fortunately, there are no marked changes in the properties of $NaClO_3$ and $KClO_3$ solutions in the region of these anomalies, at least, no known ones. As regards KCl solutions, there is a discontinuity in the derivative of the solubility with respect to temperature in the range 22-27°C [4], which in type and position coincides with the largest of the low-temperature anomalies in the growth rate. For concentrated KCl solutions near 60°C there is a round maximum in the speed of ultrasound [5], and also a peak in the partial molar volume [6, 7], these corresponding to the round crest in the growth rates. The growth-rate anomalies therefore reflect anomalies in the thermodynamic properties of the solution, and the round hump anomaly corresponds to a smooth change in properties, while the sharp discontinuity in the growth rate over a narrow temperature range reflects a higher-order phase transition. We assume that the three anomalies below 20°C correspond to three phase transitions in KCl solution that have not yet been detected by other methods; such multiple transitions in condensed systems have been reported elsewhere [8, 9].

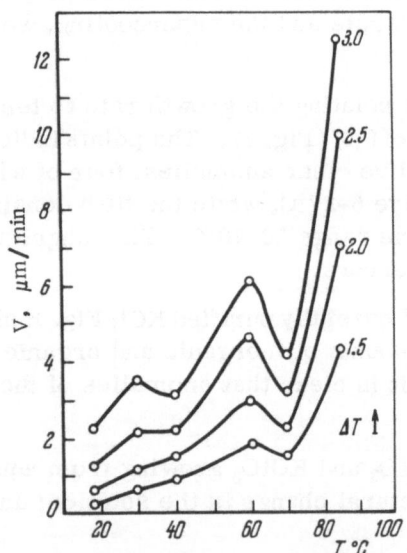

Fig. 2. Growth rate of KCl from technical solutions as a function of temperature.

One can relate the growth anomalies to anomalies in the thermodynamic parameters by taking the growth rate as a function of supercooling; we assume a linear result $v = \beta \Delta \mu_{fa}$ to simplify the treatment, and this is nearly the case in our experiments, because the seeds had very high dislocation densities [10].

The growth rate was measured on the vertical face of the seed near the lower edge; the concentration at this point differs little from that within the bulk of the solution, and so $\Delta \mu_{fa} = \Delta \mu_p$; we expressed the difference of the chemical potentials in the solution in terms of the supercooling, which gives $\Delta \mu_p = (\Delta T/T) \Delta \bar{h}$, where $\Delta \bar{h}$ is the partial molal heat of crystallization as averaged over the temperature range of the supercoolings.

The kinetic coefficient β can be derived from the theory of reaction rates as

$$\beta = \frac{kT}{h} \cdot \frac{1}{\gamma^*} \exp\left(-\frac{F^*}{RT}\right) \frac{a}{RT} .$$

Here a is the activity of the solute, and a/RT may be considered as constant if there is a linear relationship between the solubility and the temperature, as is the case for KCl; F^* is the free energy of activation in the standard state, which is not dependent on the concentration and only slightly and monotonically dependent on the temperature, while γ^* is the activity coefficient of the activated complex, which may be a hydrated molecule adsorbed at a step, and if we assume that there is no interaction between such molecules, we can assume as a first approximation that $\gamma^* = 1$. Then the growth rate is given by

$$v = \frac{k}{h} \cdot \frac{a}{RT} \exp\left(-\frac{F^*}{RT}\right) \overline{\Delta h} \Delta T.$$

The only quantity here that can vary other than monotonically with temperature is $\Delta \bar{h}$.

If the supercooling is low, $\Delta \bar{h}$ equals the partial molal heat of dissolution for the saturated solution; in the range 22-27°C, the anomaly in this quantity is due to the discontinuity in the temperature coefficient of the solubility, which results also in anomalies in the growth rate. We lack data on the heats of solution for KCl solutions at high temperatures, and we calculated the partial molal heat of solution for KBr and NaCl solutions at high temperatures and high concentrations (4 m), the calculation being based on the activity coefficients of [11] and the heats of solution at low concentrations and the partial molal specific heats at infinite dilution [12]. Figure 3 shows the partial heats of solution as functions of temperature; the quantity has a peak at ~60°C for both substances. The thermodynamic properties of KCl are intermediate between those of KBr and NaCl, so there should be an analogous maximum in the partial heats of solution, and this should be so for small supersaturations and hence the maximum growth rate around 60°C.

Fig. 3. Partial molal heats of solution of KBr and NaCl in 4 m solutions as functions of temperature.

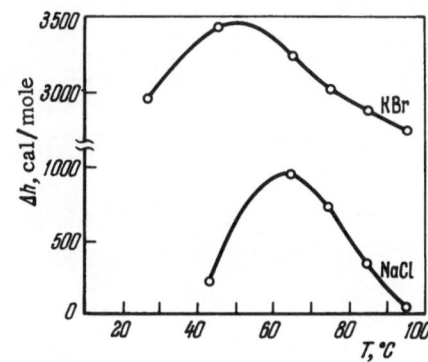

All the anomalies shift to lower temperatures as the supersaturation is increased; the higher the temperature at which an anomaly occurs, the greater this shift. We consider that this effect indicates a shift in the phase-transition point as the supersaturation increases.

Growth-rate anomalies are extremely common; in fact, all three substances for which anomalies have been observed have no exceptional properties and were selected for examination to some extent accidentally. On the other hand, many solutions have anomalies in various thermodynamic and kinetic properties, and it is assumed that higher-order phase transitions occur, so one should find growth-rate anomalies for many substances.

Literature Cited

1. Yu. O. Punin and T. G. Petrov, Kristallografiya, 13:922 (1968).
2. T. G. Petrov, Kristallografiya, 2:777 (1957).
3. V. V. Sipyagin, Kristallografiya, 12:687 (1967).
4. M. I. Shakhparonov, Introduction to the Molecular Theory of Solutions [in Russian], Gostekhizdat, Moscow (1956).
5. I. G. Mikhailov and Yu. P. Syrnikov, Vestnik LGU, ser. fiz. khim., Vol. 13 (1958).
6. L. A. Dunn, Trans. Faraday Soc., 64:2951 (1968).
7. A. J. Ellis, J. Chem. Soc., Series A, 1138 (1968).
8. A. Ubelhode, Melting and Crystal Structure [Russian translation], Mir (1969).
9. S. M. Stishov, Zh. Éksp. Teor. Fiz., 1196 (1967).
10. A. A. Chernov and B. Ya. Lyubov, in: Growth of Crystals, Vol. 5A, Consultants Bureau, New York (1968), p. 7.
11. R. R. Robinson and R. Stokes, Electrolyte Solutions [Russian translation], Mir (1963).
12. K. P. Mishchenko and G. M. Poltoratskii, Thermodynamics and Structure of Aqueous and Nonaqueous Electrolyte Solutions [in Russian], Khimiya, Leningrad (1968).

THE RELATION OF GROWTH CONDITIONS TO SECTOR BOUNDARIES IN KDP CRYSTALS

S. S. Fridman, N. S. Stepanova, and A. V. Belyustin

KDP crystals have been examined in parallel light between crossed polarizers [1]; to extend the research on this topic we also used the methods of pinhole illumination [2], ultramicroscopy [3], and etching [4].

A KDP crystal belongs to the point group $\bar{4}2m$, and it has a prismatic habit, being placed by the {101} bipyramid and the {100} prism. The crystals were grown by the static method of [5] at pH of 4.6-5.0 and temperatures from 30-50°C; the seeds were Z-cut plates 25-35 mm in size. The crystals grew mainly by displacement of the bipyramid faces, but there was sometimes also some growth on the prism faces. The growth conditions determined whether the crystals were obtained with smooth faces having no appreciable taper or with dips on the bipyramid faces and especially the prism ones, as well as some slight taper.

The crystals were cut for examination as blocks faced by prism faces and Z sections; the blocks were polished and were examined along the coordinate axes.

Division of the crystal into sectors (pyramids) can be detected from differences in the sectors, e.g., differences in color [6] or via the boundaries between them. The interfaces between the crystallographically distinct $\langle 100 \rangle$ and $\langle 101 \rangle$ growth pyramids were always visible; sometimes they could be seen even with the unaided eye. When the crystal was viewed along the Z axis between crossed polarizers, the boundaries were seen as sharp lines, and much the same result was given by pinhole illumination and etching (Fig. 1). A layered structure was characteristically seen on viewing from the side. Ultramicroscopy sometimes revealed the small point inclusions at boundaries, and those between the sectors were revealed by the pinhole method along the Z axis and by etching a Z section.

The division of the crystal into $\langle 101 \rangle$ pyramids is clearest between crossed polarizers set at 45° to the X and Y axes. If an edge is displaced strictly in the Z direction as the crystal grows, the boundary is represented by a sharp line. Any lateral shift causes the line to broaden into a band (Fig. 2a). The same is seen on considering the results from the pinhole method: a sharp line corresponds to a sharp trace of an edge, while a broad band corresponds to a diffuse picture (Fig. 2b). The sharper boundaries are seen quite clearly with the polarizers parallel, and they separate regions differing in intensity, as has been found for a number of other crystals. Figure 2b shows the boundary between two opposite (right and left) faces of the bipyramid; it is the trace of an edge perpendicular to the Z axis. Lines of this kind have been seen, as have traces of vertices receiving three or four adjacent faces, on examining crystals by the pinhole method along the X and Y axes. Such a line can bend as the crystal grows on account of displacement of the edge or vertex, and it may disappear and then reappear (Fig. 3).

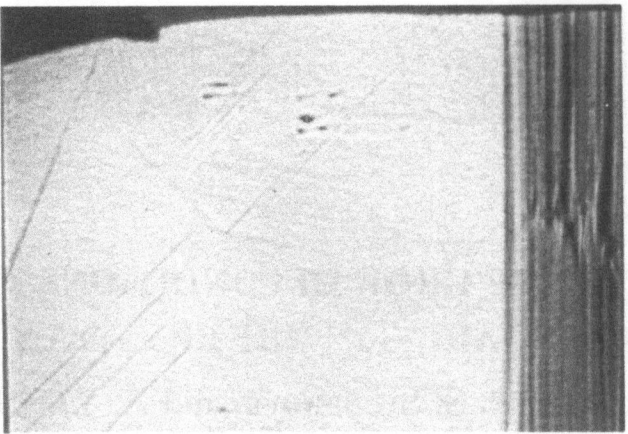

Fig. 1. Growth on (010) face and boundaries between ⟨101⟩ growth pyramids in KDP crystals
(view along Z axis, pinhole method).

The boundaries between ⟨101⟩ sectors may be visible only in part or may be absent; they become less sharp as the temperature is raised, and they are usually absent at 50°C. If such a crystal is cut into parts by height, the region with the boundaries will be of dull appearance, while a region without such boundaries will be uniformly grey.

Very often a crystal will show interfaces within sectors; Fig. 1 illustrates a case where the middle part of the growth on the (100) face has a discontinuity in the bands, which show mutual displacement, which occurs when the (100) face has dips and layers produced on opposite sides of the face that converge towards the middle. An analogous picture is seen on pyramid faces, on which the layers are displaced from the lateral edges towards the middle. The corresponding boundary is seen on viewing along the Z axis (lower part of Fig. 2b). One finds increased stresses where the layers meet, which are seen from the light transmitted between crossed polarizers. The exact intensity in the latter often varies as between the two

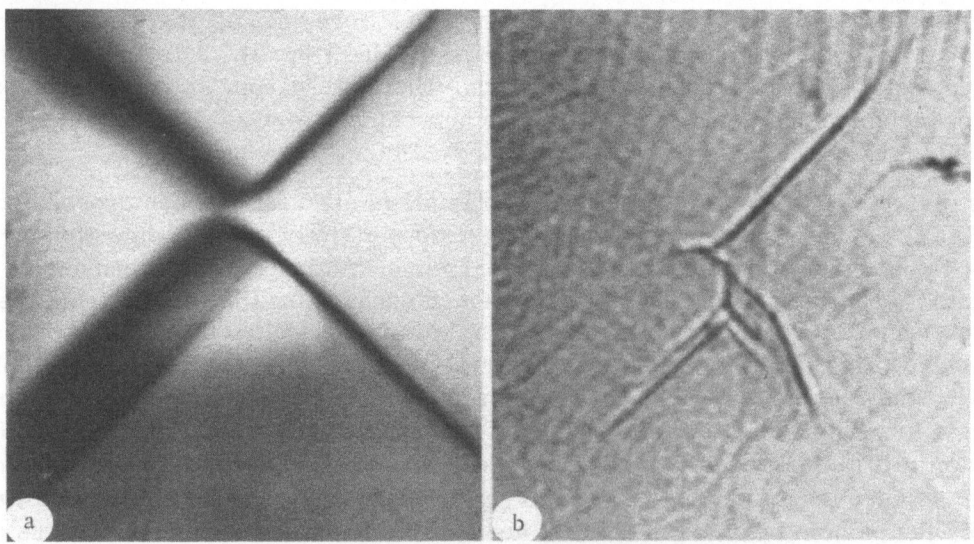

Fig. 2. Boundaries between ⟨101⟩ growth pyramids. a) Crossed
polarizers in diagonal position; b) shadow pattern.

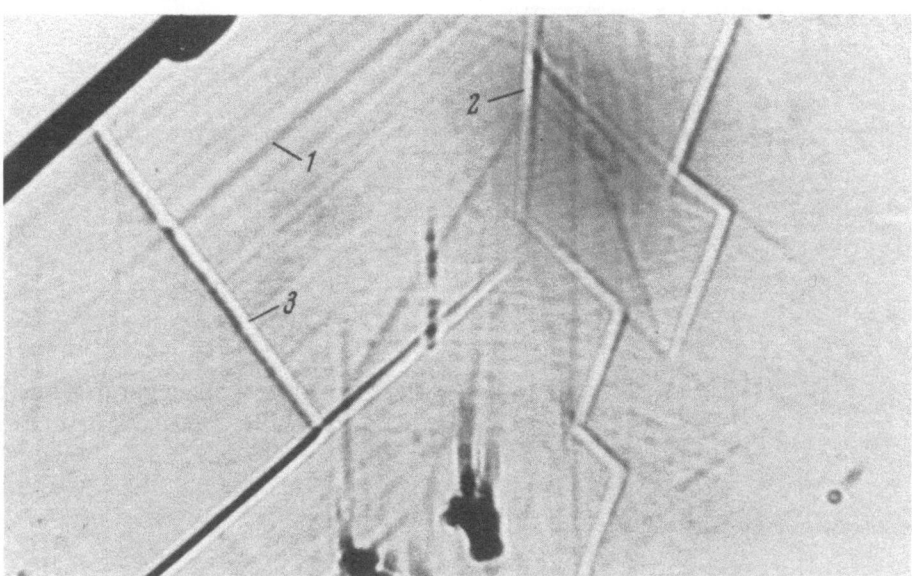

Fig. 3. Growth layers (1), trace of motion of vertex (2), and kinks (3) (view on X axis, pinhole method).

sides of a boundary; often the crystal will crack along such a boundary. Dips occur at lower pH, and they tend to disappear as the concentration is raised; the growth layers become continuous throughout the face, and the corresponding interface boundary vanishes. Lattice disorientation and displacement can occur where the growth layers of different faces meet, which results in elevated impurity and defect concentrations; also, the relationship between the interfaces and the optical anomalies between crossed polarizers indicates a major role for lattice distortions that are not directly related to the entry of impurities. Usually a crystal with $\langle 101 \rangle$ sectors will show a tendency for the plane of the optic axes to be parallel to the $\{100\}$ faces [1, 7]. As a result, the planes of the optic axes in adjacent sectors are mutually perpendicular, and hence also are the directions of elongation of the unit cell. The mutual rotation of the distorted lattices stimulates the formation of boundaries as also sectors within them.

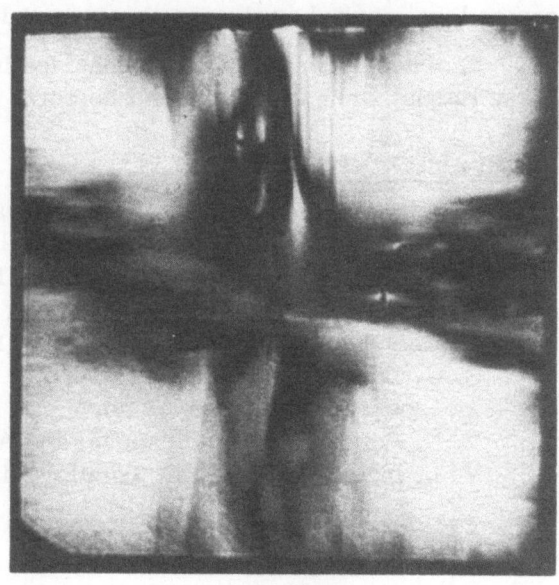

Fig. 4. Crystal with planes emerging from a regeneration pyramid (crossed polarizers parallel position).

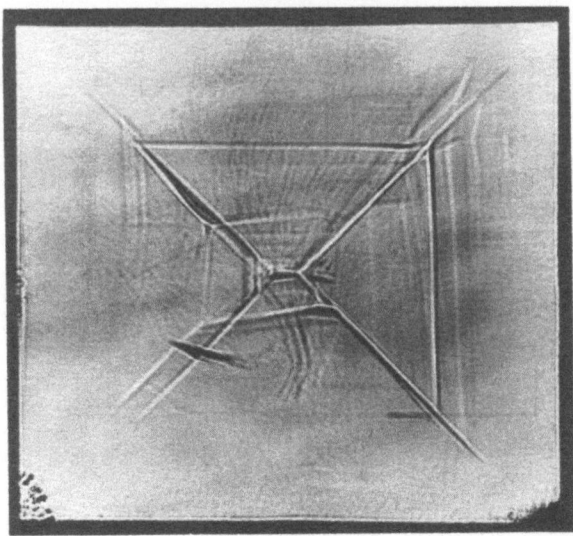

Fig. 5. Blocks in crystal (view on Z axis, pin-hole method), natural size.

Figure 3 shows a further type of boundary within a sector, namely, a fault line. On one side of such a line, the growth layers are stronger than on the other, and they are displaced one relative to another along the Z axis. Frequently, at this point there are inclusions of solution and mechanical impurities, which indicates that the fault line is a boundary between independently growing parts of the crystal.

Sometimes the pinhole method reveals planes emerging from the regeneration pyramid at various angles to the Z axis, which are visible on viewing along the X(Y) and Z axes. Between crossed polarizers, the planes give rise to rosettes of the type described by Indenbom and Tomilovskii [8] (Fig. 4). If the Z axis of the crystal is inclined to the beam, one can observe various degrees of darkening in the region between the planes; some crystals have an appreciable block structure (Fig. 5).

Conclusions

1. Boundaries are formed between and within sectors in KDP crystals on account of independent propagation of growth layers within separate parts of the surface; the tendency becomes less marked as the temperature and pH are raised.

2. The sharpness of the boundaries between growth pyramids is dependent on the extent of the biaxial anomaly in adjacent sectors.

Literature Cited

1.   V. A. Shamburov and A. V. Kucherova, Kristallografiya, 10:658 (1965).
2.   V. S. Doladugina, in: Growth of Crystals, Vol. 3, Consultants Bureau, New York (1962), p. 340.
3.   N. M. Melankholin and I. N. Guseva, Kristallografiya, 8:884 (1963).
4.   V. A. Meleshina, T. F. Chernysheva, and N. V. Russova, Kristallografiya, 12:523 (1967).
5.   A. V. Belyustin and N. S. Stepanova, Kristallografiya, 10:743 (1965).
6.   G. G. Lemmlein, Sector Structure in Crystals [in Russian], Izd. Akad. Nauk SSSR, Moscow and Leningrad (1948).
7.   V. N. Portnov, N. S. Stepanova, and A. V. Belyustin, Kristallografiya, 14:719 (1969).
8.   V. L. Indenbom and G. E. Tomilovskii, Kristallografiya, 2:190 (1957).

THE ADSORPTION LAYER AND SURFACE STRUCTURE IN THE GROWTH OF QUARTZ FROM SOLUTION

E. D. Dukova

It has been shown [1-7] that surface diffusion plays a decisive part in the growth of crystals from solution.

Here we report some experiments on the surface morphology and growth rates of crystals of β-methyl naphthalene in solution in alcohol, and these observations demonstrate the growth mechanism and the state of the adsorbed layer at temperatures approaching the melting point.

Growth Rates of Crystal Faces

Face growth kinetics may be examined in a flow of supersaturated solution when the growth mechanism is determined by surface processes [5]. The crystals took the form of thin plates and were attached to thermocouple wires to measure the temperature. The growth rate was recorded as a function of supersaturation for the side faces at temperatures of 9, 11, 12, 14.4, 16.4, 19.9, and 24.9°C (Fig. 1). The concentration C_e of the saturated solution was 15, 17.2, 18, 20, 22.5, 29, and 48%, respectively, in these cases. The melting point of β-methyl naphthalene crystals is 37°C.

A photometer system [5, 9] was used to record the shapes of the side faces at various temperatures at the point where the (110) face form (Fig. 2a) passed into a rounded one (Fig. 2c). The steps were best demonstrated by recording under fixed conditions with the lowest possible supersaturation ($\Delta c \approx 0.15\,g/100\,cm^3$), measurements being made to one unit-cell parameter.

The individual steps can be seen from the profile at 9 and 11°C (I in Fig. 3); above this temperature, the steps become smaller, but the profile reveals areas characteristic of kinematic waves (II in Fig. 3). There was no point in recording the profile above 18°C, because the steps had vanished and it was as though the surface was molten. While the steps were visible, we determined the mean light at each temperature (Fig. 4a); it is clear that this decreases as the temperature rises.

The inclination of the side faces to the plane of the plate can [9, 10] be determined approximately; Fig. 4b shows the temperature dependence of this inclination for a given supersaturation. At 9°C, the crystallographic angle is 81°, while at higher temperatures this falls rapidly, and it becomes approximately 2° at 18.5°C. This angle is close to the angle formed for ordinary kinematic waves in relation to a close-packed surface.

A previous study [9, 10] dealt with the stepped relief of the surface of β-methyl naphthalene crystals growing from a solution at a fixed temperature of about 13°C with various supersaturations; it was found that increased supersaturation breaks up the true steps into kinematic

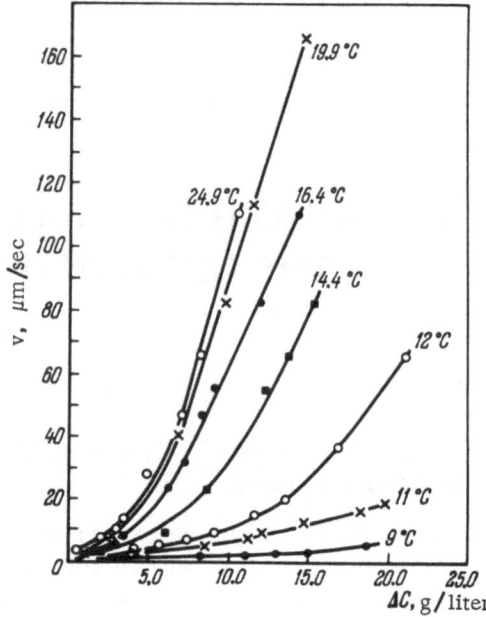

Fig. 1. Growth rate v of a side face of a crystal as a function of supersaturation ΔC at various temperatures.

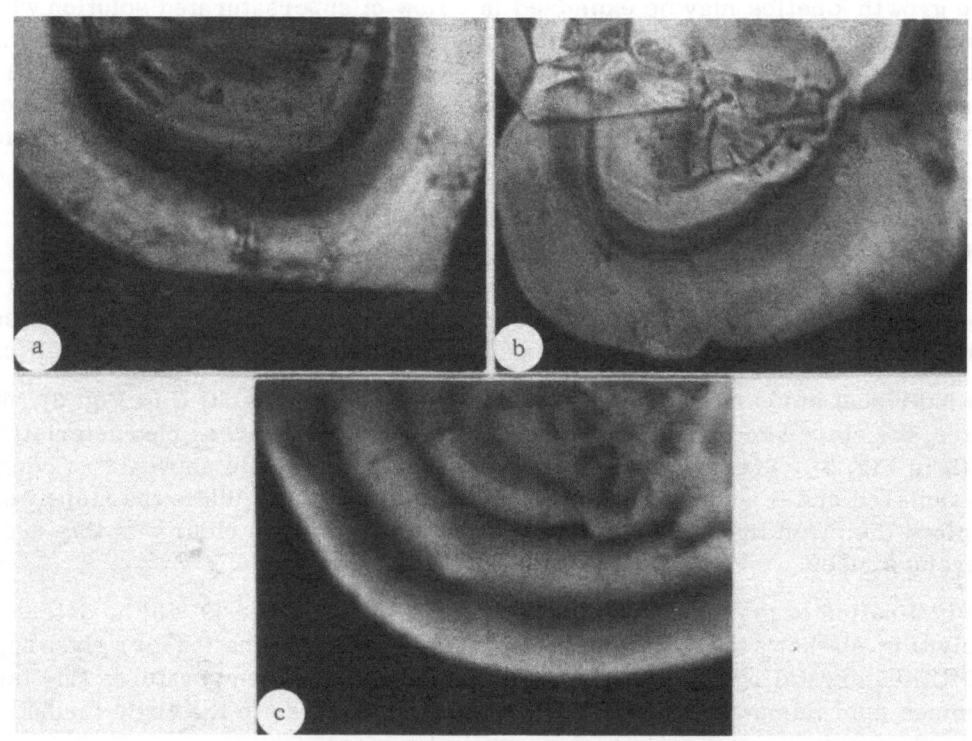

Fig. 2. Crystal shapes at various temperatures, polarized light, ×100. a) 9; b) 14.4; c) 19.9°C.

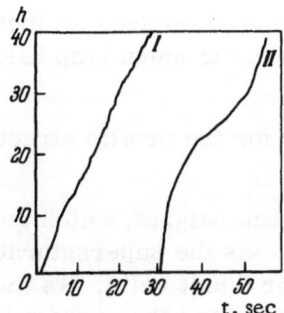

Fig. 3. Shapes of crystal sides at (I) 11°C and (II) 16°C. The ordinate is the step height h in unit-cell parameters. $\Delta C \approx 1.5$ g/liter.

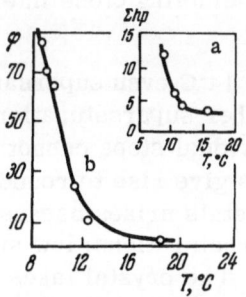

Fig. 4. Temperature dependence of (a) mean step height Σhp and (b) inclination φ of side face.

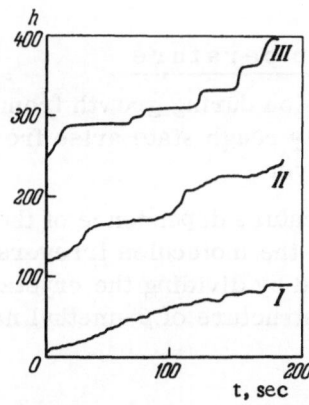

Fig. 5. Shapes of crystal sides for various supercoolings ΔT at 13°C: I) $\Delta T \approx 0.1$; II) $\Delta T \approx 0.3$; III) $\Delta T \approx 0.5$°C.

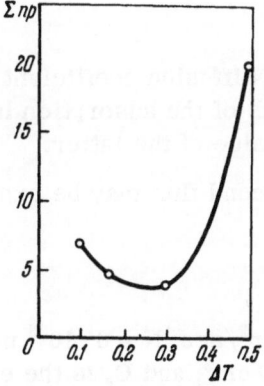

Fig. 6. Mean step height as a function of supercooling at 13°C.

waves (I and II in Fig. 5). At even higher supersaturations, one again sees an increase in the density of the macrosteps (III, Fig. 5). There is a minimum in the mean step height as a function of supersaturation (Fig. 6).

These various experiments indicate the following state for the profile structure on the sides at temperatures up to 14°C.

At low supersaturations, there is a set of steps of various heights, which tend to be fairly high (mean height 10-15 unit cells, Fig. 4a and Fig. 6). As the supersaturation increases, the steps split up into smaller ones, with mean heights of 2 or 3 unit cells. As the supersaturation is raised further, the number of high steps again increases, and the surface becomes rougher. If there is any further rise in the supersaturation, the high steps begin to behave as individual platy crystals; the layer contours cease to be parallel, as in Fig. 2, parts a and b. Crystals resembling close intergrowths are obtained [5] if the supersaturation is raised any further.

Above 14°C even supersaturations give rise to steps of mean height only 2 or 3 unit cells (Fig. 6); higher supersaturations cause the steps to break up into a molecularly rough surface, and the individual steps cannot be revealed by photometry. At 16°C and above relatively high growth rates give rise to rounded crystals with rough surfaces (Fig. 2c); the distinctive shape of these crystals arises because the inclination of the side faces is different from that of the lowest supersaturations; low supersaturations give this angle as close to 2° from the photometry data. The crystal takes the form of a very flat shell. At higher supersaturations, the flanks become steeper, and sometimes the measurements can be based on the interference colors. The inclination of a rounded face may attain 20° or more, e.g., in Fig. 2 one sees over a distance of ~2 μm from the edge a thickening of ~1 μm.

Condensation Coefficient as a Function of Temperature

The growth of facetted crystals indicates surface diffusion during growth from solution [5]; the surface structure and the transition to the molecularly rough state arise from changes in the growth mechanism and in the state of the adsorption layer.

One can deduce the growth mechanism from the temperature dependence of the growth rate or of the condensation coefficient, i.e., the proportion of the molecules irreversibly attached to the surface by collision; the number may be deduced by dividing the crystal growth rate v by the volume Ω per molecule as calculated from the structure of β-methyl naphthalene:

$$j_1 = \frac{v}{\Omega}. \tag{1}$$

The following is the number of molecules approaching the surface from the solution:

$$j_2 = \frac{D}{a}(n - n^*). \tag{2}$$

Here D is the diffusion coefficient in the solution, a is a quantity of the order of the molecular size (thickness of the adsorption layer), n is the concentration of the molecules, and n* is the equilibrium value of the latter.

This second flux may be expressed as follows:

$$j_2 = \frac{D}{a}\frac{N}{M}(C - C_e); \tag{3}$$

D $\approx 3 \cdot 10^{-6}$ cm²/sec; N = $6 \cdot 10^{23}$ mole^{-1}; $a \approx 10^{-6}$ cm; M = 142 deg/mole; C is the true concentration in deg/cm³, and C_e is the equilibrium concentration in the same units.

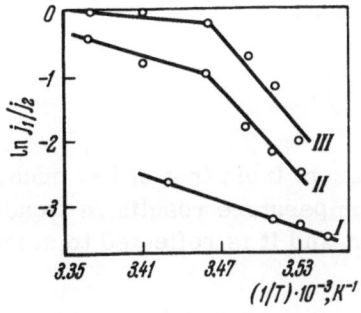

Fig. 7. Temperature dependence of the condensation coefficient for various supersaturations. I) Low; II) medium; III) high.

The condensation coefficient is

$$K = \frac{i_1}{i_2} = \frac{avM}{\Omega DN(\Delta C)}.$$ (4)

We calculated K for all our observed points (Fig. 1), and the numerical value was of the order of 0.01 for the facetted forms (layered growth at ~9°C) but approaches unity for the rounded rough forms (16-24.4°C) at the highest supersaturations.

These values for K indicate that layered growth gives rise to a layer enriched in solute molecules at the surface (only one molecule is attached out of a hundred that strike the surface), i.e., we have a surface adsorption layer. If the surface is molecularly rough, each point acts as a growth site, which explains the significance of the condensation coefficient, which here approaches unity. The results show that K is fairly constant in parts that are close to linear but alters if the supersaturation dependence of the growth rate becomes different. For this reason, it is convenient to give the mean values for low, medium, and high supersaturations for each temperature (Fig. 7).

Growth at Small Supersaturations. If the supersaturation is very low (ΔC ≈ 0.25 g/100 cm³) the curve (Fig. 1) has linear parts that may be respresented as v = βσ, where β is a kinetic coefficient and σ = ΔC is the supersaturation. Substitution here from (4) gives us K for small supersaturations:

$$K = \frac{aM}{\Omega N}\frac{\beta}{D} \approx a\frac{M}{\Omega N}\exp\frac{U_\beta - U_D^*}{kT},$$ (5)

where U_β is the activation energy for the kinetic coefficient, U_D^* being the activation energy for diffusion in the solution. The latter is small under our conditions, only 1-2 kcal/mole, and so the temperature dependence of K is determined by that of β, and under these conditions we obtained photometric evidence on the conversion of the layered structure of a thin crystal into a molecularly rough surface. Figure 7 (curve 1) shows the temperature dependence of K for this case.

The slope of the straight line gives the heat of activation as about 13.7 kcal/mole, which is close to the 14.5 kcal mole calculated from the solubility curve, which means that the adsorbed layer becomes denser as the molecularly rough surface is approached, i.e., it becomes a molecular quasiliquid film [8].

Growth at Medium Supersaturations. At any temperature above 11°C, the initial linear part is replaced by a nonlinear dependence, which may be approximated as follows, where v is growth rate and σ = ΔC:

$t°$, C	v, cm/sec	$t°$, C	v, cm/sec
11	$0.63\,(\sigma)^{1.6}$	16.4	$4.67\,(\sigma)^{1.5}$
12	$1.12\,(\sigma)^{1.5}$	19.9	$7.07\,(\sigma)^{1.5}$
14.4	$2.51\,(\sigma)^{1.5}$	24.9	$8.31\,(\sigma)^{1.5}$

One can assume generally that $v = \beta(\sigma)^{1.5}$ for this part, so we have

$$K = a \sqrt{\sigma} \frac{M}{\Omega N} \exp\left(-\frac{U_\beta - U_D^*}{kT}\right). \tag{6}$$

The crystal morphology at high temperatures is different from that at low ones; in the latter case, the crystals are defective, while raising the temperature results in steadily less defective rounded crystals. This is a complicated process, and it is reflected to some extent in the dependence of $K = j_1/j_2$ on $1/T$ (II in Fig. 7).

The curve has two points which are separated by the value 16°C ($1/T = 3.47$); at low temperatures there is a sharp change in the dependence of $\ln(j_1/j_2)$ on $1/T$ (growth of defective crystals, which we do not consider). Above 16°C, where one gets steadily less defective rounded crystals, the temperature dependence is less pronounced; deactivation energy above 16°C is about 13.2 kcal/mole, which is close to the heat of solution. Here again we get consolidation of the adsorption layer for a quasiliquid film. We can estimate the change in *a* for the two processes with equal activation energies for the kinetic coefficient, since the thickness of the adsorption layer appears in the exponential factor. Then

$$a_2 = \frac{K_2}{K_1} \frac{a_1}{\sqrt{\sigma}}, \qquad \sigma = \Delta C. \tag{7}$$

This applies for 16.4–24.9°C (Fig. 7), and the region of nonlinear growth-rate dependence on supersaturation is small and passes into a linear one as the supersaturation decreases or the temperature is raised; then [7] the adsorption layer becomes thicker. Figure 8 shows the thickening as the critical supersaturation is reduced, which denotes a transition to a linear dependence of the growth rate on the supersaturation for a rough surface. It is estimated that the adsorbed layer at 24.9°C has thickened by roughly a factor of 1000, so it is virtually a film of molten material.

Growth at High Supersaturations. Here the dependence of the growth rate on the supersaturation is linear at 12°C and above, which is characteristic of a molecularly rough surface. The parts of linear character may be represented as

$$v = \beta(\sigma - \sigma_k).$$

The initial supersaturation σ_k tends to a constant value, which indicates the establishment of a stable growth mechanism [5].

The condensation coefficient for high supersaturations is

$$K_3 = a_3 \left(1 - \frac{\sigma_k}{\sigma}\right) \frac{M}{\Omega N} \exp{-\left(\frac{U_\beta - U_D^*}{kT}\right)}. \tag{8}$$

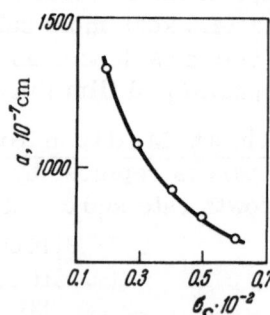

Fig. 8. Thickness *a* of adsorption layer and critical supersaturation σ_c.

The morphology of the growing crystals is even more temperature dependent in this supersaturation range; one goes from crystal intergrowths at low temperatures to crystals with molecularly rough side surfaces at high ones (Fig. 2). Figure 7 shows the temperature dependence of K; up to 16°C, this is pronounced, and one gets defective crystals, while between 16.4 and 24.9°C, where one gets rounded crystals with rough surfaces, the value changes; the heat of activation of this process is about 2.85 kcal/mole, which is very close to the 2.4 kcal/mole for the latent heat of fusion.

The result of (7) shows that the absorption layer has become a film of melt of almost macroscopic thickness, which correspondingly alters the activation energy for the kinetic coefficient, and this approaches the latent heat of fusion. The growth of a molecularly rough surface is from a melt under such conditions, not from a solution.

I am indebted to A. A. Chernov for a discussion of the results.

Literature Cited

1. W. L. McCabe and R. P. Stevens, Chem. Eng. Prog., 47:168 (1951).
2. A. Van Hook, Ind. Eng. Chem., 40:85 (1948).
3. A. W. Huxson and K. L. Knox, Ind. Eng. Chem., 43:144 (1951).
4. I. W. Mullin and I. Carside, Trans. Inst. Chem. Eng., 45:291 (1967).
5. E. D. Dukova and E. V. Gavrilenko, Kristallografiya, 14:856 (1969).
6. R. Bennema, J. Crystal Growth, 1:278 (1967).
7. A. Oberlin and M. Hucher, in: Absorption et Croissance Crystalline, Paris, Centre Nationale de Récherche Scientifique (1965), p. 407.
8. E. D. Dukova and D. Nenov, Kristallografiya, 14:106 (1969).
9. G. R. Bartini, E. D. Dukova, I. P. Korshunov, and A. A. Chernov, Kristallografiya, 8:758 (1963).
10. E. D. Dukova and A. A. Chernov, Kristallografiya, 8:765 (1963).

PART III

CRYSTAL GROWTH FROM MOLTEN SOLUTIONS

GROWTH OF SCHEELITE-GROUP TUNGSTATE AND MOLYBDATE SINGLE CRYSTALS FROM MOLTEN SOLUTIONS

L. I. Potkin

There is much interest in quantum electronics in tetragonal crystals with the structure of calcium tungstate when the dope is a transition metal; to extend the wavelength range of masers and to increase the performance, one needs to examine homologs of scheelite containing various dope materials. Here we present results on the growth of single crystals of the tungstates and molybdates of calcium, strontium, and barium containing various paramagnetic ions.

Single crystals of tungstates and molybdates of the scheelite group can be grown from pure melts (essentially via Czochralski's method) and also from solutions (under hydrothermal conditions) and also from solution in melt [1-3]. Melts have advantages such as high growth rate and scope for producing large crystals, but they represent considerable technical difficulties on account of the need to maintain high temperatures; certain substances also evolve additional difficulties due to volatility near the melting point or else incongruent melting. Also crystals grown from melts as a rule have various undesirable defects such as a block structure nonstoichiometric composition, etc.

There are also considerable technical difficulties in growing single crystals of tungstates and molybdates under hydrothermal conditions; one needs a high-pressure apparatus lined with a thermally stable and uncorroded material.

There is therefore considerable interest in making these crystals from solutions in melts at atmospheric pressure and comparatively low temperatures; it is now generally accepted that one can use crystallization from solution for this purpose, and the methods include those of temperature difference, i.e., recrystallization in a long vessel with a certain vertical temperature gradient, which causes transport of the initial material from the hot end (where it dissolves) to the cold one (where it crystallizes). This temperature-difference method provides for continuous crystallization at a constant temperature, and this involves minimal change in growth conditions during the production cycle, which facilitates production of large perfect crystals. Also, one does not need such careful temperature control as in other methods. For these reasons, we used the temperature-difference method to grow single crystals of tungstates and molybdates.

We chose the solvent on the basis of [3] and also the need to use low temperatures; one can vary the concentration of the solute in a solution within certain limits, and also the temperature coefficient of the solubility, in order to obtain the best performance. The components of the mixture are chosen so as to form a simple eutectic, which makes it possible to perform the operation at temperatures below the melting points of any of the components, which simplifies the apparatus and the control, while making the solvent less corrosive and thus expanding

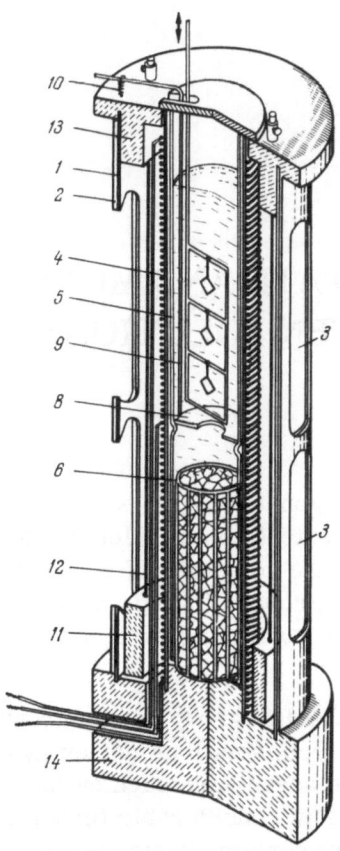

Fig. 1. Crystallization apparatus for growing single crystals from solution in melts by the temperature-gradient method.

the choice of vessel materials. In the case of these calcium, strontium, and barium tungstates and molybdates, a suitable solvent is a eutectic mixture of the chlorides of potassium and lithium, which melts at about 360°C; the solubilities of the above compounds in this LiCl + KCl mixture at 500°C are about 4 wt.% for $CaWO_4$ and 10 wt.% for $BaMoO_4$. The solubilities of the other tungstates and molybdates of Ca, Sr, and Ba lie between these values. We observed no double-decomposition reactions between the solute and solvent in the range 400-600°C. This LiCl + KCl eutectic is not very corrosive at temperatures up to 600°C, so it is possible to use silica vessels, which allows one to observe the crystal growth. Figure 1 shows the apparatus. The vessel 5 made of silica is placed in a two-zone resistance furnace consisting of the tube 4 made of pyrex glass wound with two nichrome heaters, together with the insulating jacket

Fig. 2. Characteristic shapes of crystals of tungstates and molybdates of Ca, Sr, and Ba grown from solution in LiCl and KCl.

represented by the stainless-steel tube 1, which is wrapped in asbestos tape 2, with top closure 13 and base 14, which are made from light firebrick. The same material is used for the ring 11, which can be moved vertically by the rods 12 to produce in the dissolution zone a certain temperature gradient sufficient to produce uniform dissolution of the mixture throughout the volume, which is very important in providing stable growth conditions. The insulating jacket has windows for viewing the dissolution of the initial material and the growth of the crystals. The initial material is prepared by fusing the powdered reagent in an oxyhydrogen flame or by sintering the material in a holder 6 also made of silica, which is placed in the lower part of the crystallization vessel 5. The stop 8 is placed in the middle of the vessel between the upper and lower heaters, which is intended to provide the best balance between convected motion and temperature difference. This stop is attached to the rod 9 and is pressed by spring 10 onto a ring projection on the crystallization vessel 5. The seeds are supported on a silica frame. A rod with its end brought out allows the frame to be turned or moved vertically; we used a reciprocating vertical motion. Most of the experiments were done with a temperature of 500°C in the crystallization zone and a temperature difference of 7-10°C; the growth rate increases in the same direction as the solubility in the sequence Ca, Sr, Ba, and the values for the tungstates are less than for the molybdates. The growth rate for $CaWO_4$ was 0.4 mm/day, while that for $BaMoO_4$ was 2-3 times larger.

The paramagnetic ions were introduced into the solution as the chlorides; an attempt to use the oxides directly in the initial material was unsuccessful, because the oxide of the rare-earth elements have very low solubilities in LiCl + KCl up to 600°C. We obtained the following single crystals: $CaWO_4 : Nd^{3+}$, $CaWO_4 : Gd^{3+}$, $CaWO_4 : Ho^{3+}$, $CaMoO_4 : Nd^{3+}$, $CaMoO_4 : Mn^{2+}$, $CaMoO_4 : Yb^{3+}$, $CaMoO_4 : Gd^{3+}$, $CaMoO_4 : Ce^{3+}$, $SrWO_4 : Er^{3+}$, $SrWO_4 : Nd^{3+}$, $SrMoO_4 : Gd^{3+}$, $SrMoO_4 : Mn^{2+}$, $SrMoO_4 : Er^{3+}$, $BaMoO_4 : Mn^{2+}$, $BaMoO_4 : Gd^{3+}$, $BaMoO_4 : Nd^{3+}$, $BaMoO_4 : Yb^{3+}$, $BaMoO_4 : Ho^{3+}$, $BaMoO_4 : Ce^{3+}$, $BaWO_4 : Nd^{3+}$, $BaWO_4 : Gd^{3+}$, $BaWO_4 : Yb^{3+}$.

These crystals were 5-10 mm in size, which was quite sufficient for examination of their paramagnetic behavior. The habit of the crystals was the usual one for this group of substances, namely, dipyramidal. The main faces were provided by the (011) dipyramid (Fig. 2).

At low supersaturations, one also found (013) and (014) dipyramids; the exact habit of the crystals was also dependent on the dope. For instance, $BaMoO_4 : Nb$ crystals had only (013) dipyramids, while $BaMoO_4 : Yb$ crystals had (011) and (013) dipyramids.

The physical studies have been made on these single crystals by ESR at the All-Union Mineral Raw Material Synthesis Research Institute and at Kazan University.

Literature Cited

1.   L. G. Van-Vitert, J. J. Rubin, and W. A. Bonner, J. Am. Ceram. Soc., 46:512 (1963).
2.   L. N. Dem'yanets, in: Hydrothermal Synthesis of Crystals, Consultants Bureau, New York (1971), p. 65.
3.   I. N. Anikin, in: Growth of Crystals, Vol. 1, Consultants Bureau, New York (1959), p. 259.

GROWTH OF SINGLE CRYSTALS OF β-EUCRYPTITE AND β-SPODUMENE

V. A. Ioffe and Z. N. Zonn

Lithium aluminosilicates have become important because of their unusual thermal parameters, which are due to structural features. High-temperature eucryptite or spodumene has a three-dimensional silicon−oxygen framework with the Li ions in the holes [1, 2]; β-eucryptite and β-spodumene are also the basic crystalline phases in Li-aluminosilicate glass sinters.

The Li aluminosilicates in the $Li_2O-Al_2O_3-SiO_2$ system were examined by making single crystals of β-eucryptite and β-spodumene incorporating various paramagnetic ions.

The position of the Li was examined in this laboratory by NMR, while the paramagnetic ions were examined by ESR [3]. The exact amounts of dope were monitored by spectral analysis (M. P. Semov).

The single crystals were made by crystallization from fluxes in platinum crucibles by programmed temperature reduction in furnaces with silite and platinum heaters. The $LiAlSiO_4$ was crystallized from a mixture of eucryptite composition with an excess of LiF by Winkler's method [4]. The vertical temperature difference ΔT in the melt was 15-20°C.

The faces of the crystals were affected by the temperatures used with a fixed temperature difference. Cooling at 3 deg/hr from 1200 to 1100°C was followed by 24 hr at the lower temperature and then cooling to 700°C; this led to expansion of the ab plane and considerable height reduction in the hexagonal c pyramid, with the vertex becoming blunted.

At 5 deg/hr between 1200 and 700°C, we obtained single crystals with hexagonal-pyramid and bipyramid faces, the c axis being prominent.

The crystals formed several layers near the surface. The top surface was a mosaic pattern of crystals with their vertices turned inwards; the crystals in a clump were not intergrown but were cemented together by thin layers of glassy material of eucryptite composition. The crystals had mainly a hexagonal-prism habit in the upper layers, whereas the lower ones had more perfect single crystals with hexagonal pyramid and bipyramid faces, as well as composite crystals with parallel intergrowth along the c axis. The crystals are isostructural with β-quartz, and they were 6-10 mm long along the c axis and weighed 0.6-0.8 g. There was a very reproducible 90-95% yield of β-eucryptite single crystals.

When seeds were used, we were able to make larger single crystals (up to 15 mm long along the c axis). The weight of the largest crystal was 1.66 g.

X-ray and chemical analyses were accompanied by measurement of the density, refractive indices, and microhardness, all of which agreed well with published values [1, 4, 5].

Fig. 1. Single crystals of β-spodumene, $\times$ 5

Single crystals of β-spodumene were grown from a $LiAlSi_2O_6$ melt in $LiVO_3$ [7].

The Li vanadate was made by fusing Li_2CO_3 and V_2O_5 of AR and reagent grades, with subsequent crystallization of the $LiVO_3$ melt; the composition was checked by x-ray and chemical methods.

We found that $LiVO_3$ dissolves 19 wt.% $LiAlSi_2O_6$ at 1220°C. The best ratio between the compound and the flux was 1:10, and crystallization then began at 1190–1195°C. The cooling rate was 3–5 deg/hr from 1220 to 800°C with ΔT of 15–20°C.

We obtained transparent β-spodumene single crystals with regular octahedron faces; the size along a major diagonal was up to 6 mm (Fig. 1).

Fig. 2. Single crystals of γ-spodumene, $\times$ 5

X-ray and chemical analyses were accompanied by measurement of the density, refractive indices, and microhardness, all of which agreed well with published values [1, 6, 9].

We examined the effects of the temperatures on the size and perfection of the β-spodumene single crystals; we used various ΔT and cooling rates.

At 1 deg/hr from 1220 to 700°C with ΔT of 25-30°C we obtained single crystals of γ-spodumene, which were transparent and had good hexagonal pyramid and prism faces. The material is isostructural with β-eucryptite and hence β-quartz. The yield was 80%, with sizes up to 5 mm along the c axis (Fig. 2).

This is the first time that γ-spodumene single crystals of this size have been made by the flux method. The analysis was SiO_2, 61.99; Al_2O_3, 29.60; Li_2O, 8.51, so there was a certain deficiency in silica, as has been reported [7] for small crystals made from spodumene glass.

Our x-ray results agreed with those of Li [8] for γ-spodumene from glass; d = 2.39 g/cm^3, γ 1.527, α 1.521, mp 1510°C.

Differential thermal analysis between room temperature and 1100°C revealed no phase transitions in the γ-spodumene, and this was confirmed by light microscopy up to the melting point.

We were able to make single crystals of β-eucryptite containing Fe^{3+}, Co^{2+}, Ni^{2+}, Mn^{2+}, Cr^{3+}, and Ta, as well as ones of β- and γ-spodumenes containing Fe^{3+}. It was found that there was restricted entry of the dope into these crystals; for instance, spectral analysis indicated that not more than 0.5% Fe^{3+} enters β-eucryptite.

Also, β- and γ-spodumenes accepted only 0.01 wt.% Fe^{3+}; dope in excess of 1% in the initial mixtures for all three crystals led to formation of a second phase involving the dope components.

The β- and γ-spodumene crystals were grown in Li vanadate, and spectral analysis showed that the β-spodumene contained 0.2-0.3 wt.% vanadium (0.4-0.5 wt.% for γ-spodumene). The difficulty in adding dope may have been due to the presence of much vanadium.

Literature Cited

1. R. Roy, D. Roy, and E. F. Osborn, J. Amer. Ceram. Soc., 33:152 (1950).
2. R. Roy and E. F. Osborn, J. Amer. Chem. Soc., 71:2036 (1949).
3. V. A. Ioffe and L. V. Dmitrieva, Abstracts for the Fedorov Jubilee Session [in Russian], Leningrad (1969), p. 81.
4. H. G. F. Winkler, Acta Crystallogr., 1:28 (1948).
5. R. Roy, Z. Kristallogr., 111:185 (1959).
6. W. E. Kramer, J. Amer. Ceram. Soc., 49:5 (1966).
7. N. A. Toropov, Structural Transitions in Glasses at Elevated Temperatures [in Russian], Nauka, Moscow (1966).
8. Chi-Tang-Li, Z. Kristallogr., 127:327 (1968).
9. Chi-Tang-Li and D. R. Reacor, Z. Kristallogr., 126:46 (1968).

PREPARATION AND PROPERTIES OF CRYSTALS OF
THE $LiLn(WO_4)_2$ DOUBLE TUNGSTATES FOR
Ln = RARE EARTH, Y, or Fe

P. V. Klevtsov, C. P. Kozeeva, R. F. Klevtsova,
and N. A. Novgorodtseva

Much attention has been given in recent years to the production of double tungstates and molybdates of the alkali metals and rare earths or yttrium and other trivalent metals, on account of the possible technical uses, especially in quantum electronics. Similar compounds containing iron may be of interest on account of their magnetic properties.

We have made detailed studies of the sodium−rare earth tungstates and molybdates that crystallize with a scheelite structure; some single crystals were grown from solution in a melt and from melts by Czochralski's method. With $NaGd(WO_4)_2$ and $NaLa(MoO_4)_2$ single crystals activated by Nd^{3+} we obtained induced emission [1, 2], but the class of compounds as a whole has been little studied as regards chemical crystallography or physical properties, and not much is known about possible practical uses. Many of the compounds have not even been made, especially not in the single crystal form.

Proper choice of the best method for producing single crystals involves a knowledge of the physicochemical properties and crystallographic characteristics, as well as the solubility, melting point, type of melting, former crystal structure, any high-temperature polymorphic transitions, etc. We have examined the crystallization of these compounds from solutions in melts in order to obtain such data for the double tungstates of lithium and the rare earths, and also yttrium and Fe(III).

The solvent was Li_2WO_4 or $Li_2W_2O_7$, because the melting points of these are 730 and 740°C, respectively, and they introduce no additional ions into the system. The initial mixture for the double tungstates was a mechanical mixture of WO_3 with the oxide of the trivalent metal Ln_2O_3 together with Li_2WO_4 in the stoichiometric proportions: $Ln_2O_3 + 3WO_3 + Li_2WO_4$.

The crystals were made by spontaneous crystallization on slowly reducing the temperature of the platinum crucible, which had a capacity of 50-100 cm^3; the uniform mixture of the charge and solvent was heated in the crucible to 1100-1200°C, and it was kept in the molten state for several hours before cooling at 3 deg/hr to 700 or 600°C. The crystals were extracted from the product by boiling in water.

In this way we made the double lithium tungstates of iron, yttrium, and all the rare-earth elements except cerium. The crystals were up to 5 mm in size.

The crystals of $LiLn(WO_4)_2$ for the rare earths form two morphologically distinct types; from lanthanum to gadolinium the crystals take the form of tetragonal dipyramids, while for

107

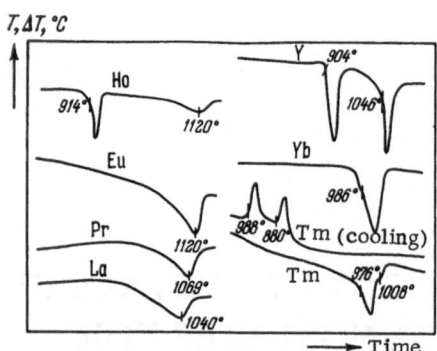

Fig. 1. Differential thermal analysis (dta) curves for LiLn(WO$_4$)$_2$ with Ln = La, Pr, Eu, Ho, Tm, Yb, and Y.

the heavy rare-earth elements from erbium to lutetium and also for yttrium they take the form of monoclinic plates. We obtained both modifications in the case of the middle elements terbium, dysprosium, and holmium. The yield of the monoclinic modification increased rapidly with the atomic number of the element. In the case of terbium, the number of platy monoclinic crystals was very small, while for holmium they formed about half of the total.

The reasons for this behavior were examined via x-ray and differential thermal analysis (dta) studies on the materials.

We found that the dipyramidal form corresponds to the scheelite structure; the platy crystals containing the elements from Tb to Lu and Y have not previously been made; they are isostructural and monoclinic, their structure being related to that of wolframite (Fe, Mn)WO$_4$, but a difference from the latter is that the unit cell of LiLn(WO$_4$)$_2$ has a doubled a parameter. For instance, LiYb(WO$_4$)$_2$ has $a = 9.89$; $b = 5.77$; $c = 4.98$ Å; and $\beta = 93.4°$, space group P2/n. For comparison, MnWO$_4$ has $a = 4.83$; $b = 5.76$; $c = 5.00$ Å, $\beta = 91.16°$; space group P2/c [3]. The monoclinic LiLn(WO$_4$)$_2$ crystals have the Li$^+$ and Ln^{3+} cations in crystallographically independent positions. These differences between the monoclinic structures of the double tungstates and of wolframite itself mean that the powder patterns are rather dissimilar.

The dta studies revealed no special features for the double tungstates with the scheelite structure, whereas the monoclinic crystals containing yttrium and the rare earths from terbium to thulium undergo a polymorphic transition on heating (Fig. 1), the temperature range between the points of polymorphic transition and melting falling rapidly as the atomic number of the element increases (Table 1). This means that the last element (lutetium) has its double tungstate LiLu(WO$_4$)$_2$ without a polymorphic transition, while LiYb(WO$_4$)$_2$ has its transition directly before the melting point.

We made high-temperature x-ray diffraction studies with a URS-50U diffractometer having a KRV-1200 high-temperature attachment, which showed that the polymorphic transition is related to transition from the wolframite structure type to the scheelite one, which oc-

TABLE 1. Polymorphic
Transition Temperatures
for LiLn(WO$_4$)$_2$ Single
Crystals

Ln	T_{pt}, °C	$T_{mp} - T_{pt}$, °C
Tb	774	346
Dy	840	268
Ho	882	204
Er	938	131
Tu	976	32

curs readily, but the reverse process requires considerable supercooling, which ranges from ~100° for $LiTm(WO_4)_2$ and increases in the series from Tm to Tb, with the transition often being incomplete (there are additional lines on the power patterns after complete cooling, which relate to the scheelite phase). The difficulty of the inverse transition increases as the atomic number of the rare earth decreases, i.e., as the polymorphic transition point falls.

The x-ray and dta evidence indicates that the $LiLn(WO_4)_2$ compounds can be made; the results can be extrapolated to show that a monoclinic form should exist for the double tungstate of lithium with adjacent lighter elements below Tb, at Gd and perhaps Eu. The scheelite modification is metastable for these elements at room temperature. It is not possible to produce a low-temperature monoclinic modification for two reasons: the polymorphic transition has a very low temperature, so one is unable to crystallize the LiGd tungstate directly in the monoclinic form, and it is difficult to convert the scheelite modification to the wolframite one, especially at low temperatures.

Crystals of $LiFe(WO_4)_2$ take the form of flattened prisms and are monoclinic; their powder patterns are similar to that of wolframite, but the x-ray studies showed that there are strong pseudo-repeat distances along the a and b directions, which makes the values of a and b twice those for wolframite. The space group of $LiFe(WO_4)_2$ is C2/c. The lithium-iron tungstate is a new variety with the wolframite structure; the Li^+ and Fe^{3+} cations have an ordered array as in monoclinic lithium—rare earth tungstates.

The dta data and x-ray data show that $LiFe(WO_4)_2$ does not have high-temperature polymorphic transitions and melts without decomposition at 1035°C.

These syntheses show that one can use molten salts to make single crystals of the $LiLn(WO_4)_2$ of appropriate size; the melting points are relatively low (all the products melt in the range 975-1135°C), so they are convenient for growing crystals from the melt, but the latter method is very difficult to apply for $LiLn(WO_4)_2$, where Ln is a rare earth from Tb to Tm, on account of the polymorphic transition. A study has been made [4] of hydrothermal synthesis of double lithium—rare earth tungstates from La to Nd, which gave a triclinic modification, which on heating in air passes irreversibly into the scheelite one around 725°C.

Literature Cited

1. G. E. Peterson and P. M. Bridenbaugh, Appl. Phys. Letters, 4:173 (1964).
2. S. A. Fedulov, Z. I. Tatarov, L. P. Shklober, N. I. Sergeeva, G. N. Antonov, and M. Z. Gurevich, Izv. Akad. Nauk SSSR, ser. neorg. mat., 2:1905 (1966).
3. Powder Diffraction File, ASTM, 13-434.
4. P. V. Klevtsov and L. Yu. Kharchenko, in: Growth of Crystals, Vol. 7, Consultants Bureau, New York (1969), p. 293.

PREPARATION AND EXAMINATION OF SOME DOUBLE TUNGSTATES AND MOLYBDATES

I. G. Avaeva, V. B. Kravchenko, T. N. Kobyreva, and B. I. Kryuchkov

There are several reports [1-5] on the production of single crystals of double tungstates and molybdates of the alkalis and rare earths of type $M^I Ln^{III}(WO_4)_2$, where M^I = Li, Na, K; some of them have been used as laser materials [2]. Double tungstates and molybdates of the alkali metals can be made with the lighter elements of group III, such as Al and Ga, but no systematic study appears to have been made on these. The compound $NaAl(WO_4)_2$ has been made as a ceramic [6], while single crystals of $NaIn(WO_4)_2$ have been grown from solution in a melt as source materials for masers [7].

We have made a systematic study of the scope for spontaneous nucleation from solutions in melts for the compounds $M^I M^{III}(M^{VI}O_4)_2$, where M^I = Li, Na; M^{III} = Al, Sc, Ga, Y, In; M^{VI} = Mo, W; the solvents were tungstates, ditungstates, and dimolybdates of lithium and sodium. We used platinum crucibles of volume from 30 to 250 ml, with the ratio of solute to solvent varying from 1:3 to 1:4.5. In most cases the mixture also contained traces of the paramagnetic ions Cr^{3+}, Fe^{3+}, Gd^{3+}.

Table 1 gives the results on preliminary melting of the charge at temperatures below the working point actually employed.

The temperature was reduced at 1-5 deg/hr; in some cases, the different initial temperatures resulted in different polymorphic modifications, e.g., for $LiSc(WO_4)_2$, which occurred on superheating the melt by 40°.

The paramagnetic ions had various effects on the results; Gd^{3+} greatly reduced the size of the crystals and made the faces less perfect, while the analogous effect from Fe^{3+} was considerably less, while Cr^{3+} in most cases did not alter the size or faces of the crystals, producing only a change in color. For instance, the crystals of $NaIn(WO_4)_2$ containing Gd^{3+} were about 0.1 × 1.5 mm in cross section, while crystals made under analogous conditions containing Cr^{3+} were 1 × 4 mm. The colors stated in Table 1 relate to crystals containing Cr^{3+}. Chromium produces practically no color in compounds containing Y apart from lithium-yttrium tungstate, which has the wolframite structure type, while $LiY(MoO_4)_2$, $NaY(MoO_4)_2$, and $NaY(WO_4)_2$ have the scheelite structure, which is probably due to the tendency of Cr^{3+} to take up octahedral coordination.

Some of the compounds were submitted for chemical analysis, which confirmed the stoichiometric relations between the components.

TABLE 1. Crystallization Conditions and Products

Compound	Solvent	Temp., °C	Density	Color
$LiAl(MoO_4)_2$	$Li_2Mo_2O_7$	1100—600	3.90	Pink needles
$LiSc(MoO_4)_2$	$Li_2Mo_2O_7$	1000—600	3.50	" "
$LiGa(MoO_4)_2$	$Li_2Mo_2O_7$	900—600	4.13	" "
$LiY(MoO_4)_2$	$Li_2Mo_2O_7$	1000—600	4.67	Colorless dipyramids
$LiIn(MoO_4)_2$	$Li_2Mo_2O_7$	1100—600	4.21	Pink needles
$NaSc(MoO_4)_2$	$Na_2Mo_4O_{13}$	980—650	—	Pink platelets
$NaY(MoO_4)_2$	$Na_2Mo_2O_7$	1100—650	—	Colorless grains
$LiAl(WO_4)_2$	$Li_2W_2O_7$	1200—750	—	Green needles
$LiSc(WO_4)_2$	Li_2WO_4	940—750	—	Bluish needles, green fragments
$LiGa(WO_4)_2$	Li_2WO_4	950—750	—	Colorless needles, pink platelets
$LiY(WO_4)_2$	Li_2WO_4	1100—750	6.73	Pink platelets
$LiIn(WO_4)_2$	Li_2WO_4	950—750	7.31	" "
$NaSc(WO_4)_2$	$Na_2W_2O_7$	950—750	6.34	Bluish needles
$NaGa(WO_4)_2$	$Na_2W_2O_7$	950—750	6.18	Yellow needles
$NaY(WO_4)_2$	$Na_2W_2O_7$	1100—750	—	Colorless grains
$NaIn(WO_4)_2$	$Na_2W_2O_7$	1100—750	7.01	Bluish needles

Most of the crystals took the form of needles or plates, except for $LiY(WO_4)_2$, which gave isometric grains or dipyramids. Sometimes compounds other than the double tungstates were formed as well, e.g., dark-green crystals of unknown composition in the runs with $LiSc(WO_4)_2$.

The crystal morphology was also affected by the temperature, the cooling rate, and the degree of evaporation, for instance, crystals of $LiY(MoO_4)_2$ on cooling at 5-6 deg/hr were often elongated and irregular in shape, whereas at 1-2 deg/hr they were regular dipyramids with glistening flat faces. Also, $NaIn(WO_4)_2$ at a high cooling rate crystallized as very much elongated needles, while at lower rates the crystals were larger and of more platy appearance.

The powder diffraction patterns and the crystal morphology indicated that most of these double tungstates have the structure of wolframite $(Fe, Mn)WO_4$ [8, 9] with various types of ordering in the univalent and trivalent cations; Table 2 gives the lattice parameters of some single crystals, together with data for $ZnWO_4$. It is likely that the molybdates as a rule have a more complicated structure.

Goniometry showed that $NaIn(WO_4)_2$ crystals have faces composed of the {100}, {010}, {401} pinacoids and {210} and {052} rhombic prisms (Fig. 1a); the crystals are often flattened

TABLE 2. Goniometric and X-Ray Data on Crystals

Compound	a	b	c	β	z	Space group
$ZnWO_4$	4.68	5.79	4.95	90°30	2	P 2/c
$NaIn(WO_4)_2$	10.07	5.80	5.03	91.2°	2	P 2/c
$LiIn(WO_4)_2$	9.65	11.45	4.95	~90°	4	—
$LiSc(WO_4)_2$	9.60	11.62	4.97	91°	4	C 2/c
$NaSc(WO_4)_2$	5.08	11.20	5.08	~90°	2	—

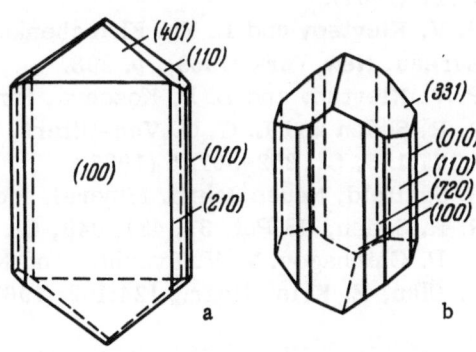

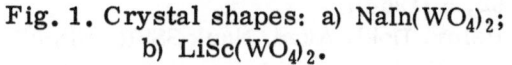

Fig. 1. Crystal shapes: a) $NaIn(WO_4)_2$;
b) $LiSc(WO_4)_2$.

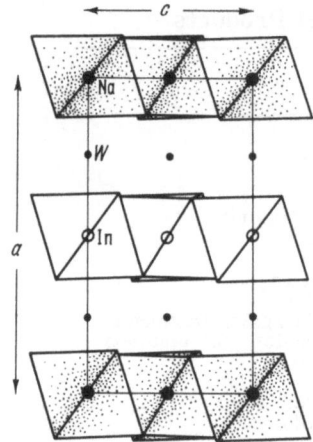

Fig. 2. Projection of the structure of $NaIn(WO_4)_2$ on (010) showing chains of octahedra along the c axis.

on the a axis and have perfect mica-type cleavage on (100). Crystals of $LiSc(WO_4)_2$ have $\{100\}$, $\{010\}$ pinacoids and $\{110\}$, $\{720\}$, $\{331\}$ prisms (Fig. 1b).

Our x-ray results show that $NaIn(WO_4)_2$ has a structure of wolframite type, with the Na and In ions separately populating infinite columns of octahedra along the c axis (Fig. 2); this results in doubling of the a parameter, and the perfect cleavage runs on the (100) planes, which contain the chains of Na octahedra. The C2/c space group for $LiSc(WO_4)_2$ leads to alternation of the different columns of all the octahedra in the (100) and (010) planes, and there is no mica-type cleavage.

All the double tungstates and molybdates were also made by ceramic methods; powder patterns showed that the solid-state reactions produce other modifications of almost all the compounds. These substances are now under detailed examination.

Conclusions

1. Spontaneous crystallization of solution in melts has given single crystals of double tungstates and molybdates of lithium or sodium with cations of group III; most of the compounds have not been made before.

2. It has been found that the growth conditions and doping materials affect the crystal morphology.

3. Most of the double tungstates have structures of wolframite type; the molybdates as a rule have more complex structures.

Literature Cited

1.   S. Preziosi, R. R. Soden, and L. G. G. Van-Uitert, J. Appl. Phys., 33:1893 (1962).
2.   A. A. Kaminskii and V. V. Osiko, Izv. Akad. Nauk SSSR, ser. neorg. mat., 1:2049 (1965); 3:417 (1967).
3.   P. V. Klevtsov and L. Yu. Kharchenko, in: Growth of Crystals, Vol. 7, Consultants Bureau, New York (1969), p. 293.
4.   P. V. Klevtsov and L. P. Kozeeva, Izv. Akad. Nauk SSSR, ser. neorg. mat., 4:1379 (1968).
5.   R. R. Soden and L. G. G. Van-Uitert, US Pat., 3,177,154, Cl. 252-301, 5 (1965); US Pat. 3,177,156, Cl. 252-301, 5 (1965).
6.   H. Saalfeld, Neues Jahrb. Mineral. Monatsh., No. 9, 207 (1955).
7.   R. R. Soden, US Pat. 31, 481, 049, Cl. 252-62, 5 (1964).
8.   A. P. Chichagov, V. V. Ilyukhin, and N. V. Belov, Dokl. Akad. Nauk SSSR, 166:87 (1966).
9.   D. Ülku, Z. Kristallogr., 124:192 (1967).

GROWTH OF Fe–Y GARNET SINGLE CRYSTALS BY SPONTANEOUS CRYSTALLIZATION WITH A SPECIFIED NUMBER OF NUCLEI

L. N. Averina, B. I. Birman, O. M. Konovalov, and T. R. Mnatsakanova

A theoretical calculation is presented for the necessary rate of temperature reduction in spontaneous crystallization with a fixed number of crystallization centers, and some results are given from the growth of single crystals of yttrium–iron garnet.

The following assumptions are made in the theoretical calculation.

1. The number of nuclei is set and does not alter during the growth.*
2. The nuclei and the growing crystals are hemispheres; the initial sizes of the nuclei are represented by a set radius R_0.
3. The growth rate v of all the crystals is the same and is constant; the effective diffusion coefficient in the melt is D, while convective mixing in the melt is taken into account in the boundary-layer approximation, that layer having thickness δ.

The components of the melt have the relationship

$$C_Y(t) + C_{Fe}(t) + C_P(t) = 1, \tag{1}$$

where C_Y, C_{Fe}, and C_P are the weight concentrations of the yttrium oxide, iron oxide, and some of all the other components in the solution. There is the following relation between the temperature and the concentrations at the crystallization front, which are denoted by the denoted by the index ϕ:

$$T_\phi(t) = T_n - m[C_n - C_Y^\phi(t)] - n[1 - C_Y^\phi(t) - C_{Fe}^\phi(t)], \tag{2}$$

where T_n and C_n are the peritectic temperature and concentration, m is the slope in the temperature fall for the liquidus due to the excess over stoichiometric in the proportion of iron oxide (m > 0), and n is the slope of the liquidus temperature reduction due to the solvent (n > 0).

The masses of all components of the system are conserved, so we can relate the concentrations in the melt at an arbitrary time t and the corresponding concentrations in the initial'

---

*This is possibly only when there is no bifurcation (Editor).

mixture, which are denoted by superscript 0:

$$C_Y(t) = \frac{C_Y^0 G_0 - C_Y^* G_1(t)}{G_0 - G_1(t) - G_2(t)}, \qquad C_{Fe}(t) = \frac{C_{Fe}^0 G_0 - (1 - C_Y^*) G(t_1)}{G_0 - G_1(t) - G_2(t)}, \tag{3}$$

where G_0 is the initial weight of the mixture, $G_1(t)$ is the weight of the material crystallized out, $G_2(t)$ is the weight of evaporated solvent, and C_Y^* is the yttrium oxide concentration in the solid.

The following equations relate the concentrations at the crystallization front to those in the main body of the melt:

$$C_Y^\Phi(t) = \frac{C_Y(t)}{1 - \psi(t)} - C_Y^* \frac{\psi(t)}{1 - \psi(t)}, \qquad C_{Fe}^\Phi(t) = \frac{C_{Fe}(t)}{1 - \psi(t)} - (1 - C_Y^*) \frac{\psi(t)}{1 - \psi(t)}, \tag{4}$$

where the growth function $\psi(t)$ takes the following form for low growth rates:

$$\psi(t) = \frac{v\delta}{D} \frac{R_0 + vt}{\delta + R_0 + vt}. \tag{5}$$

We substitute (3) and (4) into (2), and then subtract the result for the crystallization front temperature at time t from the value of this temperature at t = 0 to get the following expression for the temperature to be followed during cooling in order to get a constant value for v:

$$\frac{T_\Phi(0) - T_\Phi(t)}{\theta} = \frac{\psi(t) - \psi_0}{(1 - \psi_0)[1 - \psi(t)]} + \frac{G_1(t) - \eta G_2(t)}{[1 - \psi(t)][G_0 - G_1(t) - G_2(t)]}, \tag{6}$$

where for brevity we have

$$\theta = n(1 - C_{Fe}^0 - C_Y^0) + m(C_Y^* - C_Y^0), \qquad \eta = \frac{m(C_Y^0 + C_{Fe}^0 + nC_Y^0)}{\theta},$$

and ψ_0 is the value of ψ at t = 0.

Fig. 1. Crystals in crucible, natural size.

TABLE 1. Crystallization Results

Composition, wt.% [1, 2]	Charge weight, g	Crystal yield, %	Number of crystals				Crust weight, g	Evaporation loss, %	Run length, hr
			20 mm	10 mm	5 mm	5 mm			
Y_2O_3—11.1 Fe_2O_3—15.6 PbO—32.2 B_2O_3—1.8 PbF_2—39.2	1048	18	2	4	1	—	144	—	336
The same	1080	18	—	4	2	—	155	17	408
" "	4200	8	4	—	—	—	158	—	601
" "	2855	8.5	5	18	30	—	41.6	15	312
" "	2930	7	3	35	11	—	46.5	10	312
Y_2O_3—11.1 Fe_2O_3—15.6 PbO—38.7 PbF_2—31.7 CaO—0.5 B_2O_3—2.8	2500	18	—	1	38	5	315	5.5	408
Y_2O_3—11.1 Fe_2O_3—15.6 PbO—39.2 PbF_2—32.2 B_2O_3—1.8 CaO—0.1	300	12.6	3	—	10	—	204.7	14	673
Mean	2500	28.5						28.8	

We differentiate (6) with $\psi \ll 1$; $G_1 \ll G_0$; $G_2 \ll G_0$ to get an expression for the cooling rate needed to get a constant growth rate:

$$\left(\frac{dT}{dt}\right)_{req} \approx \theta \left\{ \frac{v^2\delta^2}{D(R_0+\delta+vt)^2} + \frac{2\pi\rho Nv(R_0+vt)^2}{G_0} - \frac{\eta}{G_0}\frac{dG_2}{dt} \right\},$$ (7)

where ρ is the density of the crystals. The yttrium−iron garnet single crystals were grown from a $PbO-PbF_2-B_2O_3$ solvent in platinum crucibles 100 mm in diameter and 150 mm high; the crucible was placed in a vertical oven heated by resistance windings, and the temperature difference over the height of the crucible in the working space was ~50°C.

The crucible contents were heated to 1280° for 10 hr and were kept at that temperature until the melts had become uniform. The crystals grew on reducing the temperature from 1280 to 1050° at a mean rate of 0.9 deg/hr, and then the melt was poured off and the crucible with the crystals was allowed to cool along with the oven to room temperature.

These tests usually gave a small number of relatively large single crystals roughly equal in size (Fig. 1).

Table 1 gives some results on the growth of these crystals under these conditions.

The cooling rate has to increase gradually as time passes in order to provide a constant growth rate; the value is estimated as 0.75-1.45 deg/hr, which agrees satisfactorily with the mean rate of 0.9 deg/hr actually employed.

Literature Cited

1.   W. H. Grodkiewicz and L. G. Van Uitert, French Patent. 1,419,723.
2.   W. H. Grodkiewicz, E. F. Dearborn, and L. G. Van Uitert, J. Phys. Chem. Solids, S8 (Suppl.):441 (1967).

EFFECTS OF THE PHASE BOUNDARIES OF THE COMPOUNDS FORMED IN MOLTEN SOLUTIONS USED TO GROW GARNETS

V. A. Timofeeva and N. I. Luk'yanova

Controlled crystallizati on involves problems for all the crystalline phases that accompany the desired one; previously [1] we have considered the phase boundaries for the compounds $YFeO_3$, $Y_3Fe_5O_{12}$, $PbFe_{12}O_{19}$, YBO_3, Fe_2O_3 formed in systems with the ternary solvent (PbO, B_2O_3, PbF_2). Here we report some further results on the crystallization ranges in such systems, which have been used for growing garnets of various compositions. We used a crystallization apparatus with a cover having a long neck [1], which reliably retains the volatile lead compounds in the solution, the evaporation loss not exceeding 1-2 wt.%. Curves 1 and 2 of Fig. 1 show the solubilities of $Y_3Fe_5O_{12}$ in two different solvents; increase in the solubility of $Y_3Fe_5O_{12}$ to 40% at 1380°C is facilitated by raising the molar proportion of lead oxide in the solution; incorporation of a fourth component, BaO (curve 3 of Fig. 1), increases the solubility of $Y_3Fe_5O_{12}$ to 50% at 1330°C, but the increased complexity of the solvent composition results in more phases in the system.

As the system is cooled from 1450 to 750°C, the various phases* have their appropriate temperature ranges of crystallization under equilibrium conditions. The highest-temperature phase is a perovskite one; as the temperature is reduced, the cubes of the perovskites begin to dissolve slowly, and at their vertices and edges there appear crystals of garnet (Fig. 2). The phases $PbFe_{12}O_{19}$ and Pb_2OF_2 then arise on further temperature reduction. Change in the molar relationships of the solvent components and introduction of BaO enable one to widen the temperature range of garnet crystallization and to increase the concentration of the garnet component in a certain temperature range; with $Y_2O_3 : Fe_2O_3 = 1:1$, the growth of the $Y_3Fe_5O_{12}$ crystals occurs without production of the phase $PbFe_{12}O_{19}$ down to 750°C, i.e., the garnet part crystallized most completely from the melt. At 750°C, the phase Pb_2OF_2 rapidly appears, which halts the growth of the garnet. A certain excess of Fe_2O_3 (from 20 to 60%) in the garnet mixture causes [1] the phase $PbFe_{12}O_{19}$ to arise, which coexists over a certain temperature range with the garnet phase (from 1105 to 1080°C).

The growth rate of the yttrium—iron garnet is represented by a curve with a broad peak, which relates to the onset of formation of the plumbite phase. Further cooling causes the garnets to dissolve on account of transition of the unstable part (Fe_2O_3) from the garnet to the plumbite. At even lower temperatures, the $Y_3Fe_5O_{12}$ crystals no longer dissolve, because the

---

* The phases were identified by microscopic examination and x-ray phase analysis from drops of the quenched solution or from samples of the crystals extracted from the solution at various crystallization temperatures.

116

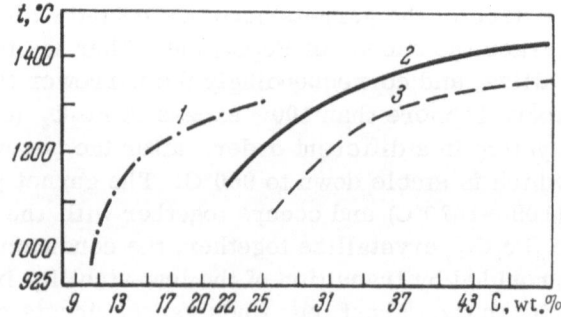

Fig. 1. Solubility curves for $Y_3Fe_5O_{12}$ in solvents: 1) 0.8 $PbO-0.3$ B_2O_3-1 $PbFe_2$, 2) 1.3 $PbO-0.3$ B_2O_3-1 PbF_2, 3) 1.1 $PbO-0.5$ B_2O_3- 1 $PbF_2-1.5$ BaO.

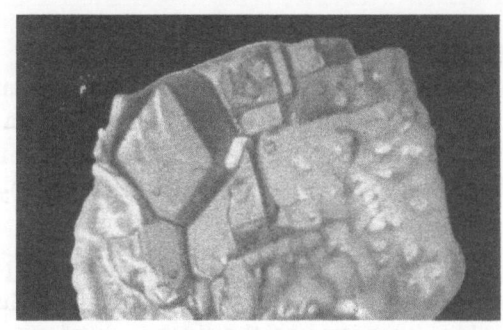

Fig. 2. Small YIG crystals on perovskite, ×10.

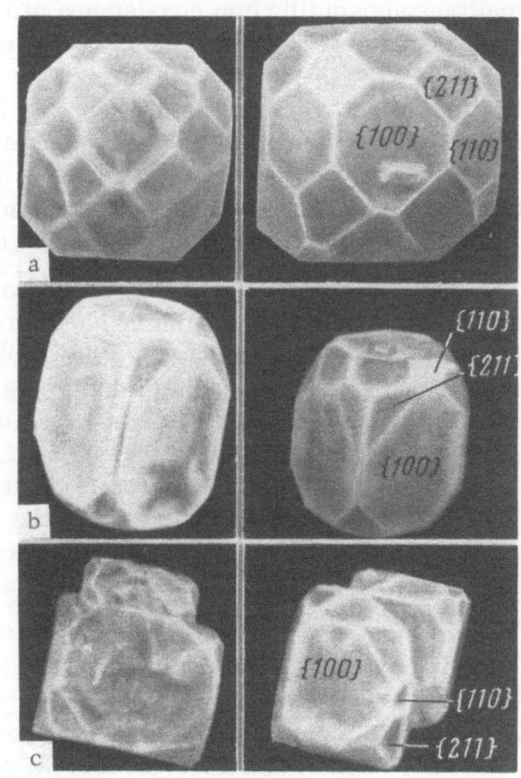

Fig. 3. Crystal habit in relation to face growth rate, ×3. a) Isometric, with identically developed $\{110\}$, $\{211\}$, $\{100\}$; b) elongated, $\{100\}$ face predominant, $v\{110\} > v\{211\} > v\{100\}$; c) cubic habit, $\{100\}$ face predominant, $\{110\}$ almost lost, $v\{110\} > v\{211\} > v\{100\}$.

surface of the garnets becomes coated with several layers of foliated plumbite crystals. The larger the excess of Fe_2O_3, the higher the temperature at which the plumbite begins to crystallize, and correspondingly the narrower the temperature region for the growth of garnet. If there is more than 100% excess of Fe_2O_3 (case IV of Fig. 2), the phases are produced in the system in a different order: after the perovskite and iron oxide there crystallizes plumbite, which is stable down to 960°C. The garnet phase occurs only over a narrow temperature range (1090-1020°C) and occurs together with the plumbite phase in the melt. When $YFeO_3$ and $Y_3Fe_5O_{12}$ crystallize together, the conditions for equilibrium of $PbFe_{12}O_{19}$ and $Y_3Fe_5O_{12}$ are provided by transition of the low-stability iron oxide from one to the other, here from the plumbite to the garnet; one finds garnet crystals at the corners and edges of the plumbite.

There are several phases also when one grows aluminum garnets in systems with analogous solvents; the first material to be produced is corundum, which covers only the range 1230-1220°C. Garnet crystallization covers the range 1230-1130°C, and below this one gets the spinel $BaAl_2O_4$.

In the case of the gallium garnets, the range 1400 to 870°C produces only the garnet phase; it has previously been shown [2] that the rare earths from Dy to Lu do not form compounds of perovskite type with gallium. We examined the phase composition of the solution for the system $Yb_2O_3-Er_2O_3-Ga_2O_3-Nd_2O_3-PbO-B_2O_3-PbF_2$; although this mixture has a very complicated composition, we found no phases other than the garnet of composition $(Yb, Er)_3 \cdot Ga_5O_{12} : Nd$.

A multiplicity of phases is particularly common when one grows crystals from small seeds under dynamic conditions. If there is a hole in the cover, the composition of the solution can change, which can influence the equilibrium, also, one can use the presence of a second phase to alter the supersaturation in the desired phase and to produce the most favorable conditions for growth of this on seeds. We used seed crystals of composition $(Yb, Er)_3 \cdot Ga_5O_{12}$ (lattice parameter 12.27 Å) in a system with a large excess of Fe_2O_3 in the region of simultaneous equilibrium coexistence of the garnet and plumbite phases to make uniform layers of yttrium-iron garnet; epitaxial overgrowth begins at the boundary or a little earlier (by about 10°C) for production of the garnet phase, when the melt contains only the phase $PbFe_{12}O_{19}$. Heterometry occurs not only in the growth of $Y_3Fe_5O_{12}$ on $Gd_3Ga_5O_{12}$, which has a lattice parameter of 12.37 Å which equals the parameter of $Y_3Fe_5O_{12}$ [3]. In our case, heterometry was also not observed in the case of the composition $(Yb, Er)_3Ga_5O_{12}$. The use of seed crystals of yttrium-aluminum garnet composition does not allow one to produce layers free from defects on account of heterometry (the lattice parameter of $Y_3Al_5O_{12}$ is 12.01 Å).

The fourth component BaO in the solution affects the morphology of the garnets; in the system used to grow the yttrium-aluminum garnet we found changes in the face growth rates (v_{hkl}) of the garnet. In this case, the {100} fast-growing face is transformed to a slowly-growing one and participates in the facets of the $Y_3Al_5O_{12}$ crystal. Parts a-c of Fig. 3 show crystals of $Y_3Al_5O_{12}$ bearing three types of faces common in garnet; the habit of the crystals indicates how the face growth rates vary. Initially, the face growth rates are approximately equal, which results in an isometric habit and identical development of all faces (Fig. 3a), while subsequently the growth rate of the {100} face falls, and this then begins to predominate in area over the other faces. The {110} face acquires the highest growth rate, and this gradually becomes smaller, so the crystal as a whole acquires a cubic appearance. The {211} face undergoes the same changes.

Literature Cited

1.   V. A. Timofeeva and N. I. Luk'yanova, Kristallografiya, 14:884 (1969).
2.   V. A. Timofeeva, T. V. Lebedeva, and T. S. Kon'kova, Kristallografiya, 10:92 (1965).
3.   R. C. Linares, J. Crystal Growth, 3/4:443 (1968).

THE EFFECTS OF GROWTH CONDITIONS ON THE MORPHOLOGY OF Y–Al BORATE CRYSTALS

I. É. Gerasimova, T. I. Timchenko, and O. G. Kozlova

The yttrium–aluminum borate $YAl_3(BO_3)_4$ is [1] in the class of insular borates with the general formula $RX_3(BO_3)_4$, where R is a rare earth and X is Al or Cr; this group of compounds has piezoelectric response and also fluorescence, and the crystals can be used as matrices for quantum oscillators.

These crystals were first made by Ballman [2] in 1962, and Knox and Mills [3] made x-ray measurements that showed that the borate is isostructural with the mineral $CaMg_3(CO_3)_4$.

We made crystals of this borate in the system $K_2SO_4 - MoO_3 - B_2O_3 - Al_2O_3 - Y_2O_3$ from solution in a melt by temperature reduction involving spontaneous crystallization; the shapes were fairly simple, and the habit was dominated by the faces of the prism and rhombohedron, with a pinacoid not present on all crystals.

The habit and surface sculpturing of the faces are dependent mainly on the cooling rate and the presence of certain low-melting impurities.

When the rate is ~1 deg/hr, the crystals are largely isometric; the prism and rhombohedron faces are almost equal in size, and the pinacoid is absent (Fig. 1).

The rhombohedron faces are smooth, with thin growth layers and single large subindividuals (Fig. 2); the prism faces have vicinals in the form of cones.

Introduction of PbF_2 in the proportion of 0.2 mole part per 1 mole of borate causes the crystals to become prismatic when the cooling rate is 1 deg/hr; the pinacoid face is still absent, while the rhombohedron faces are much smaller in area (Fig. 3).

If the cooling rate is nearly tripled, the habit and face structure are appreciably affected; the crystals become prismatic and the pinacoid appears sometimes.

The most prominent prism faces are covered with fine striations; the rhombohedron faces become rougher on account of humps of height from 6-10 μm.

Addition of LiF at the rate of 0.5 mole per mole of borate under the same conditions substantially improves the crystal quality and accelerates the growth; the crystals are up to 4-4.5 mm in size. The pinacoid face is quite prominent, and the habit is prismatic, but the difference in growth rate between the rhombohedron and pinacoid on the one hand and the prism on the other tends to disappear and the crystals become less elongated (Fig. 4).

The rhombohedron and pinacoid faces are equally covered by large vicinals of height from 20-30 μm; the number of vicinals on the rhombohedron faces is very much less than that for crystals obtained without doping.

119

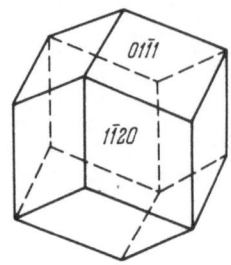

Fig. 1. Shape of $YAl_3(BO_3)_4$ crystals produced
by slow cooling.

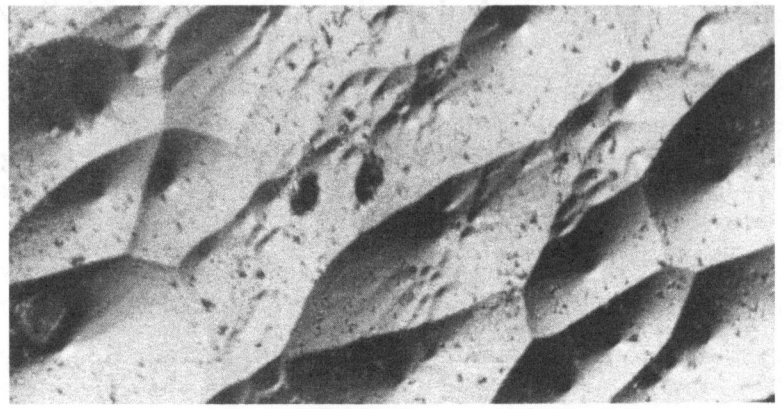

Fig. 2. Growth figures on a prism of Y−Al borate
produced by slow cooling, ×450.

Fig. 3. Shape of $YAl_3(BO_3)_4$ crystals produced
by a trace of PbF_2.

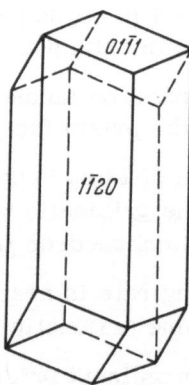

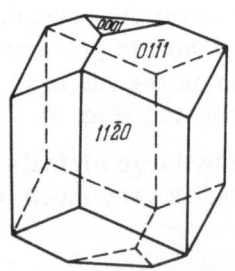

Fig. 4. Shape of $YAl_3(BO_3)_4$ crystals produced
by LiF in mixture.

The isometric crystals become elongated as the supersaturation is increased.

Literature Cited

1. V. B. Kravchenko, "Crystallochemical studies on some borates and borosilicates," Thesis, Institute of General and Inorganic Chemistry, Novosibirsk (1963).
2. A. Ballman, Amer. Mineralogist, 47:1380 (1962).
3. A. Mills, Inorg. Chemistry, 1:960 (1962).

PREPARATION AND EXAMINATION OF EPITAXIAL
LAYERS OF GaSb-GaAs SOLID SOLUTIONS

Yu. M. Kozlov, N. N. Sheftal', L. S. Garashina, and D. M. Kheiker

A^3B^5 solid solutions provide extensive scope for varying the properties of semiconductors. The GaSb−GaAs system is an interesting one in this class but has been largely neglected [1, 2]. It is difficult to make GaSb−GaAs solid solutions as single crystals on account of homogenization problems, because the A^3B^5 have oriented chemical bonds, which hinder diffusion [2]. These difficulties are largely overcome if one uses thin films. GaSb−GaAs solid solutions have been made as films [3] by vacuum evaporation, but it is difficult to make high-quality films in this way. There are also considerable difficulties in using gas transport for the purpose. We have grown GaSb−GaAs epitaxial films from solution in gallium metal of 000 grade by the isothermal static method in a vertical apparatus [5]. The initial materials were SU-0000 antimony and n- and p-type GaAs with various carrier concentrations.

The substrates were (111) single-crystal plates. The process was operated in pure dry hydrogen. A solution in gallium was prepared [6] saturated at 650°C with gallium antimonide and arsenide, which was kept at 670°C for 6-8 hr to homogenize before cooling to 650°C for inserting the substrate, after which the temperature was further reduced by 3-5°. The temperature near the substrate was kept constant to 0.5°. The density of the solvent is very different from that of the solutes, so there was vertical concentration and temperature gradients, with the latter not exceeding 5 deg/cm for isothermal growth.

The growth rate on the gallium side of the substrate was much higher than that on the arsenic side with these gradients of 5-10 deg/cm: 20-30 μm films were produced on B(111) but 70-80 μm on A(111). The thicknesses were reproduced to ± 5 μm. The layers also differed considerably in morphology: these on A(111) were smooth and had dislocation densities less than 10^5 cm^{-2}; there were isolated packing defects and twin lamellae. The B(111) layer had many truncated and hollow growth pyramids, with dislocation densities up to 10^6 cm^{-2}.

The Hall mobility and carrier concentration were measured* at 300°K by van der Pau's method as 1 cm^2/V-sec and $1.21 \cdot 10^{18}$ cm^{-3}. The electrical insulation of the film was provided by the GaAs (111) high-resistance substrate, which had $\rho_1 = 10^8$ ohm-cm (Hall factor taken as unity).

These films were examined with x-rays (Toshiba AFV-201 diffractometer, CuK$_\alpha$).

The patterns showed (111) reflections from GaSb$_{1-y}$As$_y$ and GaAs$_{1-x}$Sb$_x$ (Fig. 1), which indicate that the (111) planes in the films were parallel to those in the substrate. Published

---

* Measurements made in collaboration with V. G. Sidorov.

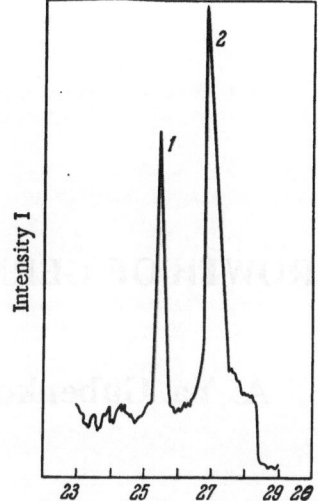

Fig. 1. Part of the x-ray pattern of a GaSb—
GaAs specimen. 1) 111 peak for $GaSb_{(1-y)}As_y$;
2) 111 peak for $GaAs_{(1-x)}Sb_x$ (epitaxial GaSb—
GaAs solid solution produced at 650-645°C).

values [3, 4] are 6.095 and 5.653 Å for the cubic cells of the substrate and standard GaSb as
deduced from the (111) peaks, which agreed with those found:

Phase	Lattice constant. Å
GaAs	5.65
$GaAs_{1-x}Sb_x$	5.73
$GaSb_{1-y}As_y$	6.03
GaSb	6.09

These results show that replacement of Sb (radius 1.36 Å) by As (1.18 Å) reduces the
cell parameter.

We get x = 0.17 and y = 0.14 if we assume [3, 4] that the cell parameter varies linearly
with composition.

We are indebted to G. G. Matwear for assistance.

Conclusions

Liquid epitaxy produced GaSb—GaAs solid solutions, which are difficult to make in
other ways. X-ray tests show that the films actually are GaSb—GaAs.

Literature Cited

1. J. C. Wooley, in: "Preparation of III—V Compounds," R. K. Williardson and H. L. Goering, eds., Reinhold, New York (1963).
2. N. A. Goryunova, Chemistry of Diamond-Type Semiconductors [in Russian], Izd. LGU
   (1963); Complex Diamond-Type Semiconductors [in Russian], Izd. Sov. Radio, Moscow
   (1968).
3. E. K. Muller and J. L. Richards, J. Appl. Phys., 35:1233 (1964).
4. M. E. Straumanis and C. D. Kim, J. Electrochem. Soc., 112:112 (1965).
5. Yu. M. Kozlov, Proceedings of the Second All-Union Symposium on the Growth of Crystals and Films of Semiconductor Compounds [in Russian], Izd. SO AN SSSR, Novosibirsk
   (1969) (in press).
6. R. N. Hall, J. Electrochem. Soc., 110:385 (1963).

GROWTH OF GERMANIUM CRYSTALS FROM SOLUTIONS IN METALS

A. Ya. Gubenko, N. A. Kononykhina, and M. B. Miller

There is interest in the growth of crystals from solution in molten metals as regards crystallization kinetics and mechanism and also as regards applications [1-3]. As regards semiconductors, the method is usually used with seeds in Czochralski's method or by means of zone melting with a temperature gradient.

The usual problems in the use of molten metals arise in studying the growth kinetics in relation to the properties of the solvents and addition of dope materials; we consider these aspects for the crystallization of germanium.

The germanium crystals were made by Czochralski's method and by zone melting with a temperature gradient from solutions in the sections Ge−[Au + (0.1−1) wt.% Sb], Ge−[Sn + (1−5.5) wt.% Sb], and Ge−[Au +1 wt.% Ga] in the corresponding systems, and also from binary solutions of germanium containing Au, Ag, Sn, or Pb. The crystallization temperatures were 600-890°C. The crystals were grown on seeds oriented usually along [111], except for some special tests mentioned specifically below in the zone method. In Czochralski's method, the growth rate was varied over the range 0.01-1 mm/min; the crystal rotation speed was 80 rpm and the crystal diameter was 8-9 mm, which caused the effects from accumulation of solvent components at the front to be negligible [4, 5]. In the zone-melting tests, the growth rate was varied within the same limits, while the temperature was varied from 600 to 850°C; the gradient was kept constant. The points used in constructing the graphs are mean values from 3 or 4 runs each.

We sought to elucidate the speed limitations on the zone in the Ge−Au and Ge−Sn experiments by measuring the crystallization rate under identical conditions with various orientations of the seed plate or dissolving germanium plate; we found that a plate oriented on [100] increased the rate by a factor 1.8 relative to one oriented on [111]. However, this change did not alter the zone speed, so the processes at the front where the material dissolves do not in this case limit the zone speed.

We examine the effects of small amounts of antimony on the crystallization of Ge in the systems Ge−Au and Ge−Sn by zone melting; the speed of the molten zone is substantially dependent on the antimony concentration in the solution (Fig. 1).

The zone speed v_{zo} is dependent on the antimony concentration to an extent that increases with the crystallization temperature; the antimony dependence becomes weak at low temperatures, and at 600°C vanishes completely, so the main tests on the effects of additives on the zone speed were made at 850°C. It has been shown previously [3, 4] that antimony and gallium up to 1 wt.% have practically no effect on the slope of the liquidus curve for binary solutions of

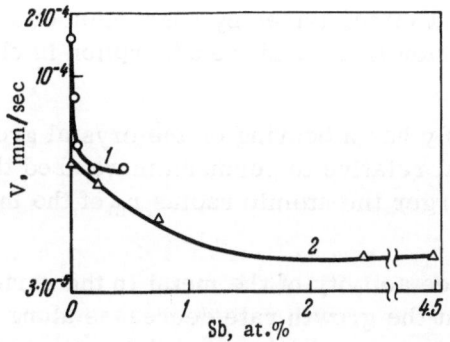

Fig. 1. Speed of molten zone at 850°C as a function of Sb concentration in the melt for (1) Ge−(Au + Sb) and (2) Ge−(S + Sb).

Ge−Au and Ge−Sn. We compared the curves of Fig. 1 with Langmuir absorption isotherms and with results of antimony uptake in both methods in order to elucidate the effects of antimony on the growth rate; the results from the various series were similar.

Figure 2 shows the dependence of the antimony concentration in the crystal on the concentration in the Ge−Au solution: 1) the curves are similar to the Langmuir isotherms, 2) the antimony concentration that produces a plateau in Fig. 3 (at 850°C) is the same as in Fig. 1. We also found that addition of the gallium did not influence the growth rate of the germanium crystals in either method, while there was no plateau on the relation between crystal concentration and solution concentration.

The effects were less clear in the tin system, and the plateau region was attained at higher antimony concentrations in the liquid, while the growth-rate curve in Fig. 1 was smoother. In a Ge−Sn solution, the antimony is relatively weakly adsorbed at the crystallization front [4, 5], and so the surface activity of antimony in Ge−Sn is less than that in Ge−Au, which may be due to difference in the surface activities of the solvents on account of differences in the atomic radii: $r_{Au} < r_{Sn}$.

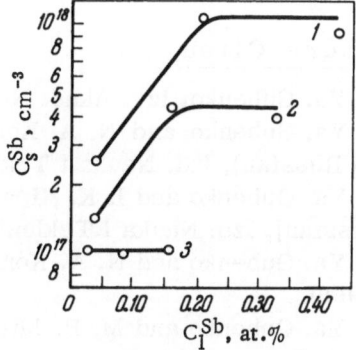

Fig. 2. Sb concentration in Ge crystals as a function of concentration in the liquid at (1) 800°C, (2) 850°C, and (3) 890°C.

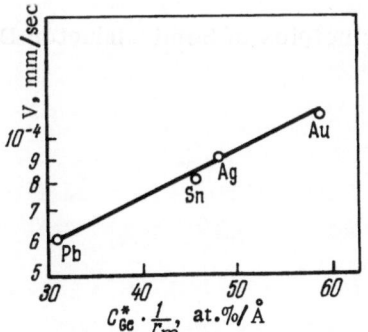

Fig. 3. Speed of molten zone at 850°C in binary systems of Ge with metals.

The results thus indicate that the third-component effect arises by adsorption of a surface-active material (e.g., antimony) at the crystallization front, and the adsorption is close to Langmuir type.

The surface activity of the solvent metal obviously has a bearing on the crystal growth rate. As a measure of the surface activity of the metal relative to germanium we used the values of the atomic radii in the liquid state [6]; the larger the atomic radius r_m of the metal relative to r_{Ge}, the greater the surface activity.

This criterion indicates an increase in the surface activity of the metal in the series Ge−Au, Ge−Ag, Ge−Sn, Ge−Pb; our tests showed that the growth rate decreases along this series in binary systems when the crystals are grown by zone melting, and the zone speed is then dependent on the concentration of the solute [7]. Therefore, the observed growth rates are plotted as ln V versus C_{Ge}/r_m, where C_{Ge} is the solubility of germanium in the metal [8]. Figure 3 shows that crystal growth rate decreases logarithmically as the surface activity of the solvent increases.

Conclusions

1. The seed orientation has an appreciable effect on the zone speed when germanium is crystallized by zone melting with a temperature gradient from solutions in gold and tin, while the orientation of the dissolving crystal has no such effect.

2. We have examined the effects of small amounts of a third component on the germanium crystallization rate in Ge−Au and Ge−Sn; a trace of antimony considerably reduces the crystallization rate, and a saturation effect occurs at high antimony concentrations. It is concluded that this effect arises by adsorption of antimony as a surface-active material at a crystallization front, the adsorption being close to Langmuir type.

3. We have examined the crystallization of germanium in binary solutions of the compositions Ge−Pb, Ge−Sn, Ge−Ag, Ge−Au; the crystal growth rate decreases logarithmically as the surface activity of the solvent increases.

Literature Cited

1. A. Ya. Gubenko, Izv. Akad. Nauk SSSR, ser. neorg. mat., 4:440 (1968).
2. A. Ya. Gubenko and N. A. Kononykhina, in: Crystallization Mechanism and Kinetics [in Russian], Izd. Nauka i Tekhnika, Minsk (1969), p. 459.
3. A. Ya. Gubenko and I. K. Kiparisova, in: Crystallization Mechanism and Kinetics [in Russian], Izd. Nauka i Tekhnika, Minsk (1969), p. 187.
4. A. Ya. Gubenko and N. A. Kononykhina, Izv. Akad. Nauk SSSR, ser. neorg. mat., 4:1787 (1968).
5. A. Ya. Gubenko and M. B. Miller, Izv. Akad. Nauk SSSR, ser. neorg. mat., 6:471 (1970).
6. V. K. Grigorovich, Izv. Akad. Nauk SSSR, ser. met. i toplivo, 6:98 (1960).
7. W. A. Tiller, J. Appl. Phys., 34:2757 (1963).
8. V. M. Glazov and V. S. Zemskov, Physicochemical Principles of Semiconductor Doping [in Russian], Nauka, Moscow (1967).

BASIC TRENDS IN CRYSTAL GROWTH BY THE
MOVING-SOLVENT METHOD

V. N. Lozovskii, V. P. Popov, G. S. Konstantinova,
and V. A. Ivkov

The two commonest forms of crystallization from the moving solvent are the growth of whisker crystals from a vapor via a liquid phase (VLS process) and zone melting with a temperature gradient [1-3]. Here we survey results on the trends in the latter case.

Here the crystallization is determined by processes not only at the crystallization front and in the adjacent layers of liquid but also in the volume of melts and at the dissolving boundary; there is a combination of dissolution and crystallization at the boundaries of the zone, with diffusion of the dissolved atoms in the melts, which leads to a characteristic dependence of the crystal growth rate v on the zone length l (with linear dimension in the displacement direction). We find that v increases as l decreases, and tends to a maximal value v_{max}; the region of most marked variation of v with l occurs in various ranges in l in accordance with the mechanism of the kinetic processes at the zone boundaries. Figure 1 shows v/v_{max} as a function of l for normal growth or dissolution of a crystal A', and the same for growth at screw dislocations (B and B'), and also via mechanisms involving two-dimensional nuclei (C and C'). These curves have been constructed from Tiller's theory for the possible limiting values of the kinetic coefficients [4]; the diffusion parameter is the product of the diffusion coefficient for the melt and the reciprocal slope of the liquidus line; for the binary system in use we took it as $1.46 \cdot 10^{-6}$ cm^2/sec–deg, which corresponds to zone melting processes in the Al–Si system at 1000°C. Variation of the diffusion parameter within reasonable limits for systems based on Si and Ge results in some shift and displacement of the A, B, and C curves along the l axis, but the general picture remains the same.

The saturation regions on the curves in Fig. 1 represent the diffusion state in zone melting; the region of rapid increase in v corresponds to the limiting action of kinetic processes at the zone boundaries. If it is found that v is largely independent of l, one can say in the absence of component evaporation [5] that atom diffusion in the zone limits the crystal growth rate. If v is substantially dependent on l, it remains an open question whether the kinetic processes at the boundaries have the rate-limiting effect, and we do not know at which boundary (the melting or crystallizing one) the processes are more difficult and what the mechanisms might be.

Figure 1 shows points obtained from various sources for v as a function of l in various systems [8-11]; the points fall in the region characteristic of the layer mechanisms BC', and the region BB', which corresponds to the dislocation mechanism, fits better to the experimental points than does the region representing two-dimensional nuclei. The curve shapes show also that the kinetic effects in zone melting in the InSb – In system correspond well with the screw-dislocation mechanism, better than for the two-dimensional nuclei one.

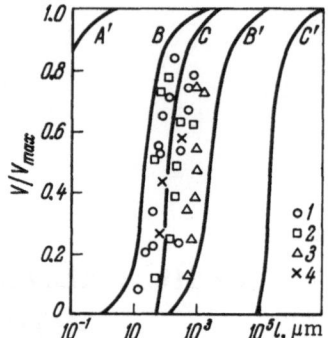

Fig. 1. Relative zone speed as a function of thickness. The observed points 1-4 are from [8-11] respectively.

There is no convincing proof of a limiting action of the zone boundary during zone melting in the kinetic state; Tiller [4] supposed from general considerations that the processes at the low-temperature boundary are the slower, while Wernick [11] established that the activation energy for the process in the Ge—Al system is fairly large at about 31 kcal/mole, which has been interpreted as meaning that dissolution is the rate-limiting step, as in [6], where the deduction was based on the formation of microscopic faces at the dissolution boundary for linear Ge—Au zones moving in germanium. However, these results are insufficient to indicate unambiguously that the dissolution is the rate-limiting step. It has been shown [8, 9] for Si—Al and Si—Ag systems that large activation energies may correspond to diffusion conditions. Facets are formed on the dissolution face on account of instability there arising from concentration superheating [12]. Only the most slowly growing faces should produce a stable configuration on the dissolution face, and these are {111} faces for Si and Ge.

We have examined the motion of linear zones for various solvent metals (Al, Au, Ag, Pt, Pd, Ni, Cu, Fe, and Sn) with silicon in zone melting for various mutual orientations of the crystal, the zone, and the direction of the temperature gradient. We used temperatures from 850 to 1250°C and temperature gradients from 20 to 100 deg/cm. In all cases the solution front was bounded by {111} planes, while the crystallization front had smooth outlines. If the {111} planes on the dissolution front are not parallel to the zone axis, the instability arising at this boundary results in a breakup of the zone into separate parts [14].

The main experiments were done under conditions such that the growing layers of the crystal formed in doped growth bands, while the dissolution front was indicated by a p—n junction arising on freezing the zone. The shape of the growth bands in the p—n junction gives the true morphology of the crystal—melt interface, not the one produced at the crystal — eutectic boundary after stopping the zone melting. This is the first time that such a method has been applied to this form of zone melting.

A decisive experiment for revealing the limiting boundary is examination of the migration speed for a planar zone when the dissolving and growing crystals have various orientations, as we did for Si—Al at 1100°C for zone thicknesses of 15, 35, and 66 μm using silicon plates oriented on {111}, {100}, and {110} planes. Change in the orientation from {111} to {100} in the dissolving crystal caused the speeds of the zone to alter on average by 8, 20, and 80%, respectively. When the orientation of the growing crystal was changed, we found no change in zone speed, and analogous results were obtained for {111} and {110} orientations. Then for Si—Al we have established unambiguously that the dissolution in zone melting is more difficult than crystallization if the dissolving and crystallizing pieces of silicon are oriented on {111} planes. One expects this conclusion to be correct for all the above systems based on silicon, since the morphologic features of the zone boundaries are similar to those for the Si—Al system, but one cannot rule out that other systems may have the crystallizing boundary as the

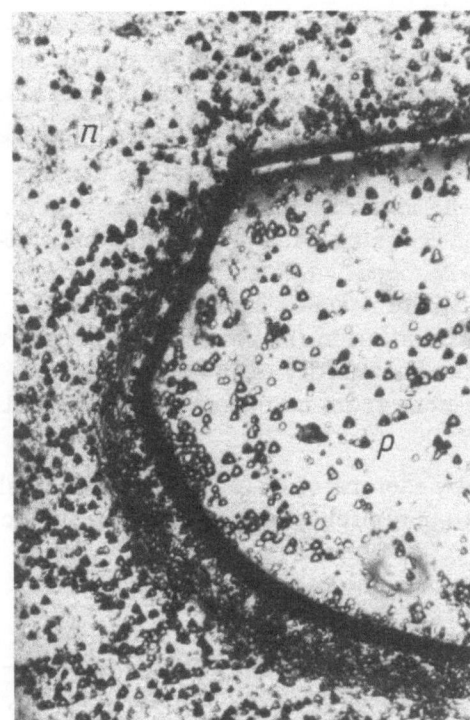

Fig. 2. Section of a silicon specimen with dislocation etch pits, ×300. The p region is a layer
of silicon recrystallized with Al as solvent,
while the n region is the initial crystal.

Fig. 3. Cross section showing movement of
linear zone and growth bands, ×300.

rate-limiting one, especially when the crystallization is hindered by the introduction of surface-active dopes [15].

A basic feature of the crystallization from a thin zone of moving solvent is that the concentration supercooling is substantially less than in other methods of growing a crystal at the rate, so crystallization from a moving solvent may be used to produce perfect single crystals.

We examined the defects in silicon films grown by zone melting; we found that inclusions of the melt may occur only at the stage when comparatively thick zones enter the crystal ($l > 100$ μm) without the use of a seed, when these zones penetrate into the crystal and new layers of solid state initially start to grow in from the sides. During such growth, the crystallization front is drawn out and is a source of branching microzones. As the zone moves onwards, the curvature of the crystallization front falls and the microzones are not produced. There is a reduction in the temperature gradient and a stabilization in the general temperature conditions as the zone enters the material, which reduces the probability of inclusions of a second phase [16]. When there is stable motion of a planar zone from a seed crystal, one finds no inclusions of the parent solution.

We found that the dislocation densities were reduced by factors of 2-10, as has previously been observed for the GaAs−Ga system. Figure 2 shows a photomicrograph of an etched specimen revealing the recrystallized and initial parts of the crystal. There is an accumulation of dislocations at the boundary of the deposited layer probably because of a large difference in the dope content in the growth and initial crystals.

Under certain conditions we found that dope bands occurred; Fig. 3 shows a photomicrograph of a specimen where selective etching reveals the recrystallized layer and these bands, which are related to temperature variations during the zone melting. If the mean temperature was stabilized to 2°C during zone melting, one obtained electrically uniform crystals with not more than 10% spread in specific resistance.

Conclusions

1. One finds layerwise crystallization and dissolution in all the systems where the mechanism has been established for the boundaries of the liquid zone in zone melting with a temperature gradient.

2. In the kinetic state, the crystallization rate in zone melting may be limited not only by processes at the growing front but also at the dissolving boundary, and the latter are the rate-limiting ones for many binary systems based on silicon.

3. The method allows one to grow perfect crystals; defects such as inclusions of a second phase and dope bands are comparatively easily eliminated by choice of appropriate growth conditions.

Literature Cited

1. D. T. I. Hurle, J. B. Mullin, and E. R. Pike, J. Mater. Sci., 2:46 (1967).
2. A. I. Mlavsky and M. Weinstein, J. Appl. Phys., 34:2885 (1963).
3. W. G. Pfann, Zone Melting [Russian translation], Metallurgizdat, Moscow (1960).
4. W. A. Tiller, J. Appl. Phys., 34:1757 (1963).
5. W. A. Tiller, J. Appl. Phys., 36:261 (1965).
6. R. G. Seidensticker, J. Electrochem. Soc., 113:152 (1966).
7. J. H. Wernick, J. Chem. Phys., 25:47 (1956).
8. A. I. Udyanskaya, Trudy Novocherkassk. Politekh. Inst., 170:33 (1967).
9. E. A. Nikolaeva, Trudy Novocherkassk. Politekh. Inst., 170:40 (1967).

10. R. W. Hamaker and W. B. White, J. Appl. Phys., 39:1758 (1968).

11. J. H. Wernick, J. Metals, 9:1169 (1957).

12. R. G. Seidensticker, in: Crystal Growth Problems [Russian translation], Mir, Moscow (1969), p. 197.

13. R. G. Rhodes, Imperfections and Active Centers in Semiconductors [Russian translation], Metallurgiya (1968).

14. V. P. Popov and V. N. Lozovskii, Proceedings of the 4th All-Union Conference of Young Scientists on Production and Use of Semiconductor Materials [in Russian], GIREDMET, Moscow (1969).

15. A. Ya. Gubenko et al., this volume, p. 124.

16. V. N. Lozovskii, V. P. Popov, and N. I. Dabrovskii, Proceedings of the Second Conference of Young Scientists from the Rostov Area, Natural Sciences Section [in Russian], Rostov-on-Don (1968).

GROWTH OF GERMANIUM SINGLE CRYSTALS FROM SOLUTION IN MERCURY

A. P. Cherkasov, R. Sh. Enikeev, and I. S. Aver'yanov

Mercury-doped germanium single crystals are of considerable interest because they have a photoelectric response in the far infrared. The basic requirements are high structural perfection, maximal mercury concentration, and uniform distribution.

Mercury-doped germanium is made by zone melting in mercury vapor [1, 2] and by pulling from the melt in the vapor. The product has a dope concentration of about $5 \cdot 10^{14}$ cm^{-3} and a dislocation density of 10^3-10^4 cm^{-2}.

We examined the growth of Ge crystals from solution in mercury to improve the structure and raise the mercury concentration. We assumed a retrograde solubility and a maximum content in the range 700-900°C, so the tests were done in that range.

Figure 1 shows the apparatus. The quartz tube contains a piece of Ge; mercury is poured in, and the tube is evacuated and sealed before being placed in a thick-walled steel tube. The necessary counterpressure was provided by pouring some mercury into the tube, which was then sealed up before being placed in an oven. The run temperature varied from 700 to 900°C, which was maintained for 24 hr and followed by cooling to 400°C over two or three days.

The dissolved germanium was deposited. Mercury is denser than Ge, so the latter forms a surface crust of small platy crystals in the form of regular triangles, hexagons, and rhombs (Figs. 2 and 3). The largest crystals were of 5-6 mm side and thickness 1 mm.

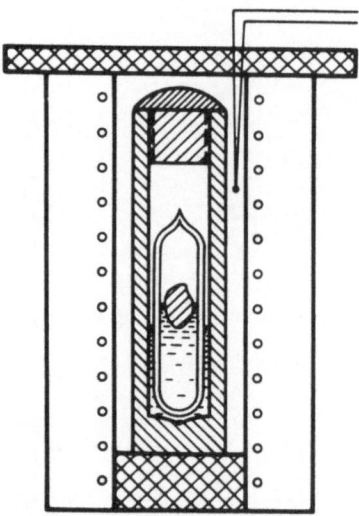

Fig. 1. The crystal growth apparatus.

132

Fig. 2. Hexagonal platy germanium single crystal, ×25.

Fig. 3. Platy rhombodial germanium single crystal, ×25.

Fig. 4. Octahedral germanium crystal, ×10.

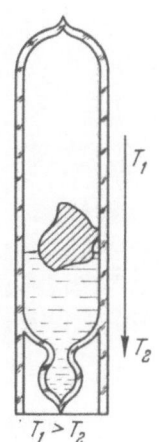

Fig. 5. Tube for making isometric germanium crystals.

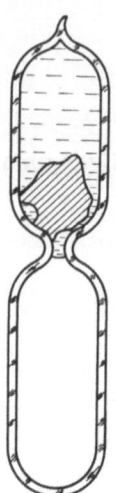

Fig. 6. Tube for making the largest crystals.

Under the crust were crystals in the form of dendrites, octahedra, and tetrahedra (Fig. 4) which were up to 20 mm long. The octahedra and tetrahedra were mostly very small, although individual ones were up to 2.5 mm in size.

Figure 5 shows a modified tube for producing larger crystals, in which there was hardly any free surface for the mercury.

The largest isometric crystals were grown in a special tube (Fig. 6) without temperature reduction; instead, the tube was gradually lowered into a cooler part of the oven; and crystallization began in the lower part. The small crystals floated up and redissolved, while the large ones were retained by the constriction and continued to grow.

We examined the structure and electrical characteristics. There were no dislocations, and the crystals had p-type conduction. The specific resistance was 0.2 ohm-cm for crystals grown at 800°C, which represents a mercury concentration of $2 \cdot 10^{16}$ cm^{-3} if the conductivity is determined by the mercury.

This method gives mercury concentrations raised by nearly two orders of magnitude with improved structure.

Literature Cited

1.   S. B. Borello and H. Levinstein, J. Appl. Phys., 33:2947 (1962).
2.   J. K. Lennard, Proc. Nat. Electronics Conf., 8:816 (1962).

PART IV

CRYSTAL GROWTH FROM MELTS

MORPHOLOGY OF CRYSTALS PULLED FROM THE MELT

M. D. Lyubalin

There is no specific theory of the morphology of rounded crystals grown from a melt; one cannot apply to these the concepts and methods of classical morphology, which were formulated from freely grown polyhedral crystals, and which relate the shape to the internal structure and conditions of formation.

The main features of a rounded crystal as distinct from a polyhedral one are due to the motion of the growing crystal through the melt, as during pulling, and to the considerable flow of heat from the melt into the crystal through the growth surface.

Effects of Pulling

We examined these for clarity by reference to a freely growing polyhedral crystal, neglecting the effects of capillary forces and melt nonuniformity (Fig. 1). It is clear that pulling produces face elements of two types: true faces, edges, and corners and also pseudofaces, pseudoedges, and pseudocorners.

A pseudoface is a surface that is formed on pulling by a set of lines of intersections between the melts and a true crystallographic plane; a pseudoedge is the line composed of a set of pseudovertices, which are points of intersection of the melt with the true edge.

True facet elements belong to the interface between a crystal and a melt, i.e., they occur at the growing end of the crystal; the pseudo elements occur at the part of the growth front where there are three phases: crystal, melt, and gas or vacuum. During pulling, the set of these gives rise to a side surface, and a pseudoface differs from a true face in almost never participating in the growth of the bulk. The shape and position of a pseudoface on a crystal are to some extent arbitrary, because they are determined not only by the growth rates of the true faces v_n but also by the orientation and speed of pulling v_b. If all these quantities are known, the position of a pseudoface can be found from the formula*

$$\alpha = \tan^{-1} \left(\frac{v_n}{v_b} \operatorname{cosec} \varphi \pm \cot \varphi \right), \tag{1}$$

where α is the angle between the normal to a pseudoface and the pulling direction, while φ is the angle between the normal to the true face in that direction.

A pseudoface can coincide with a true face for a particular relation between v_n and v_b (III in Fig. 1b).

Not all possible true faces, edges, and corners participate in producing the volume of a pulled crystal; Fig. 1d shows an immobile crystal growing at the surface of a melt ($v_b = 0$);

---

*The plus sign applies for $\varphi < 90°$.

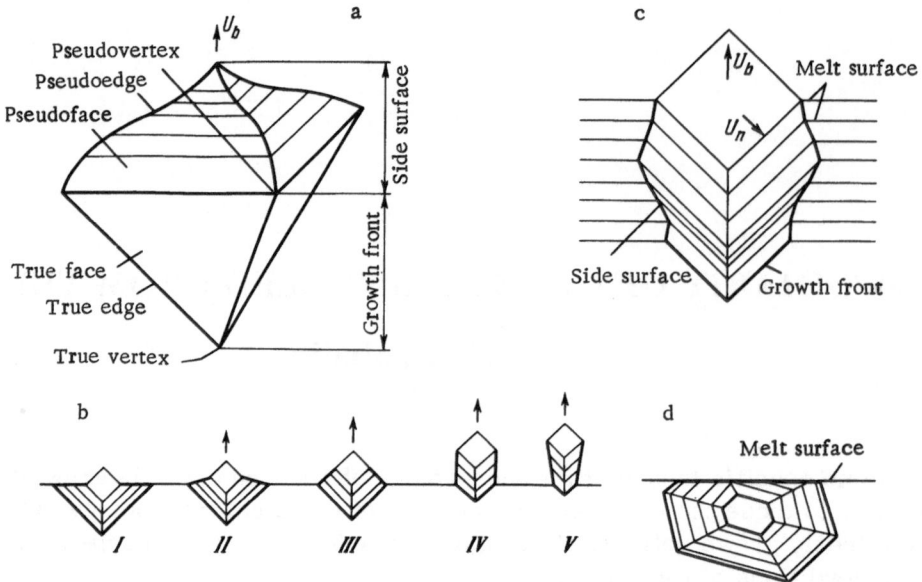

Fig. 1. Effects of intersection: a) Three-dimensional crystal; b) shape change with change in v_B, $v_n =$ const; c) shape change for $v_B =$ const, $v_n \neq$ const (level of melt reduced instead of crystals pulled); d) growth of a fired crystal ($v_B = 0$) side surface coinciding with surface of melt.

that it is clear that the growth of a vertex ceases at once after through the surface of the melt, while an edge ceases to grow when both of the vertices belonging to it intersect the melt surface. A face is lost completely if the direction of displacement of all of its vertices constitutes an angle less than 90° with the surface of the melt. If the crystal moves upwards ($v_b > 0$), all vertices, edges, and faces vanish, if their velocities of displacement in the pulling direction are less than v_b. Then pulling results in geometrical selection, with survival of only those true facial elements that are best oriented relative to the surface of the melt, or, in other words, make large angles with the normal to it.

This argument shows that pulling results in two types of face surface: side ones and a growth front, whose roles in crystal production are different.

The side surface reflects the change in the growth from the start to end and therefore bears at least a partial chronological record of the process, this information usually being derived from the internal morphology via special sections [1].

Effects of Heat Fluxes

We consider the effects on crystal shape from heat fluxes via the growth surface via the following assumptions:

(1) the melt directly at the growing surface is supercooled;
(2) the degree of cooling of the melt is the greater the higher the crystallization rate.

The following equation gives the speed of an isotherm for a constant temperature at the growth front:

$$\frac{dr(t)}{dt} \approx \sqrt[3]{\frac{\varkappa}{L\gamma} r_{\text{cool}} (T_{\text{cf}} - T_{\text{cool}}) \frac{1}{9t^2}} , \tag{2}$$

where r_{cool} is the radius of the spherical cooler, T_{cool} is the temperature of this, L is the latent heat of crystallization, γ is the crystal density, $\varkappa$ is the thermal conductivity, and T_{cf} is the temperature at the crystallization front.

We have derived this expression for growth of crystals in a field with spherical symmetry by the method [2], which involves joint solution of the equations of thermal conduction and crystallization kinetics in the field with the symmetry of a cylinder; (2) implies that the growing surface is displaced parallel to itself in the steady state, when the temperature gradients are constant. In this case, one should have equal rates for crystal growth and displacement of the melting isotherm; this condition is met when the front lags far enough behind the melting isotherm for the supercooling to provide the necessary growth rate. High pulling rates demand high supercoolings.

The shape of a pulled crystal is dependent not only on the supercooling but also on the relation between the temperature gradients in the melt and the crystal; in the steady state, and with a planar growth front, the pulling rate is defined by the equation for heat balance:

$$v_{\text{B}} = \frac{\varkappa\,(G_{cr} - G_m)}{\gamma L}. \tag{3}$$

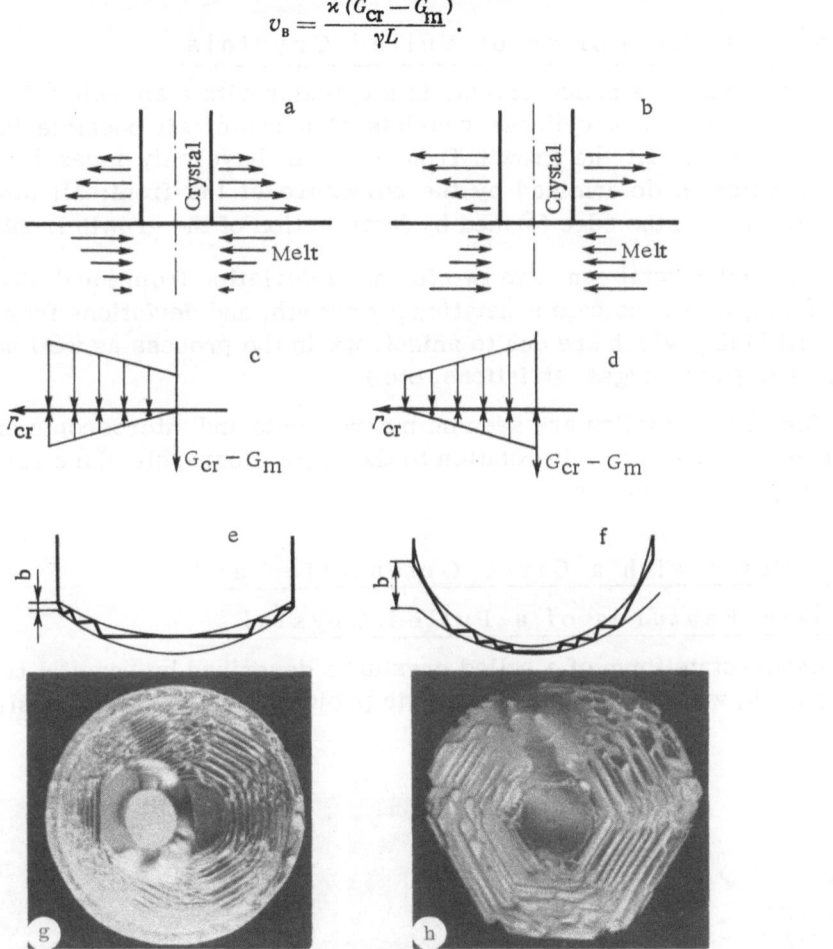

Fig. 2. Crystal shape in relation to heat-flux geometry in crystal-melt system. a, b) Heat flux patterns; c, d) axial gradients in melt and crystal at interface; e, f) different shapes for germanium crystals (b is the gap between the growth surface and the instantaneous position of the melting isotherm), g, h) photographs of growth fronts; ×5.

It is assumed here that the thermal conductivity $\varkappa_m$ of the melt is the same as that of the crystal $\varkappa_{cr}$ at the interface ($\varkappa_m = \varkappa_{cr} = \varkappa$); while G_{cr} and G_m are the temperature gradients in the crystal and melt at the interface.

It follows from (3) that pulling at a given speed can be performed with various G_{cr} and G_m if the difference between these is unaltered; the crystal grows its faces better when there is a small positive gradient in the melt, but crystals that do not differ macroscopically can be produced at different pulling rates, as can be seen if the region of the supercooled melt is represented as a gap between the instantaneous position of the growth surface and the melting isotherm [3]. Now this approach enables one to explain also why one gets crystals with different cross sections and growth-front shapes for identical mean diameters and pulling rates. Figure 2 shows two cases of crystal growth under different conditions; at each point on the front, the total axial gradients ($G_{cr} - G_m$) are identical, but in Fig. 2a and c, the positive gradient G_m at the periphery is greater than that at the center, and one gets a crystal with a rounded cross section. Parts b and d of Fig. 2 show the converse case; faces can arise at the edge of the front, but this is difficult at the center.

Ideal Steady-State Forms of Pulled Crystals

The ideal form of a pulled crystal is a cylinder with a smooth side face and base; the side surface of such a cylinder consists of a set of all possible faces parallel to the pulling direction. At the growth front one can have only faces lying in a small circle of radius ρ, which is determined by the curvature of the front; all other possible faces may be represented in the edge formed by intersection of the growth front with the side surface.

A real pulled crystal can have two forms of deviation from ideal shape: deviations from cylindrical form, which indicate nonstationary growth, and deviations from smoothness due to inflections and kinks, which are due to anisotropy in the process as well as to transient conditions (the planar parts, edges, striations, etc.).

Both forms of deviation are seen on macroscopic and microscopic examination; the relief of the growth front varies in relation to the crystallographic plane representing a particular microscopic area (Fig. 3).

Path of a Point with a Given Orientation and
Morphologic Features of a Pulled Crystal

The steady-state form of a pulled crystal is described by parallel transfer, a distinction from free growth, which can be described via projection from center of similitude. Under

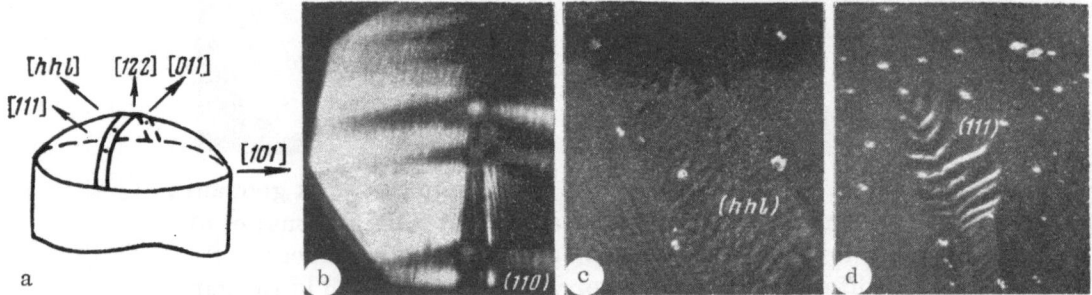

Fig. 3. Differences in fine structure of growth fronts: a) schematic; b-d) photographs; ×120.

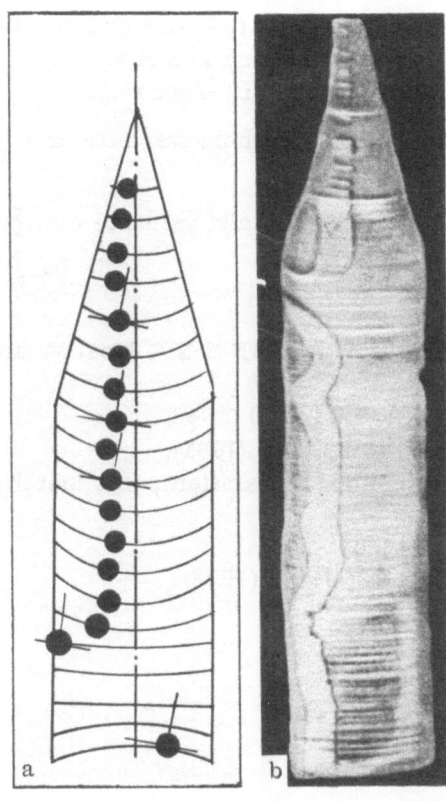

Fig. 4. Path of a point with a given orientation at the growth front. a) Schematic; b) channel in the crystal [4] related to (111) face.

steady-state conditions during pulling, the path of any point with a given orientation is a straight line parallel to the pulling direction; changing growth conditions lead to curvature of the path. The positions of a point with a given orientation move relative to the geometrical axis of the crystal when the curvature of the growth front alters, as is clearly seen by etching longitudinal sections of doped single crystals of certain semiconductors such as germanium when there is a macroscopic (111) face at the front (Fig. 4).

Facetting and habit are terms used to describe the shape of a polyhedral crystal; to describe a circular crystal, one needs to (1) establish the crystallographic symbol for the most prominent elements of the surface, (2) determine the visible symmetry of the crystal by comparison of parts of the surface identical in crystallographic symbol, and (3) determine the ratio of the dimensions of parts of the surface that differ in crystallographic symbol (to define the surface curvature). Steps 1 and 2 have been solved by photogoniometry [5], while the curvature may be determined by photogoniometry or from the distortion [6].

Conclusions

1. One gets geometrical selection on pulling crystals; out of all the possible face elements, the crystals retain only those that are best oriented relative to the free surface of the melt. One gets two types of surfaces: a side surface and a growth front, which differ in role and in crystal formation.

2. One can get constant temperature gradients in crystal and in the melt at the interface if the growing surface is displaced parallel to itself; steady-state growth of a pulled crystal is described geometrically by parallel transfer, a distinction from free growth, which is described by translation from a center of symmetry.

3. The shape of a pulled crystal is determined by the ratio of the temperature gradient in the melt and crystal; one can get crystal; one can get crystals differing in shape with a given linear crystallization rate, and also ones identical in shape with different rates.

4. The curvature at the surfaces is an important morphological characteristic of a rounded crystal.

I am indebted to Professor N. N. Sheftal' for interest in the work and valuable criticism.

Literature Cited

1. G. G. Lemmlein, Sector Structure in Crystals [in Russian], Izd. AN SSSR, Moscow and Leningrad (1948).
2. V. V. Gerasimenko and B. Ya. Lyubov, Kristallografiya, 13(4):750 (1968).
3. A. Trainor and B. E. Bartlett, Solid State Electronics, 20:23, 166 (1961).
4. J. Dykov, Imperfections in Semiconductor Crystals [Russian translation], Izd. Metallurgiya (1964), p. 279.
5. V. A. Mokievskii and Ch. D. Dzhafarov, ÉVMO, 91:1 (1956).
6. M. D. Lyubalin and V. A. Mokievskii, Kristallografiya, 13(4):735 (1968).

OBSERVATIONS ON THE SOLIDIFICATION OF
A DROP OF NaCl MELT

E. Hartmann

In the 1930s, Gyulai [1] began to examine the solidification activity of molten NaCl; here I consider the problem in relation to the morphology of the resulting grains.

The thermal conductivity of the NaCl melt (melting point 804°C) was observed by the simple method described by Gyulai [1, 2]. A drop of NaCl was melted in a platinum loop heated by alternating current; the loop diameter was 2 mm, while the platinum wire had a diameter of 0.3 mm and the current was 6 A. The solidification on reducing the current was observed visually or photographically via a microscope operated at ×5-80. The drop was 2-3 cm from the lens, and the axis of the microscope was horizontal to avoid condensation of NaCl vapor on the lens. The platinum loop containing the drop was mounted on a micromanipulator. Another current-heated loop was placed beyond the molten drop, so the rear side did not solidify and interfere with observations on the last face of the drop. The loop also served as a light source. For reflection use we employed a device with a projector lamp, a lens, and a right-angled prism.

This simple apparatus allows one to vary the heating rate and to repeat experiments rapidly many times. Observations in transmission or reflection facilitate distinction of processes occurring at the surface and in the bulk of the transparent melt. The drop takes the form of a lens, which considerably complicates focusing the microscope and sometimes results in poor photographs.

The solidification was examined with rapid cooling (sudden switching off of current) or with slow cooling.

In the first case, the drop solidifies in a few seconds, the exact time being dependent on the size and temperature, the shape of the platinum loop, the distance between the drop and the second wire, the temperature of the second loop, and other factors, i.e., on the heat transfer and degree of supercooling. If cooling was rapid, one could distinguish three types of crystallization, although there were no sharp boundaries between these.

In the first type, the outer surface of the drop rapidly becomes covered with a thin layer of grains with mirror-smooth faces. Under this transparent layer, the melt contained a multitude of spherical crystals, which grow rapidly to the size of 50 μm (Fig. 1). At the end of the solidification, spherical stars are also present. The crystallization ends when all the spherical stars, which become facetted, are in contact with one another, while the outside of the drop remains covered with flat platelets.

In the second type, the surface of the drop is formed by stellate crystals (Fig. 2), whose number is less than that of the small plates in the previous type; these stars intergrow, while

143

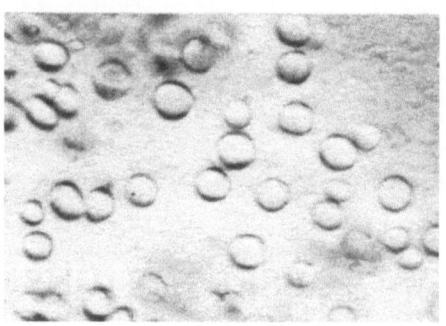

Fig. 1. Spherodial crystals formed in molten
NaCl by rapid cooling, ×90.

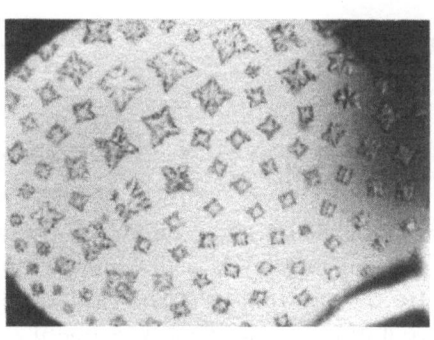

Fig. 2. Dendritic crystals formed on a rapidly
cooled drop of NaCl, ×25.

under them there are produced three- or four-ray dendrites (these are rare, Fig. 3), whose contours are initially very much rounded, but which subsequently become more complicated and sharper in the four-rayed case along the ⟨100⟩ direction. The dendrites grow into the gaps of the drop, but the outer surface of the drops have crystals that are usually smooth, in spite of the row of surface holes along ⟨110⟩ direction (Fig. 4). Observations in reflected light and in transmission indicate that the surface holes are traces of narrow channels formed by growth that gradually fill up.

In the third type, one gets several large dendrites; reflected light shows that the solidified drop has small hummocks composed of parallel terraces, while in the vertices there are flat parts with rows of holes in the form of a cross.

There is a correlation between the types of rapid solidification and the drop temperature; the higher the temperature, the fewer the nuclei, the larger grain sizes, and the more pronounced the steps on the outer surfaces of the grains. If a drop solidifies instantly after melting, one gets the first type; if the melt is kept for 2-3 seconds at high temperature before rapid cooling, one gets the second type; and one gets the third type when the superheating is maintained for about 10 seconds.

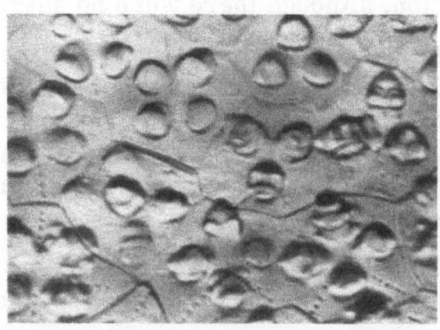

Fig. 3. Cloverleaf dendrites in molten NaCl,
×90.

Fig. 4. Smooth surfaces on a frozen granule, ×80.

When the cooling is slow, the crystallization starts at a single point; during melting, the residue of the solid phase vanishes at the same point. One observes an interesting effect on cooling: the crystal grows, volutes rapidly, and stops doing so only when much of the melt has cooled. The stroboscope indicates that the speed of rotation may be as high as 1200–1400 rpm. If the crystallization is slow, one usually gets only one grain; one gets either platy forms (Fig. 5 a, b) or dendritic ones (Fig. 5c). The rows of surface holes always lie along ⟨110⟩ directions.

We determined the orientations of the grains from the disposition of the cleavage planes and from etch figures (we used acetic acid). The crystals were cleaved by heating to 600° and plunging in paraffin oil; the resulting cracks lay at 45° to the rows of surface holes, so the orientation of the rows was ⟨110⟩.

A line film enabled us to determine the crystal growth rates; and the shift in the boundaries between the grains gave the growth rates at intervals of 100–1000 μsec; the values were several times larger than those reported in [3]. The crystals produced in the melt had a spherical shape, probably on account of surface tension; it is considered [4, 5] that this can be effective even below the melting point, where the strength of the crystal is small, i.e., the crystal becomes spherical, and this would explain the rounded dendrites seen at the start of growth, which subsequently become sharper.

The ⟨110⟩ surface holes probably indicate that the surface growth is blocked in these directions; it has been observed [6] that such blocking occurs in these directions when NaCl

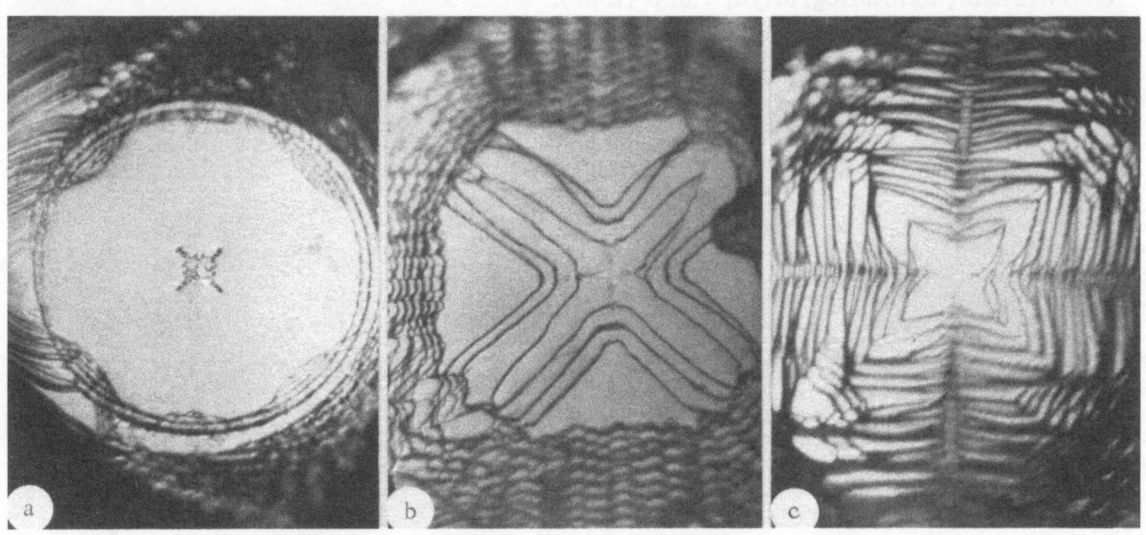

Fig. 5. Surfaces of crystals formed by slow cooling, ×50.

crystals grow from the vapor state, and the effect was ascribed to impurities. One can explain as follows the effects of melt temperature on crystallization type: the higher the melt temperature, the less the rate of loss of heat as the melt approaches the onset of crystallization, and the number of crystallization centers is reduced when the cooling rate is lowered. If the number of nuclei is small, each grows for a fairly long time before it meets another, and therefore it becomes large and more complicated in shape. A similar effect has been reported for metals. It has been found that a low casting temperature facilitates increase in the number of crystals, whereas high pouring temperatures reduce the number of crystals, and the crystal size increases [7]. It may be that the superheating time affects the type of crystallization not only by the heat-loss rate but also via its effects on catalytic particles, because the latter can [8] reflect the dependence of the nucleation rate on the thermal history of the liquid.

The rotation of the crystal during growth is not related to crystallization; the nonuniformity in the temperature and surface tension in the melt give rise to flows, which occurred around the growing crystal and thus the latter revolves along with the melt in the drop.

Conclusions

1. One gets a multitude of spherical nuclei when a drop of NaCl melt cools rapidly.

2. The crystallization type and morphology are dependent on the melt temperature and cooling rate.

3. The crystals rotate at 1200-1400 rpm as the solidification proceeds.

I am indebted for advice to the late Professor Z. Gyulai and to Professor M. P. Shaskol'skaya.

Literature Cited

1.   Z. Gyulai, Z. Kristallogr., A91:142 (1935).
2.   Z. Gyulai, Z. Phys., 125:1 (1948).
3.   Crystallization Mechanism and Kinetics [in Russian], Izd. Nauka i Tekhnika, Minsk (1964), p. 71.
4.   L. Graf, Z. Metallkunde, 42:401 (1951).
5.   L. Graf, Z. Metallkunde, 45:36 (1954).
6.   G. Turchani, Kristallografiya, 7:290 (1963).
7.   B. Chalmers, Solidification Theory [Russian translation], Metallurgiya (1968), p. 245.
8.   D. Turnbull, J. Appl. Phys., 21:1022 (1950).

PREPARATION AND EXAMINATION OF SINGLE CRYSTALS OF SOME FERROELECTRICS WITH THE STRUCTURE OF TETRAGONAL K–W BRONZE

O. F. Dudnik, V. B. Kravchenko, A. K. Gromov, and Yu. L. Kopylov

Single crystals have been made recently for various ferroelectrics employed in electronic devices [1, 2]; many of these have the structure type of tetragonal potassium–tungsten bronze [3-8], and particular interest attaches to solid solutions in the system $BaNb_2O_6-SrNb_2O_6$ [4] on account of the very high electrooptic constants found there.

There are several papers [9-11] on the $(Ba, Sr)Nb_2O_6$ system for ceramic specimens; it is found [9] that solid solutions with the tetragonal-bronze structure are formed in this system in the range 15-80 mol.% $SrNb_2O_6$. Rather different solubility limits have been stated [10] for this system: 20-60 mol.% $SrNb_2O_6$. Smolenskii et al. [10] also found that the $BaNb_2O_6-CaNb_2O_6$ system gives rise to solid solutions in the range 15 to 40 mol.% $CaNb_2O_6$, and they assumed that analogous solid solutions would occur in the $(Ba, Sr, Ca)Nb_2O_6$ ternary system.

So far, single crystals have been obtained only for $Ba_xSr_{1-x}Nb_2O_6$ with $x = 0.25 - 0.75$ [4]; they were grown by Czochralski's method from iridium crucibles and were of a dark amber color, but they could be lightened to yellow or pale amber by prolonged annealing at 1400°C in oxygen.

We have examined the scope for making single crystals in the $(Ba, Sr, Ca)Nb_2O_6$ system in the three initial binary systems, and we have also examined some of their properties.

The crystals were grown by Czochralski's method in an induction oven in iridium and platinum crucibles in air usually at pulling rates v of 10-12 mm/hr and rotation speeds n of 2 rpm. The automatic temperature control was provided by a modified Redmet-201 system, which provided a constant temperature at the middle point of the bottom of the crucible and gradual temperature change with a set rate. The raw materials were the carbonates of Ba, Sr, and Ca together with niobium hexoxide, all of special purity grade.

The results for $Ba_xSr_{1-x}Nb_2O_6$ with $x = 0.25 - 0.75$ in iridium crucibles were as in [4]; with platinum crucibles we obtained single crystals ranging from dark blue with rapid cooling to colorless on slower cooling. The short-wave transmission of the colorless crystals was better than that of the dark-amber ones (Fig. 1). The reason for the blue color has not yet been established.

X-ray evidence indicates that barium–strontium niobate has point groups 4 mm [12]; if the c axis coincides closely with the pulling direction, one gets good faces on the crystal, and perfect edges where adjacent faces intersect, and the perfection increases with x. The faces

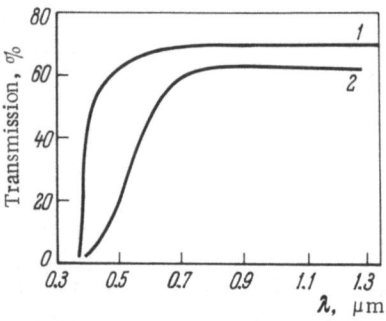

Fig. 1. Short-wave transmission of $Ba_{0.5}Sr_{0.5} \cdot Nb_2O_6$ crystals grown in crucibles of (1) platinum and (2) iridium.

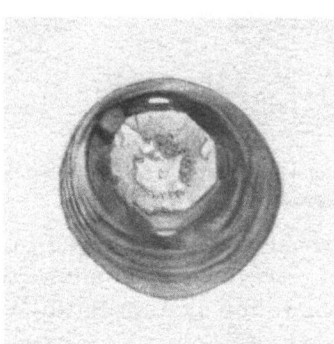

Fig. 2. Lower part of a crystal with slow detachment from the melt, × 4.

are mirror smooth, and no microrelief is observed at magnifications up to 200 times. If $x \geqslant 0.5$, the edges are usually sharp; the faces often have striations with distances between striae equal to v/n.

The iridium and platinum crucibles gave rise to patches of metal on the faces in the form of hexagons with various side lengths, and also triangles, and the {130} faces had much smaller islands of metal than the others. The surfaces of these iridium patches were rough, with a certain appearance of bulk, whereas the platinum ones were mirror-smooth.

Goniometry showed that the habit is determined by 24 faces of four prisms: {110}, {120}, {100}, and {130}; if x = 0.25, all faces are roughly equally developed. If x is increased, the four faces of the {100} prism become smaller and degenerate into very narrow strips, which is the origin of the statement [1] that the crystals have 20 faces. The crystallization front, as defined by rapid removal, was basically a more or less perfect face of the (001) pinacoid, while on spontaneous relatively slow detachment one obtained clear faces of the {201} and {111} pyramids also (Fig. 2). Figure 3 shows the idealized form of such a crystal.

We consider that this morphology arises from specific features of the growth, in particular stabilization of the cross section of the facetted crystal; if the facetting is complete (sharp edges), it persists during growth even when the melt temperature change is considerable (10°C or more), provided that this is reasonably slow.

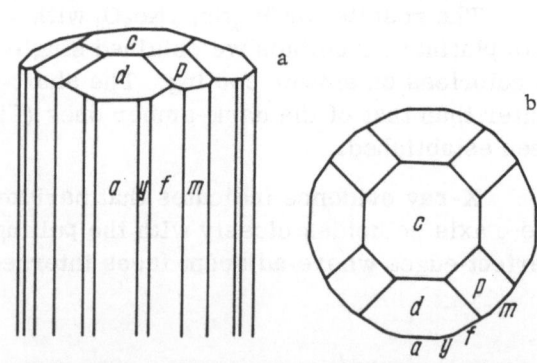

Fig. 3. Idealized $Ba_xSr_{1-x}Nb_2O_6$ crystal. a) General view; b) lower part of a crystal with slow detachment from the melt.

When crystals of other compounds such as $LiNbO_3$ and $ZnWO_4$ are grown under similar conditions, the cross sections change in response to temperature variation by as much as 1-2 orders of magnitude; gradual temperature reduction leads at some point to spontaneous and very rapid expansion of the crystal via the striated faces with high indices, and the tangential growth rate becomes much higher than normal. This gives interest to preliminary tests that show that these crystals can be grown at elevated rates (up to 40 mm/hr).

The crystal expands not only after stopping the temperature reduction as soon as expansion is noted but also when the temperature is raised slightly at this moment; before the diameter stabilizes again, the crystal expands to two-thirds or three-quarters of the diameter of the crucible, which is 30 mm. The initial pattern may have high symmetry (fourfold axis), which is one of the basic conditions for producing crystals without cracks, in which case the pattern is generally retained; then a considerable temperature rise is needed to restore the original cross section.

There is a tendency for the cracking to increase with x, i.e., as the tendency to produce faces increases, which may explain why the actual cross section does not coincide with the equilibrium one for a given growth temperature.

Conversely, for x = 0.25, where there is least tendency for the above to occur, it has never been possible to make long crystals if the temperature remains constant. Crystals up to 7-9 mm in diameter and up to 2 cm long have been made only by constant temperature reduction during gorwth, which ultimately always leads to the above spontaneous expansion, while repeated attempts to prevent expansion by stopping the temperature reduction have led to rapid detachment of the crystal from the melt. From this we have drawn the preliminary conclusion that the greater the cross section of the crystal the shorter the length that can be pulled before the onset of expansion, and thus there is an optimal rate of temperature reduction for each cross section, i.e., the one giving the greatest length. However, to restore the original cross section after expansion requires a considerable temperature rise, which enables one to continue the growth up to the next expansion and to obtain crystals of the necessary length.

To explain these effects requires additional tests, especially determination of the supercooling, which here may well exceed 10°, and one also needs to examine the composition shift in the region under the crystal, particularly for x = 0.25.

We found that crystals with x = 0.25 show least tendency to crack under given conditions of growth and cooling, which was the basis for even larger reduction in x, which is desirable on the basis of the dielectric constant ε_c' at room temperature as a function of composition (Fig. 4), since the electrooptic coefficient is [13] directly proportional to ε.

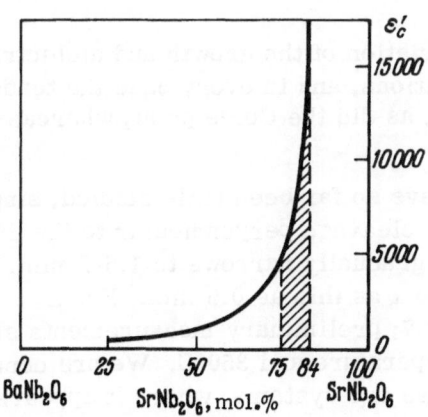

Fig. 4. Composition dependence of ε_c' (polar axis) at 20°C for $Ba_xSr_{1-x}Nb_2O_6$.

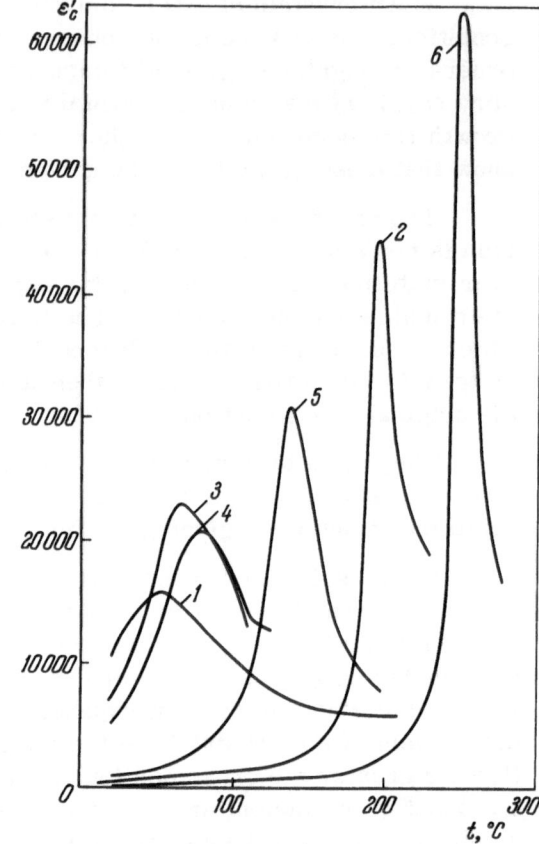

Fig. 5. Temperature dependence of ε_c' for niobate single crystals. 1) $Ba_{0.84}Sr_{0.16}Nb_2O_6$; 2) $Ba_{0.6}Sr_{0.2}Ca_{0.2}Nb_2O_6$; 3) $Ba_{0.25}Sr_{0.65}Ca_{0.1}Nb_2O_6$; 4) $Ba_{0.25}Sr_{0.65}Ca_{0.1}Nb_2O_6 : CuO$; 5) $Ba_{0.5}Ca_{0.5}Nb_2O_6$; 6) $Ba_{0.75}Ca_{0.25}Nb_2O_6$.

Crystals with the potassium—tungsten bronze structure grow down to x = 0.16; at this point, a certain excess of niobium oxide may be of value. The hatched region in Fig. 4 represents compositions for which single crystals have not previously been obtained.

In the $Ba_xCa_{1-x}Nb_2O_6$ system we grew single crystals with x of 0.5-0.84, and for x = 0.55 to 0.78 they were technically satisfactory. The Curie points are, respectively, 140, 250 and 350°C for x of 0.5, 0.75, and 0.84, respectively. The dielectric constant is largely independent of temperature up to 10-15°C away from the Curie point (Fig. 5). The crystals have a morphology similar to that of $Sr_{0.75}Ba_{0.25}Nb_2O_6$.

In the ternary system we made crystals of compositions $Ba_{0.25}Sr_{0.65}Ca_{0.1}Nb_2O_6$ and $Ba_{0.6}Sr_{0.2}Ca_{0.2}Nb_2O_6$, whose Curie points were 62 and 194°C, respectively; crystals containing Ca show less tendency to become colored.

All these solid solutions showed a monotonic variation of the growth and dielectric properties in relation to the mean radius of the divalent cations, and in every case the tendency to produce faces and to crack increased with that radius, as did the Curie point, whereas the room-temperature dielectric constant fell.

The systems $(Ca, Sr)Nb_2O_6$ and $(Mg, Ba)Nb_2O_6$ have so far been little studied; single crystals of $Sr_xCa_{1-x}Nb_2O_6$ with x of 0.5-0.8 had perfect cleavage perpendicular to the direction of spontaneous nucleation as the polycrystalline boule gradually narrows to 1.5-2 mm. These crystals are readily controlled; one can easily produce x as thin as 0.5 mm. From $Ba_xMg_{1-x}Nb_2O_6$ melts one gets crystals only near x = 0.7; preliminary measurements show that ε' for this composition varies little between room temperature and 350°C. We are continuing the dielectric measurements and x-ray studies on these two systems, where it appears that the solid solutions belong to a different structural type.

Literature Cited

1. E. G. Spencer, P. V. Lenzo, and A. A. Ballman, Proc. IEEE, 55:2074 (1967).
2. I. S. Rez, Usp. Fiz. Nauk, 93:633 (1967).
3. W. A. Bonner, W. H. Grodkiewicz, and L. G. Van Uitert, J. Crystal Growth, 1:318 (1967).
4. A. A. Ballman and H. Brown, J. Crystal Growth, 1:311 (1967).
5. J. J. Rubin, L. G. Van Uitert, and H. J. Levinstein, J. Crystal Growth, 1:315 (1967).
6. L. G. Van Uitert, H. S. Levinstein, J. J. Rubin, C. D. Capio, E. F. Dearborn, and W. A. Bouner, Mat. Res. Bull., 3:47 (1968).
7. E. A. Giess, G. Burns, D. F. Kane, and A. W. Smith, Appl. Phys. Lett., 11:233 (1967).
8. D. F. O. Kane, G. Burns, B. A. Scott, and E. A. Giess, J. Electrochem. Soc., 115:1018 (1968).
9. M. H. Francombe, Acta Crystallogr., 13:131 (1960).
10. G. A. Smolenskii, Ya. I. Ksendzov, A. I. Agranovskaya, and S. N. Popov, in: Solid–State Physics [in Russian], Vol. 2, Izd. AN SSSR (1959), p. 244.
11. I. G. Ismailzade, Kristallografiya, 5:268 (1960).
12. P. B. Jamieson, S. C. Abrahams, and J. L. Bernstein, J. Chem. Phys., 48:5048 (1968).
13. S. K. Kurtz and F. N. H. Robinson, Appl. Phys. Lett., 10:62 (1967).

CRYSTALLIZATION KINETICS OF SODIAN
FLUOROTREMOLITE FROM A MELT

G. A. Rozhnova, G. I. Kosulina, and L. F. Grigor'eva

There are purely scientific and also practical aspects of the physicochemical conditions for production of fluoramphibole from a melt, in particular the kinetics of the crystallization, on account of the need for new materials with specific technical properties.

We used a melt close in composition to sodian fluorotremolite ($0.5Na_2O \cdot 2CaO \cdot 5MgO \cdot 8SiO_2 \cdot F_2$); complete homogenization of the melt was provided by fusing the mixture in platinum crucibles at 1360-1400°C, which certainly exceeds the liquidus of 1190°C. The melt crystallized on cooling at various rates from the maximum temperature down to set points (1150, 1100, 1050, and 1000°C) with halts of 10 to 60 min at each temperature, and then quenching in ice. The mode of crystallization of the specimens was examined on sections.

The exact temperature and time had little effect on the nucleation frequency; the individual nuclei generally arise at the walls of the crucible in contact with the melt, and from these there grow out spherulitic clumps consisting of radially diverging amphibole needles (Fig. 1). The melt tends to produce a glass, and the crystallization is spherulitic, and this behavior is due to the marked increase in the viscosity as the temperature is reduced, for it is generally recognized that viscosity affects the crystallization rate [1-4]. It is therefore of interest to relate the crystallization kinetics to the viscosity, and we made measurements over the range 1360-1200°C with GOI viscometer and at 640-580°C by compression. The viscosities in the range 1200-640°C (the crystallization range) were obtained by extrapolation in accordance with the equation $\log \eta = a + b/T^2$ (Fig. 2).

The high crystallization temperature and the marked tendency to vitrification required special techniques in this work, and prevented a complete quantitative description of the process. We obtained only a qualitative evaluation of the crystallization capacity. The nucleation rate is low at 1180-1000°C, but the linear growth rate of the fluoramphibole crystals is high.

We used the method of [3] to examine the crystallization kinetics; we took pieces of doubly homogenized quenched glass, whose structure may be considered as similar to that of the initial melt, which were placed in an oven heated to a set temperature. After the appropriate time at the set temperature, the specimens were quenched in ice. The difference between the liquidus temperature for this melt (1200°C) and the working temperature corresponded to the supercooling; x-ray and electron-microscope results showed that the sodian fluorotremolite has a crystallization region of 620-1190°C, and also that the supercooling affects the shape and size of the crystals.

Short prismatic crystals with various orientations are produced within the body of the glass when the amphibole crystallizes; if the temperature is low and the supercooling is large,

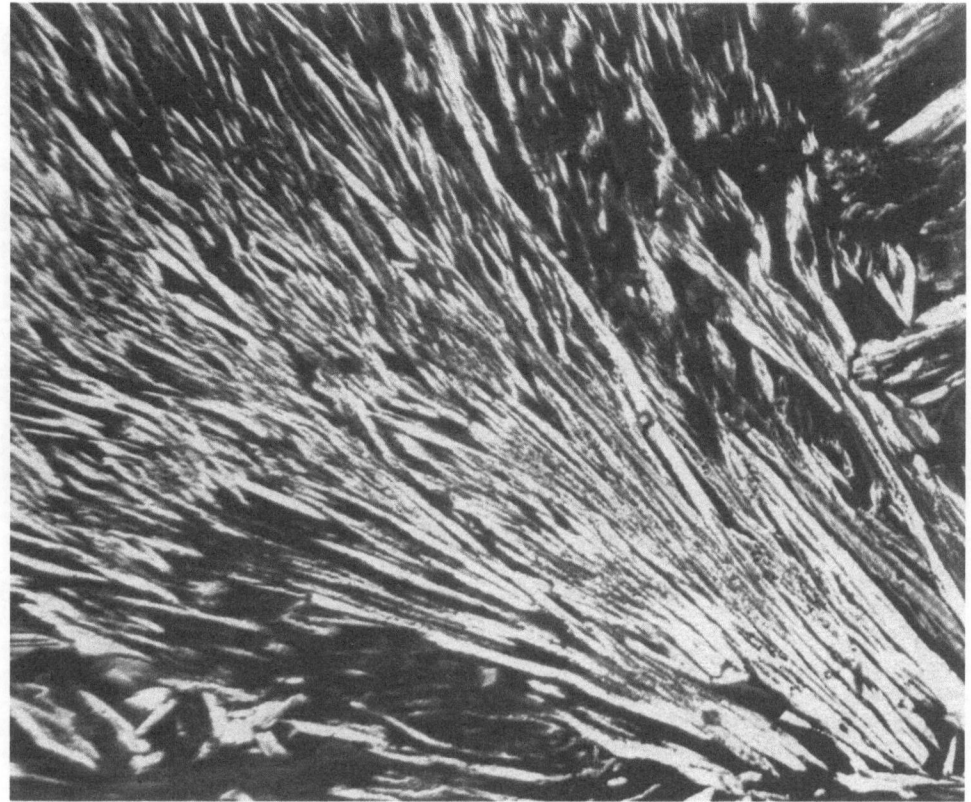

Fig. 1. Microstructure of soda fluorotremolite made by rapid cooling of melt, ×90.

the crystals are nearly isometric (Fig. 3); as the temperature is raised, the crystals become appreciably elongated along [001] (Fig. 4).

In the early stages of crystallization (Fig. 5), up to the point where the growing surfaces meet, there is a linear relation between the maximal crystal length as a function of temperature, and also such a relation for the growth rate $\Delta l/\Delta t$. Once the crystals have met, the growth rate falls, on account of change in the growth mechanism. Assuming that the growth rate of the crystals is constant, we extended the temperature curves to meet the abscissa; the resulting points (a, b) correspond to the induction time τ needed for nucleation [5]. The main main factor governing the linear growth rate is clearly the temperature.

Figure 6 shows that the linear growth rate LGR is related to the temperature and viscosity by a curve having a prominent maximum; the crystallization rate of a melt is dependent on

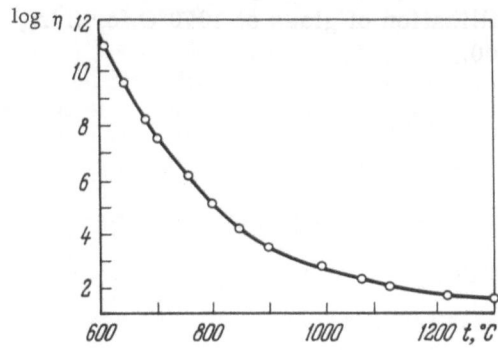

Fig. 2. Temperature dependence of log η.

Fig. 3. Electron micrograph after isothermal crystallization of glass at 700°C for 2 hr, ×16,000.

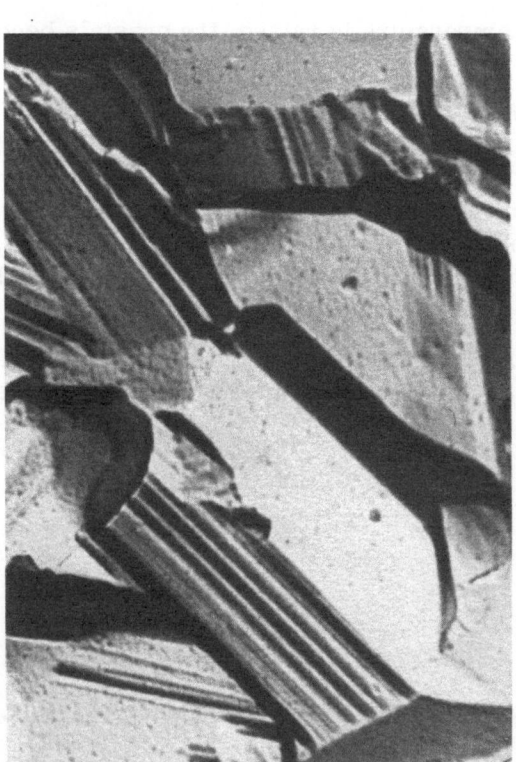

Fig. 4. Electron micrograph after isothermal crystallization of glass at 1070°C for 2 hr, ×20,000.

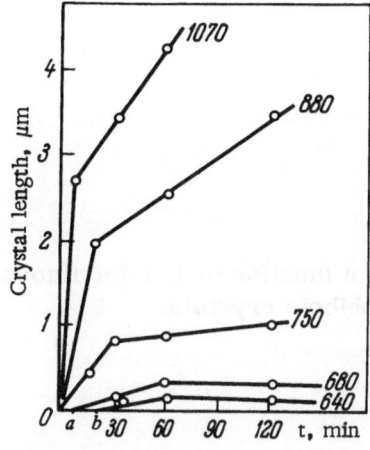

Fig. 5. Length of fluoramphibole crystals as a function of isothermal crystallization time.

the relation between the free energy of the system and the kinetic factor controlled by the particle mobility [6, 7]. To a first approximation, the growth rate of these crystals is proportional to the supercooling between the temperatures corresponding to the maximum and minimum on the right branch of the LGR curve, and the decisive features are the thermodynamic parameters. If the supercooling is of the order of 100°C, with a viscosity of $2.5 \cdot 10^2$ P(poise), one has conditions that provide for the maximal growth rate; the left branch of the growth-rate curve is similar, but the temperature dependence of the viscosity takes a different form (Fig. 2). The linear growth rate as calculated from $v = k_0/\eta + k \log \eta$ [3] agrees well with the observed values for viscosities between $2.8 \cdot 10^2$ and $1.6 \cdot 10^3$. The growth rate falls at large supercoolings on account of increased effects from the kinetic factor; when the supercooling is large and the viscosity is high, as at 900-700°C, the growth is limited by the mobility of the structural group. Figure 7 shows that the logarithm of the growth rate is linearly related to $1/T$; the activation energy for growth given by this line is 32 kcal/mole. The activation energy for viscous flow in the liquid state is 50 kcal/mole. The first value is the lower because the growth mechanism at large supercoolings involves the mobility not only of the anion complexes but also of the cation [8].

We used electron micrographs to calculate the numbers of nuclei, on the assumption that each crystal is produced from a single nucleus; it is clear (Fig. 8) that the number of nuclei arising in the glass is proportional to the time of isothermal treatment. The considerable in-

Fig. 6. Linear growth rate (LGR) and nucleation rate (NR) as functions of temperatures for fluoramphibole crystals.

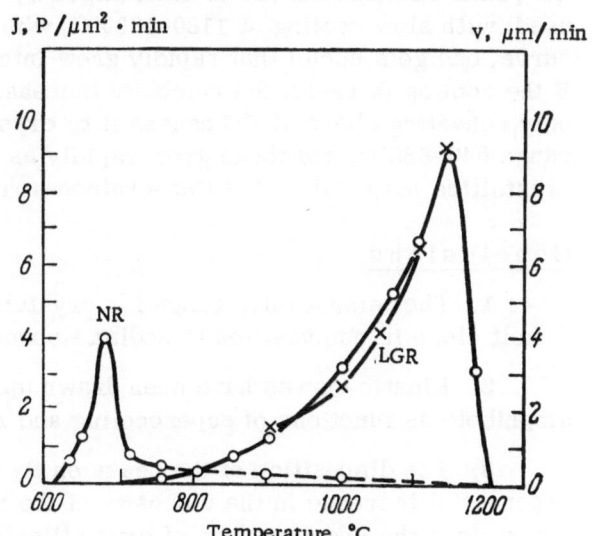

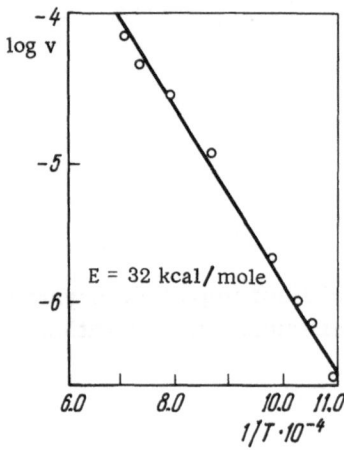

Fig. 7. Log v as a function of 1/T for fluoramphibole crystals.

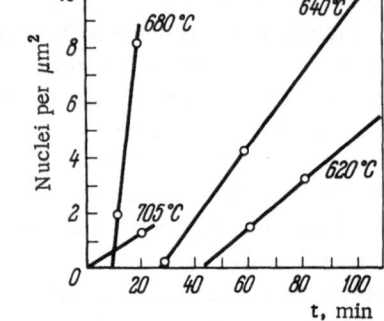

Fig. 8. Nuclei per μm^2 as a function of isothermal treatment time.

crease in the linear growth rate did not allow us to make systematic studies of nucleation at temperatures above 700°C; the slope of the temperature curve indicates, however, that the nucleation rate increases considerably.

The experiments and calculations gave the nucleation-rate curve NR, as in Fig. 6; in the region 660-680°C, where the glass has a viscosity of $1.6 \cdot 10^8$, one finds the maximum growth rate.

The NR and LGR curves have peaks in relation to the supercooling on account of the exponential dependence of the viscosity on the temperature [7, 9]. The mutual disposition of the peaks on these curves is determined by the mode of crystallization under the conditions used; with slow cooling at 1180-1000°C, which corresponds to the maximum on the growth-rate curve, one gets nuclei that rapidly grow into large radially radiated clumps of fluoramphibole. If the cooling is rapid, the viscosity increases rapidly, and the melt solidifies to glass, and only reheating above 620°C causes it to crystallize; in this case, numerous nuclei arise in the range 640-680°C, and these grow rapidly as the temperature is raised, which gives a poly-crystalline material with a fine-grained structure.

Conclusions

1. The temperature range for crystallization of fluoramphibole has been determined for a melt close in composition to sodian fluorotremolite.

2. Kinetic curves have been drawn up for the nucleation rate and linear growth rate of amphibole as functions of supercooling and melt viscosity.

3. The disposition of the peaks on these two curves in relation to supercooling is due to exponential increase in the viscosity of the melt as the temperature is reduced, which in turn determines the precise mode of crystallization at this composition.

Literature Cited

1. J. Frenkel, Phys. Z. Sowjetunion, 1:498 (1932).
2. A. Leontjewa, Acta Physicochim. URSS, 13:423 (1940).
3. A. A. Leont'eva, Zap. Vses. Min. Obshch., 72(1):62 (1943).
4. M. V. Okhotin, Zh. Fiz. Khim., 28:254 (1954).
5. V. N. Filipovich and A. M. Kalinina, Izv. Akad. Nauk SSSR, ser. neorg. mat., 4:1532 (1968).
6. V. I. Malkin, in: Problems of Metallography and Metal Physics [in Russian], Vol. 6 (1959), p. 76.
7. I. I. Kitaigorodskii and M. B. Usivitskii, Steklo, No. 4, 1 (1964).
8. F. P. H. Chen, J. Amer. Ceram. Soc., 46:467 (1963).
9. N. F. Chelishchev, in: Experimental Studies in Rarer-Element Mineralogy and Geochemistry [in Russian], Nauka, Moscow (1967), p. 84.

CRYSTAL GROWTH AND OPTICAL CHARACTERISTICS OF LANTHANIAN FERGUSONITE

G. V. Anan'eva, G. F. Bakhshieva, V. E. Karapetyan,
A. M. Morozov, and E. M. Sychev

Lanthanum orthoniobate $LaNbO_4$ crystallizes at high temperatures in tetragonal form with the scheelite structure; its melting point is 1650°C or so, and at 510-600°C there is a polymorphic transition, with the low-temperature modification isostructural with the monoclinic mineral fergusonite (space group C2), which is a rare-earth niobate, containing mainly the yttrium group of rare earths. A monoclinic $LaNbO_4$ (lanthanian fergusonite) has comparatively small deviations from tetragonal form ($a = 5.56$, $b = 11.54$, $c = 5.20$ Å, $\beta = 94°$) [1-3]. The interest in $LaNbO_4$ crystals is due to the prospects as use as laser matrices, which have been deduced from spectral and fluorescence studies of polycrystalline orthoniobates and orthotantalates of ABO_4 type, where A = Y, La, B = Nb, Ta [4, 5].

$LaNbO_4$ crystals do not occur in nature, and the first indication that they might be grown artificially is found in [4] where Verneuil's method was suggested, and $LaNbO_4$ single crystals have recently been made in this way [6].

The lanthanum niobate crystals we used were grown by pulling from the melt (Czochralski's method). We used iridium crucibles in helium or argon, the apparatus having induction heating; the temperature at the bottom of the crucible (about 1800°C) was kept constant to ±2°C during the growth by means of a thermal regulator having a sapphire light guide. The first crystals were grown on an iridium wire, as no seeds were available; the pulling rates were 5-12 mm/hr, and the lengths of the crystals attained 80 mm with a diameter of 8-15 mm, there being a perfect cleavage on (010), which is analogous to the basal plane in tetragonal crystals of scheelite type. Powder patterns showed that these crystals were monoclinic.

We examined the block structure of the $LaNbO_4$ crystals by x-ray oscillation curves and Berg—Barret deflection topography; the crystals contain mainly large fairly perfect regions a few mm in size, and the width of the oscillation curve for such parts does not exceed 8'.

The crystals show polysynthetic twinning, which is due to the polymorphism.

We examined specimens up to 0.02 mm thick, which are oriented on (010) planes, and the microscope shows that the sizes of the basic crystals were 100-150 μm or more, while the twin bands were on average a few μm thick (Fig. 1). The extinction directions of the twin crystals were turned one relative to the other around the b axis through 7°. The crystals also had regions free from these defects, which were used in the optical measurements.

The absorption spectra of unactivated $LaNbO_4$ single crystals were recorded throughout their transmission range; a plate 0.2 mm thick transmits over the range 0.27-9.5 μm (Fig. 2).

Fig. 1. Polysynthetic twins in LaNbO$_4$ in the extinction position: a) Basic crystals; b) twinned layers.

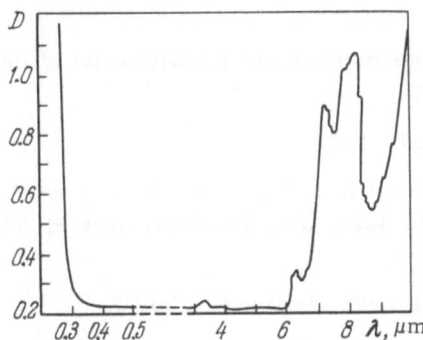

Fig. 2. Absorption spectrum (uncorrected for reflection) for LaNbO$_4$ single crystals, thickness 0.2 mm.

The absorption bands in the 6-9 μm region are found also for other oxide compounds such as CaWO$_4$, LaNa(MoO$_4$)$_2$, Ca$_5$(PO$_4$)$_3$F, which are due to lattice vibrations, including overtone and combinations of the modes of the [NbO$_4$] tetrahedra.

The refractive indices were measured by an autocollimation method on prisms of two types, which were cut in such a way that the rays were either parallel to the b axis or perpendicular to it, which enabled us to establish all three semiaxes of the optical indicatrix for the biaxial LaNbO$_4$. Table 1 gives the refractive indices at various wavelengths, and also 2V, the angle between the optic axes, which is given by

$$\tan V = \frac{\gamma}{\alpha} \sqrt{\frac{\beta^2 - \alpha^2}{\gamma^2 - \beta^2}}$$

The indicatrix has its short α axis coincident with the [010] axis, while the γ axis is close to [001]; at $\lambda = 0.5461$ μm, $\gamma - \alpha = 0.116$, while along the b axis (which corresponds to

TABLE 1. Refractive Indices of Lanthanum Niobate Crystals

λ, μm	γ	β	α	$2v$
0.4358	2.259$_9$	2.152$_5$	2.133$_5$	47°28'
0.4861	2.232$_8$	2.130$_8$	2.112$_8$	47°16'
0.5461	2.211$_6$	2.113$_8$	2.095$_8$	48°04"
0.5876	2.200$_2$	2.104$_9$	2.087$_2$	48°18'
0.6563	2.187$_4$	2.094$_5$	2.077$_1$	48°26'

the tetragonal c axis in the scheelite structure) the birefringence is also considerable ($\gamma - \beta =$ 0.1), although the monoclinic distortions in the structure are small. LaNbO$_4$ crystals are optically positive, with γ as the acute bisector.* The refractive indices of LaNbO$_4$ show high dispersion, whereas the birefringence hardly varies in the visible region; 2V increases with the wavelength. We found no dispersion in the position of the γ and β axes of the indicatrix.

LaNbO$_4$ single crystals were activated with various rare earths (Ce^{3+}, Pr^{3+}, Nd^{3+}) introduced into the melt as isostructural orthoniobates; lanthanum niobate is completely isomorphous with the niobates of the other rare earths, so one can introduce practically unlimited concentrations of the rare-earth activators without loss of lattice perfection. We obtained lasing with a LaNbO$_4$ crystal at room temperature operating at $\lambda = 1.024$ μm. Lanthanum niobate is the material of lowest symmetry that has given production of stimulated radiation.†

We are indebted to V. B. Tatarskii and A. I. Komkov for valuable advice, and also to P. P. Feofilov for interest in the work and discussion of the results.

Conclusions

1. LaNbO$_4$ single crystals are isostructural with the monoclinic modification of natural fergusonite and have been grown by pulling from the melt.

2. Characteristic growth defects have been demonstrated.

3. The basic optical characteristics of the unactivated LaNbO$_4$ crystals have been determined: the refractive indices, the angle between the optic axes, and the absorption spectra in the visible and infrared regions.

4. Induced emission from LaNbO$_4$—Nd crystals has been obtained at 300°K.

Literature Cited

1. A. I. Komkov, Dokl. Akad. Nauk SSSR, 126:853 (1959).
2. H. P. Rooksby and E. H. White, Acta Crystallogr., 16:888 (1963).
3. V. Stabican, J. Amer. Ceram. Soc., 47:55 (1964).
4. C. A. Wesselink and A. Brill, Philips. Res. Repts., 20:269 (1965).
5. N. A. Godina, M. N. Tolstoi, and P. P. Feofilov, Opt. i Spekt., 23:756 (1967).
6. L. M. Belyaev, F. I. Dmitrieva, N. M. Melankholin, A. A. Popova, and L. V. Soboleva, Kristallografiya, 14:359 (1969).
7. G. F. Bakhshieva, V. E. Karapetyan, A. M. Morozov, L. P. Morozova, and P. P. Feofilov, Opt. i Spekt., 27:936 (1969).

---

*As LaNbO$_4$ resembles scheelite, one expects that the birefringence along the c axis should be small, while the acute bisector of the angle between the optic axes should coincide with the b axis.

† See [7] on the luminescence and lasing characteristics of Nd^3 + LaNbO$_4$.

MORPHOLOGY OF InSb AND GaSb SINGLE CRYSTALS
WITH THE FACE EFFECT

M. S. Mirgalovskaya, M. R. Raukhman,
and I. A. Strel'nikova

The face effect has been described [1-11] for InSb and GaSb single crystals grown by Czochralski's method under conditions that facilitate maximum development of the channel in the billet [7, 8]. The initial material was zone-purified indium antimonide ($n \sim 3\text{-}5 \cdot 10^{14}$ cm^{-3}), together with gallium and antimony of semiconductor grade. The dopes were tellurium (donor) and zinc, cadmium, and germanium (acceptors), which were added to the melt either in elemental form or as appropriate alloys. The seed crystals were oriented on $\langle 111 \rangle$ by standard x-ray methods to 1-2°. The crystals were grown as a rule in the B[III] direction. The usual methods were employed to measure the Hall effect of the specific resistance; the thermoelectric emf was measured with a semiautomatic system at 300 and 80-85°K by a method previously described [8, 11].

The crystal was cut along the $\langle 111 \rangle$ growth axis parallel to the $\{211\}$ and $\{110\}$ planes, and also perpendicular to the $\langle 111 \rangle$ axis on the A(111) and B(111) planes in order to examine the structure.

Figure 1 shows InSb and GaSb crystals with the face effect; under certain conditions, one can obtain crystals (Fig. 1a) whose volume is almost completely formed as a result of growth on the $\{111\}$ face, which is important, because the material in the channel region may be more uniform than that outside it [7]. Figure 2 shows a longitudinal section of the upper part of the crystal, which shows the onset of channel formation; it is clear that the channel arose at the upper part of the hemisphere and subsequently increased in diameter on account of change in the curvature of the crystallization front.

Table 1 gives the parameters of the crystals; the InSb and GaSb crystals were examined by chemical etching, which showed that the channel can be detected in these crystals when they are doped either with tellurium or tellurium together with any of the above acceptor materials, the tellurium concentration playing the decisive part. The face effect is not revealed by chemical etching for crystals doped with acceptors (Fig. 3). The effect is best seen in InSb crystals of n and p types with tellurium concentrations between $8 \cdot 10^{17}$ and $8 \cdot 10^{18}$ cm^{-3}, while in GaSb specimens of n and p types it is best seen in the range from $8 \cdot 10^{17}$ to $3 \cdot 10^{18}$ cm^{-3}. We determined for these crystals the value of R, which characterizes the face effect, and the values are given in Table 1; the largest R occur for InSb crystals doped with Te and Cd. The R for Te in InSb is appreciably reduced at high Te concentrations, which is confirmed by chemical etching and measurement of the thermo-emf. The R for Te and InSb and GaSb grown in the $\langle 111 \rangle$ polar directions is somewhat larger for B[$\overline{1}\overline{1}\overline{1}$] directions than it is for A[111] ones.

161

Fig. 1. Single crystals with the face effect, ×5: a) Te-doped InSb (n ≈ 3 · 10^{18} cm^{-3}), b) Zn-doped GaSb (p ≈ 1 · 10^{19} cm^{-3}).

The measurements on the thermo-emf α substantially expand the scope for research on the face effect, especially when etching fails to reveal it; also, the thermo-emf curves allow one to determine the details of the impurity distribution at the boundaries of the channel and in the surrounding parts [7, 9, 10]. Figure 4 collects the results on the carrier distribution (p and n, together with α) for InSb and GaSb crystals of p and n types, and these can be correlated with the face effect. If the amount of dope of one sign is much greater than that of the other (p ≫ n or n ≫ p), then α is inversely proportional to the concentration n or p (Fig. 4B), which agrees with the theory of [12] and the experiments of [7, 11]. In poorly compensated

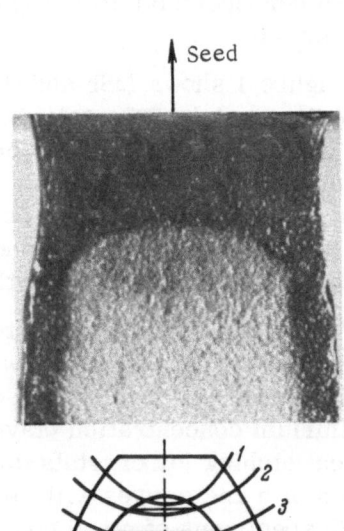

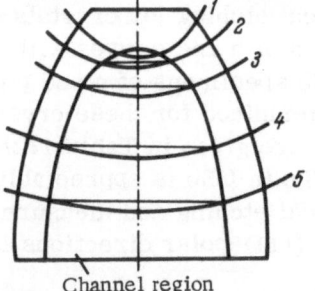

Fig. 2. Longitudinal section of the upper part of a crystal, including start of formation of a channel, and shape of crystallization front (1-5).

TABLE 1. Face Effect and Growth Conditions

Substance and growth direction	Dope	Carrier concentration range, cm^{-3}	Te concentration range showing etching channel, cm^{-3}	R; $R = \dfrac{n(p)_{\text{at fa}}}{n(p)_{\text{elsewhere}}}$	R from [2]
InSb B [$\bar{1}\bar{1}\bar{1}$]	Te	$n = 10^{14} - 5 \cdot 10^{18}$	$7 \cdot 10^{17} - 8 \cdot 10^{18}$	5—3	3—9
	Te	$n = 8 \cdot 10^{18} - 1,1 \cdot 10^{19}$		2—1.3	
	Zn	$p = 10^{14} - 2 \cdot 10^{19}$	Not detected	1.3—1.1	1.1—1.3
	Cd	$p = 10^{15} - 3 \cdot 10^{19}$	" "	2.3—2.0	2—3
	Ge	$p = 10^{15} - 5 \cdot 10^{18}$	" "	1.4—1.7	1.4—1.8
	Te+Zn ⎫ Te+Cd ⎬ Te+Ge ⎭	$p = 3 \cdot 10^{18} - 2 \cdot 10^{19}$	$7 \cdot 10^{17} - 8 \cdot 10^{18}$	1.2—1.1	—
GaSb B [$\bar{1}\bar{1}\bar{1}$]	Te	$n = 2 \cdot 10^{17} - 3 \cdot 10^{18}$	$5 \cdot 10^{17} - 3 \cdot 10^{18}$	1.8—2	1.5—2
	Zn	$p = 5 \cdot 10^{17} - 3 \cdot 10^{19}$	Not detected	1.4—1.1	1.1
	Te+Zn	$p = 3 \cdot 10^{18} - 3 \cdot 10^{19}$	$6 \cdot 10^{17} - 3 \cdot 10^{18}$	1.1—1.3	—

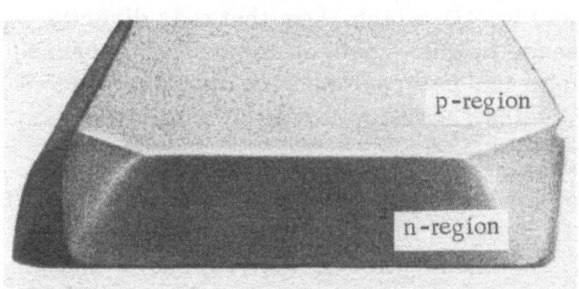

Fig. 3. Longitudinal section of an InSb single crystal doped in the upper part with Zn (p type) and in the lower part with Zn and Te (n type).

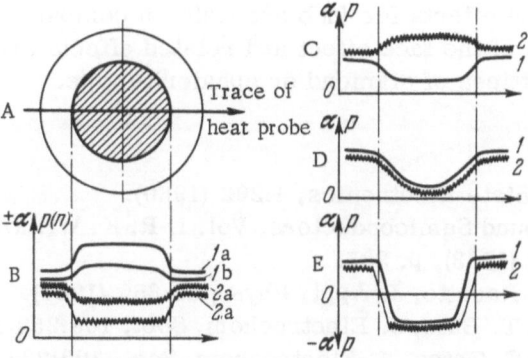

Fig. 4. Distribution of thermo-emf α and carrier concentration (n, p) in doped InSb and GaSb crystals with the face effect. A) Cross section, B-E) distribution curves for: 1) carrier concentration, 2) α; B) in n-type specimens (1a and 2a) (donor dope) and p-type ones (1b and 2b) (acceptor dope); C) weakly compensated p-type specimens; D) highly compensated p-type specimens; E) crystals with ring p—n junction.

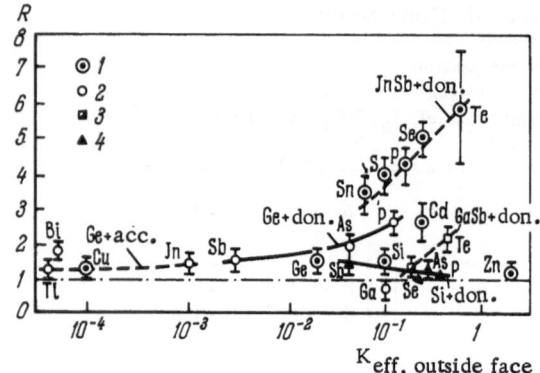

Fig. 5. Relation of R to K_{eff} outside face for donor and acceptor dopes in InSb, GaSb, and Si crystals with the face effect: 1) InSb; 2) Ge; 3) GaSb; 4) Si.

specimens, where the carrier concentrations at the face fall somewhat, while α continues to rise, we still get an inversely proportional relationship between α and p or n concentration, as Fig. 4C shows. If the compensation is nearly complete (p $\approx$ n), the concentration of three whole carriers at the face is substantially reduced, while the contribution from the n carriers increases, and then the crystal may have mixed conduction; in that case, α is substantially reduced [12]. This case is illustrated by the curves of Fig. 4D, which show that α is directly proportional to the concentration. Any further increase in the degree of compensation causes α to fall and then change sign; in that case, the crystal has two regions with opposite types of conduction (n type at the face, p type away from it), which results in a ring p—n junction (Fig. 4E).

We used our evidence and published values to plot R as a function of the partitions coefficient K for InSb, GaSb, Ge [1], and Si [6] (Fig. 5). Each donor impurity gave a definite dependence of R on K, whereas the acceptor ones did not. The adsorption model [16] shows that all the donor impurities in InSb and GaSb, which are surface-active materials, give points on the graph (Fig. 5), which lie in accordance with their surface activity [13, 14]. The substantial differences in the R for the donor dopes in InSb and GaSb are probably due, as for Ge and Si [6], to differences in adsorption in relation to the melting point of the host material [15].

We conclude from the effects for InSb and GaSb in comparison with elemental semiconductors that many aspects of the face effect and related effects are common to all semiconductor materials that have lattices of diamond or sphalerite type.

Literature Cited

1. J. A. Diknoff, Solid State Electronics, 1:202 (1960).
2. J. B. Mullin, Compound Semiconductors, Vol. I, R. K. Willardson and H. L. Goering, eds., Reinhold, New York (1963), p. 365.
3. R. N. Hall and J. H. Racette, J. Appl. Phys., 32:856 (1961).
4. W. P. Allred and P. T. Bate, J. Electrochem. Soc., 108:258 (1961).
5. M. D. Banus and H. C. Gatos, J. Electrochem. Soc., 109:829 (1962).
6. M. G. Mil'vidskii and A. V. Berkova, Fiz. Tverd. Tela, 5:13 (1963).
7. M. S. Mirgalovskaya, M. R. Raukhman, V. A. Kokoshkin, and E. I. Khodyakova, Izv. Akad. Nauk SSSR, ser. neorg. mat., 4:1857 (1968).
8. M. S. Mirgalovskaya, M. R. Raukhman, V. A. Kokoshkin, and V. V. Guzeeva, Izv. Akad. Nauk SSSR, ser. neorg. mat., 4:841 (1968).
9. M. S. Mirgalovskaya, V. A. Kokoshkin, and V. Ya. Smirnov, Izv. Akad. Nauk SSSR, ser. neorg. mat., 1:340 (1965).
10. M. S. Mirgalovskaya, G. V. Kukuladze, and V. A. Kokoshkin, Izv. Akad. Nauk SSSR, Ser. neorg. nat., 4:694 (1968).

11.   M. S. Mirgalovskaya, I. A. Strel'nikova, and V. A. Kokoshkin, Izv. Akad. Nauk SSSR, ser. neorg. mat., 4:852 (1968).

12.   L. S. Stil'bans, Semiconductor Physics [in Russian], Sov. Radio (1967).

13.   V. B. Lazarev and M. Ya. Dashevskii, Izv. Akad. Nauk SSSR, ser. neorg. mat., 1:1901 (1965).

14.   M. Ya. Dashevskii, G. V. Kukuladze, V. B. Lazarev, and M. S. Mirgalovskaya, Izv. Akad. Nauk SSSR, ser. neorg. mat., 3:1561 (1967).

15.   L. L. Kunin, Surface Phenomena in Metals [in Russian], Metallurgizdat, Moscow (1955).

16.   A. Trainor and B. E. Bartlett, Solid State Electronics, 2:106 (1961).

MORPHOLOGY OF RUBY CRYSTALS GROWN BY CZOCHRALSKI'S METHOD

G. K. Geranicheva, I. I. Afanas'ev, and T. G. Agafonova

A ruby rod grown by Czochralski's method usually has a rounded lustrous surface; the only prominent crystallographic form readily identified visually is the pinacoid.

Detailed crystallographic examination of such specimens has shown that there are plenty of simple forms, together with unusual combinations.* We have performed goniometry for five crystals with a two-circle reflecting goniometer; we examined the sawn end parts of the rods (the seed end and the growth front). For 35 ruby crystals we also made the most important [1, 3] photogoniometric measurements of the end parts and side surfaces. The resulting patterns were indexed to define the simple forms and to establish the importance of each in this method of growth; Table 1 gives the results.

The stereographic projection of Fig. 1 was drawn up from this evidence, in which the principal simple forms are shown as open circles; these are the {0001} pinacoid and the {1011} positive rhombohedron, both of which are present on all crystals without exception. Fair importance attaches also to some particular forms: the {11$\bar{2}$0} prism of the second kind and the {31$\bar{4}$5}, {32$\bar{5}$7}, {21$\bar{3}$4}, {41$\bar{5}$4} scalenohedra, which relate to the {10$\bar{1}$1} zone.

The other simple forms (positive and negative rhombohedra) are usually observed at the growth fronts, which have a dendritic or skeletal structure. The order of importance of the simple forms is {0001}, {10$\bar{1}$1}, {11$\bar{2}$0}, {31$\bar{4}$5}, {21$\bar{3}$4}, {32$\bar{5}$7}, {41$\bar{5}$4}, {11$\bar{2}$3}, {10$\bar{1}$2}, {011$\bar{2}$}, etc. The crystals characteristically have an unusual combination of these forms; similar results have been reported elsewhere [4].

For comparison, the projection shows by broken lines the simple forms on rubies grown by Verneuil's method; the belt of dipyramids so characteristic of these [1, 2] is absent on crystals grown by Czochralski's method. Apart from the {02$\bar{2}$1} negative rhombohedron and the belts of negative scalenohedra, forms always present are the {10$\bar{1}$1} positive rhombohedron and the belt of positive scalenohedra.

The mode of occurrence of a simple form is dependent on whether it forms a face on the side surface or at the growth front; the side surfaces of a ruby rod have the faces of simple forms arranged along the generator of the cylinder (Fig. 2); the orientation of the growth axis determines which simple forms appear on the side surfaces, and what are the sizes of the flat-faced parts. If the growth is on [10$\bar{1}$0], the side surface has pinacoid faces and two faces of the {11$\bar{2}$0} prism. Faces of the {10$\bar{1}$1} positive rhombohedron are represented by thin, slightly

---

*The ruby crystals were provided by A. O. Ivanov.

TABLE 1. List of Established Simple Forms

Form	Symbol	Spherical coordinates		Frequency, %	Form		Spherical coordinates		Frequency, %
		φ	ρ				φ	ρ	
Pinacoid	0001	α	0°	100	Dipyramid	$22\bar{4}3$	30°	61°12'	10
Prism	$11\bar{2}0$	30°	90°	60	"	$11\bar{2}3$	30°	42°20'	15
Rhombohed. (+)	$10\bar{1}4$	0°	18°04'	5	"	$11\bar{2}1$	30°	69°54'	5
"	$10\bar{1}2$	0°	38°14'	10	Scalenohed.	$31\bar{4}5$	26°20'	44°50'	25
"	$10\bar{1}1$	0°	57°37'	100	"	$21\bar{3}4$	19°20'	46°25'	25
Rhombohed. (−)	$01\bar{1}4$	0°	18°04'	2	"	$32\bar{5}7$	13°24'	49°00'	25
"	$01\bar{1}3$	0°	28°31'	3	"	$41\bar{5}4$	6°36'	61°21'	25
"	$01\bar{1}2$	0°	38°14'	10					

projecting ridges. If the growth axis is a direction perpendicular to (10$\bar{1}$1), the side surface of a circular rod is formed by four faces of the {10$\bar{1}$1} rhombohedron and pinacoid faces; in this orientation, the {10$\bar{1}$1} and {0001} faces are roughly equal in development (Fig. 2).

A completely different form is taken by faces that are seen near the growth front; the individual faces here usually form blunted vertices [1]. An even more unusual form is taken by faces that occur at dendritic or skeletal parts of the growth front; for instance, a pinacoid face on crystals with a 30° orientation takes the form of regular triangles in an area of skeletal growth in Fig. 3c; from the sides of the triangle there emerge belts of positive rhombohedra, while from the vertices there run belts of negative rhombohedra. The faces on the {10$\bar{1}$1} rhombohedron near the growth front form blunted corners, the area sometimes being considerable, as in Fig. 3e; from these run belts of positive scalenohedra from the [10$\bar{1}$1] zone, and often also belts of positive rhombohedra from the [11$\bar{2}$0] zone, which run out to {0001}. The faces of the {11$\bar{2}$0} prism give in parts of dendritic structure accumulations of blunted vertices of rhomboid shape, as in Fig. 3f. Figure 3 compares the main faces encountered on the side surfaces and at the growth fronts of rubies grown by Czochralski's method.

Of these observations, the most interest attaches to the occurrence of a flat (blunted corner) at the growth front; occurrence of this results in the face effect [5]. These Czochral-

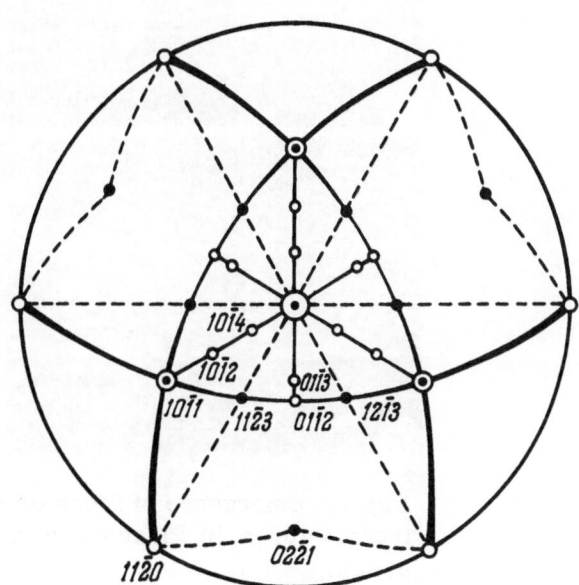

Fig. 1. Stereographic projection of simple forms for synthetic ruby crystals.

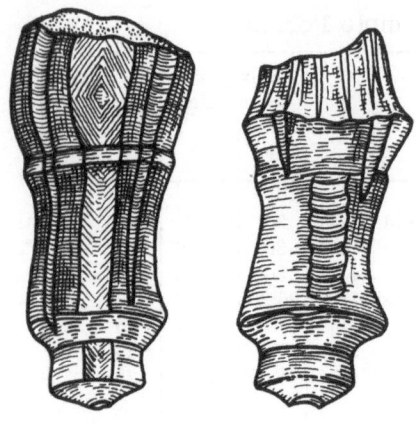

Fig. 2. General view of ruby crystal perpendicular to (10Ī1).

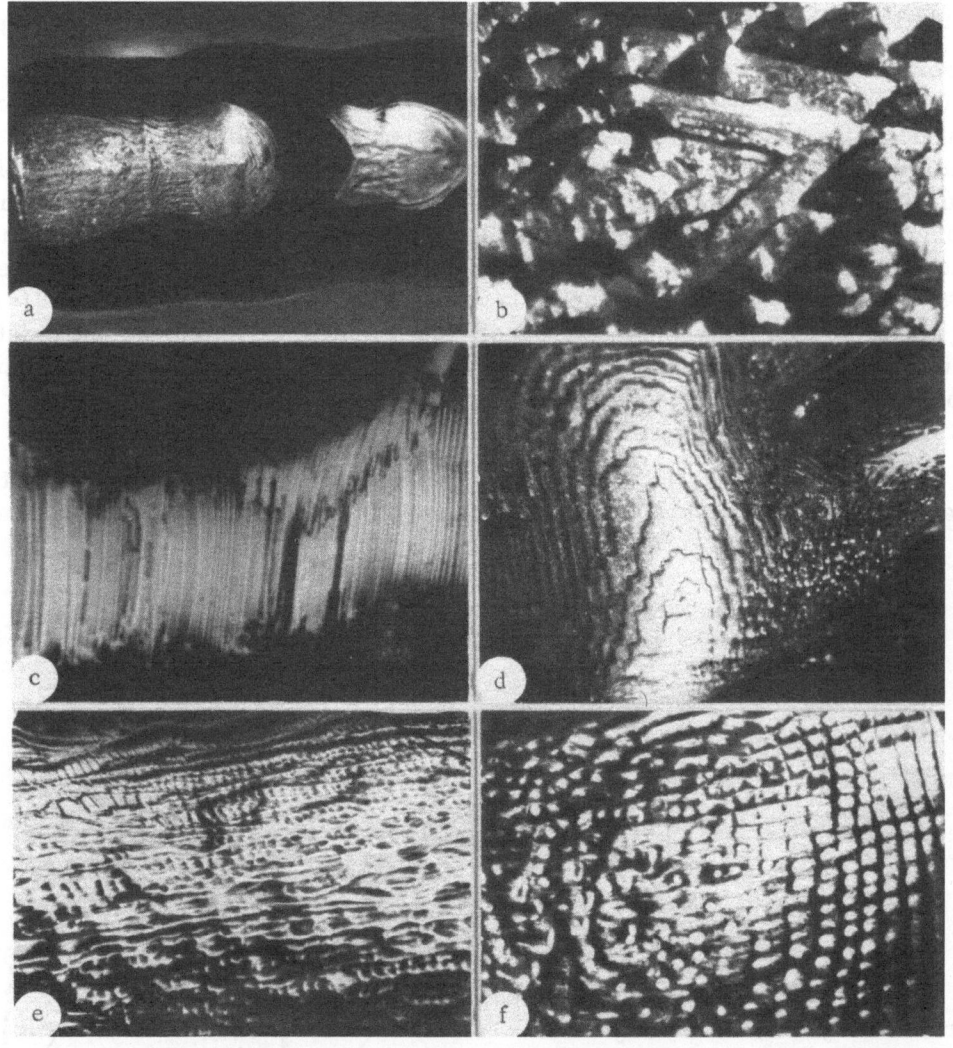

Fig. 3. Emergence of faces on the side surface and crystallization front, ×40. a, b) Pinacoid, c, d) positive rhombohedron, e, f) {11Ī0} prism.

ski rubies have small flat ($10\bar{1}1$) faces at the growth front, which results in a core to the rod with an elevated chromium content [4]. The face effect can be maximally developed or completely suppressed, in accordance with the specification for the finished crystal; the optical uniformity should be greatest in the first case, because one will obtain a component consisting of a single growth pyramid.

This morphologic study shows that there is an abundance of forms on Czochralski ruby crystals, which occur in unusual combinations, which enable one to establish the types of striation and other surface structures on the rods. The morphologic features of these surfaces facilitate rapid orientation and checking of growth conditions.

Literature Cited

1. V. N. Voitsekhovskii, Kristallografiya, 13:563 (1968).
2. N. N. Marochkin and I. I. Shafranovskii, Zap. Vses. Min. Obshch., Vol. 82, No. 1 (1953).
3. M. D. Lyubalin and M. A. Mokievskii, Proceedings of the First Conference on Crystallization by Stepanov's Method [in Russian], Izd. Fiz.-Tekh. Inst. im. Ioffe AN SSSR, Leningrad (1968), p. 155.
4. F. R. Charvot, J. C. Smith, and O. H. Nestor, J. Phys. Chem. Solids, Suppl., 2:45 (1967).
5. A. Witt and H. Geithos, in: Crystal Growth Problems [Russian translation], Mir, Moscow (1968), p. 262.

MACROSCOPIC DEFECTS IN CRYSTALS GROWN
BY CZOCHRALSKI'S METHOD

V. M. Garmash, T. P. Znikina, and V. B. Lazareva

Many modern techniques require large optically perfect crystals almost entirely free from macroscopic defects, especially for laser purposes.

Here we describe the shape, structure, and topography of crystals of yttrium-aluminum garnet (YAG), lithium niobate, barium-sodium niobate, and sodium-lanthanum molybdate, all grown by Czochralski's method, and we give details also of the topography of the microdefects (domains, negative crystals, bulk defects,* and thermal cracks).

External Morphology of Crystals

Virtually nothing has been published on the morphology of these crystals, apart from lithium niobate [1]; there has also been little published on the morphology of crystals made by Czochralski's method [2].† The morphology of such crystals includes striations on the sides and growth cones; the morphologic symmetry of the striations corresponds to the symmetry of the growth direction (Table 1).

To determine the genetic features of the crystal morphology we chose a supercooling such that the crystallization front was almost planar, while the growth fragments were well developed at it. The growth surfaces of lithium niobate crystals grown on $[000\bar{1}]$, $[10\bar{1}0]$, $[10\bar{1}2]$ are shown in Fig. 1, while those of yttrium aluminum garnet grown on [001] are shown in Fig. 2.

The growth front of lithium niobate grown on [0001] shows triangular pyramids (Fig. 1a); on the side surface the striations end in flat areas, as they do also on the growth front.

Fragments on the growth front of $LiNbO_3$ crystals arise as a result of development and intergrowth of the growth pyramids of $\{10\bar{1}2\}$ faces, as has been confirmed by x-ray methods. Tangential growth of the elementary rhombic fragments in the $\{10\bar{1}2\}$ growth pyramid leads to an edge form on $[10\bar{1}0]$ and to a vertex form on $[000\bar{1}]$ (Fig. 3b). At high growth rates, when the supercooling is several degrees, the crystals approximate to solids with three vertices that have been duplicated (Fig. 3c), while $[10\bar{1}2]$ planes occur at the growth front.

The microscope shows that the $\{10\bar{1}2\}$ faces grow by layers (the F faces are flat and grow reproducibly) [3]; sometimes the center of a face shows a growth spiral corresponding to the

---

*Bulk defects are regions within the crystal having large differences in refractive index.
† This deficiency has now been remedied to a considerable extent by studies by Yubalin (see paper in this volume) (Editor).

TABLE 1. Relation of Macroscopic Defects to Crystal Morphology

Crystal	Structural symmetry	Growth direction	Streaks on surface cyclic	Streaks on surface conical	Morphologic symmetry	Growth surface scheme	Growth surface sym.	Bulk defect scheme	Bulk defect sym.	Negative crystals scheme	Negative crystals sym.	Cleaved surfaces scheme	Cleaved surfaces sym.	Domain structure (symmetrical)
Lithium meta-niobate	$3/m$	[000$\bar{1}$]	3	6	$3\,m$	[scheme: L_3 circle]	$3\,m$			[scheme: [0001]]		[scheme: Y]	$3\,m$	$\infty:m$ / $2:m$
	↑ 1170° C ↓ 1210° C	[10$\bar{1}$0]	2	4	m	[scheme: hatched oval L_3]	m			[scheme: [10$\bar{1}$0]]	m ($3\,m$)			$2 \cdot m$
	$3\,m$	[2$\bar{1}$30]	4	6	$mm\,2$		m							
		[10$\bar{1}$2]	4	3	m	[scheme: L_3]	m			[scheme: [0001]]				— / $m\;1:m$
Al–Y garnet	$m3m$	[001]	4	4	$4/mmm$	[scheme]	$4\,mm$	[scheme: +]	$4\,mm$	[scheme]	$4\,mm$			
		[111]	3	3	$3\,m$	[scheme]	$3\,m$	[scheme: Y]	$3\,m$	[scheme]	$3\,m$			
		[110]	2	2	$mm\,2$	[scheme]	$mm\,2$	[scheme]	$mm\,2$	[scheme]	$mm\,2$			
Ba–Na niobate	$4/mmm$ ↑↓ 560° C / $4\,mm$ ↑↓ 260° C / $mm\,2$	[001]	16	8	$4/mmm$	[scheme: squares]	$4\,mm$			[scheme: dashed octagon]	$4\,mm$			$\infty:m$
		[100]	4	2	$mm\,2$									
La–Na molybdate	$4/m$	[001]	4	4	$4\,mm$	[scheme: grid]	$4\,mm$					[scheme: asterisk]	$4\,m$	
		[100]	2	2	mm									

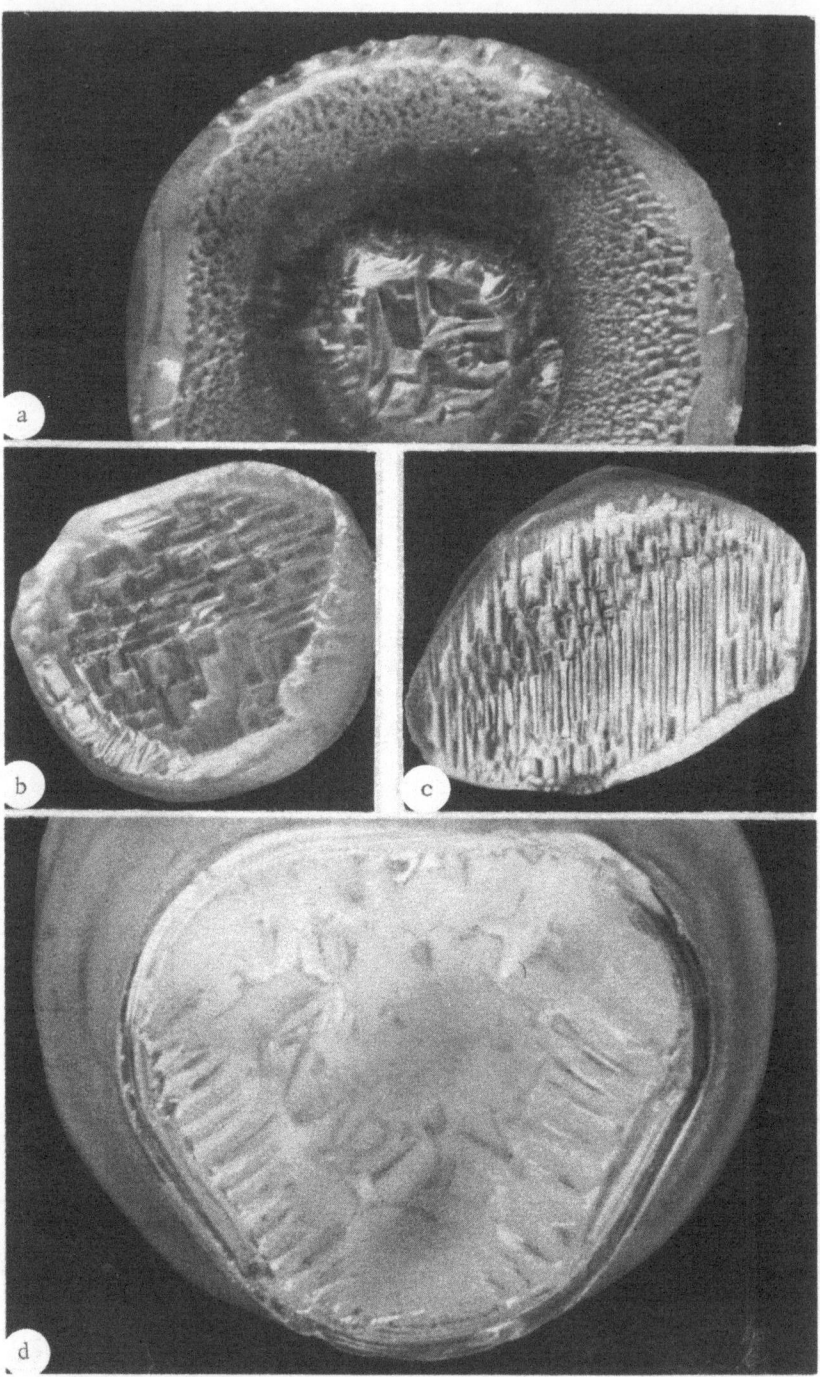

Fig. 1. Surfaces of a highly concave growth front in Li
metaniobate crystals grown in various directions, ×5:
a) $[000\bar{1}]$; b) $[10\bar{1}\bar{2}]$; c) $[10\bar{1}0]$; d) $[000\bar{1}]$.

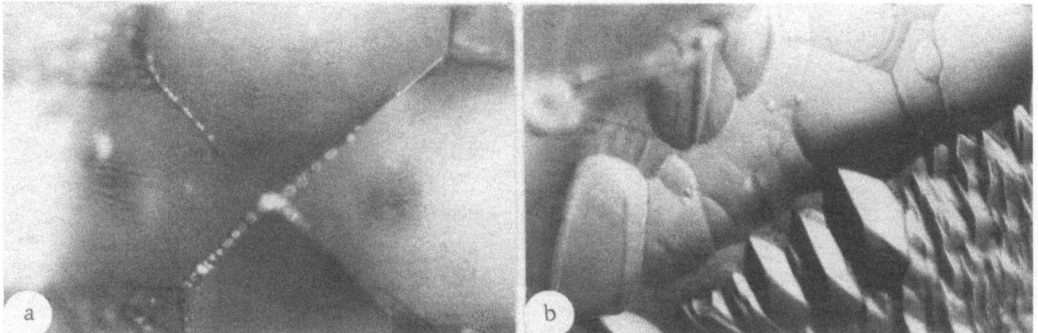

Fig. 2. Surface of growth front in YAG, ×42. a) Grown along [001]; b) difference in etching of F and K faces.

symmetry (Fig. 4). Then the striations on Czochralski crystals should be considered as irregular development of growth pyramids.

The morphology of the side surface of a crystal is dependent on the growth direction, while the morphology of the growth front is dependent on the curvature of the latter; Fig. 1 shows the growth surface of a crystal grown on [000$\bar{1}$] with a highly concave crystallization front; at the center there are three-faced pyramid projections, as on crystals of the same direction having planar crystallization fronts. As one passes from the center to the edge, there is a change in morphology, and the picture seen in the intermediate region is similar to that of a crystal grown on [10$\bar{1}$2], while on the side surfaces the picture is as for crystals grown on the [10$\bar{1}$0].

The growth front of a (YAG) crystal has flat faces (Fig. 2a); the external appearance and the etching (Fig. 2b) indicate that this is an F face, whereas the main part of the crystal grows by development of K faces.

Table 1 gives the disposition of the faces in relation to growth direction; the morphologic symmetry of the crystal is then described via the symmetry elements of the structural symmetry in the growth direction.

The smooth faces at the growth front of yttrium aluminum garnet correspond to development of a bulk defect, which is a result of differences in mode of inclusion of impurities by the F and K face pyramids; the K faces take up impurities more uniformly. Table 1 gives the morphologic symmetry of a bulk defect.

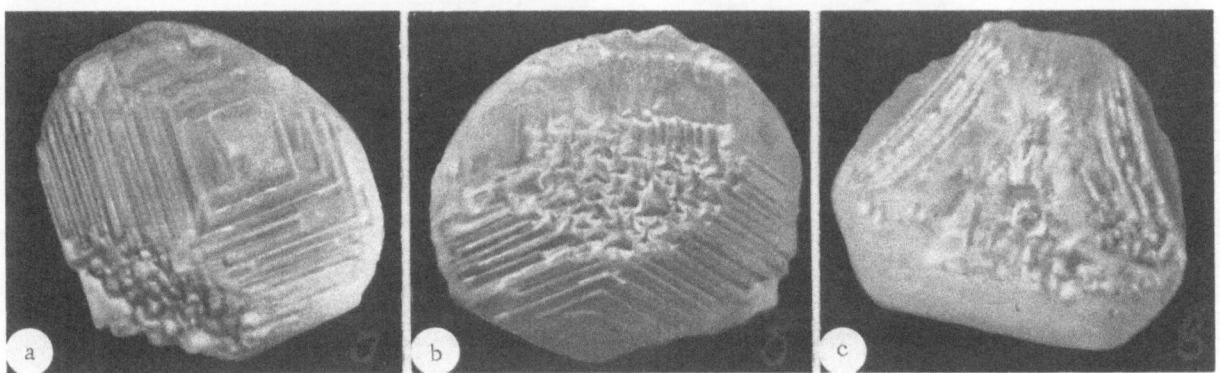

Fig. 3. Surfaces of growth fronts in Li niobate crystals with various growth rates and directions. a) Slow growth on [10$\bar{1}$2]; b) the same but [000$\bar{1}$]; c) high rate on [000$\bar{1}$].

Fig. 4. Spiral growth form on [10$\bar{1}\bar{2}$] face, ×120.

Negative Crystals

The central part of a Czochralski crystal contains macroscopic holes, which may be either rounded pores of size 1–100 μm (Fig. 5a) or negative crystals in the form of dendrites; the cross section of the latter is a closed planar edge form with reentrant angles (Fig. 5b and Table 1).

The holes arise usually on account of changes in growth conditions; negative crystals are produced as they expand (Fig. 5b). When there is pronounced narrowing of the crystal, one gets pores with rounded surfaces; when one passes from a broad part to a narrow one, both types of hole occur together. If the crystallization material contains 0.05–0.1% impurity, negative crystals can form even without change in growth conditions, while dendrites develop only very slowly at some distance from the start of expansion of the crystal.

The origin of the negative crystals is as follows. As the growth pyramids develop, the impurity rejected into the melt accumulates in the central (most supercooled) zone under the

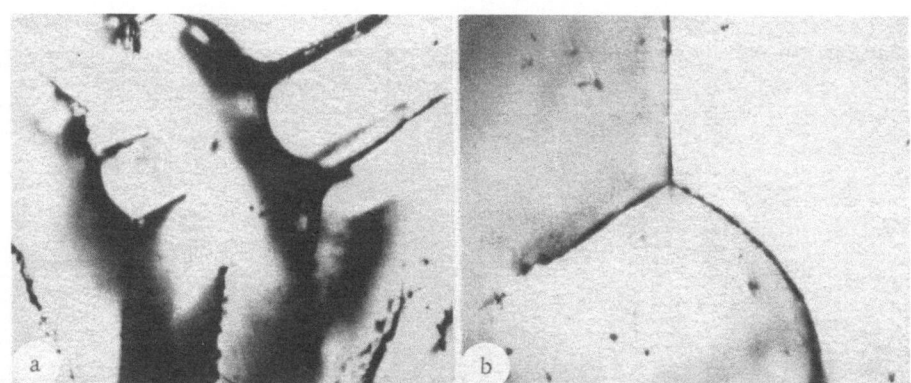

Fig. 5. Macroscopic pores (negative crystals) in crystals grown on [111] and [000$\bar{1}$]. a) Pores with rounded structure at center of dendrite; b) dendrites of negative crystals.

Fig. 6. Domain structures in Li metaniobate crystals, ×3, grown on   (a) [1010]
and (b, c) [21$\bar{3}$0].

crystal; as it accumulates, the melt temperature may rise above the melting point, so the
growth pyramids do not fuse, and this gives rise to a hole. Dendrite formation occurs when
the impurity removal rate (by diffusion or convection) is less than the rate of accumulation.

Negative crystals of rounded shape may arise as follows. As the crystal grows, the
central part becomes especially rich in impurities, and so its melting point falls below that of
the pure crystal; when crystal narrows, the rise in melt temperature may cause the central
part to melt out and leave rounded pores. This is confirmed by the correlation between the
growth conditions and the genetic features of the negative crystals.

The shape of the negative crystals is dependent on the growth direction; we find a 3m
symmetry for lithium niobate crystals along the threefold axis when the negative crystals are
formed by either of the above mechanisms. Single crystals grown on other directions have
correspondingly different morphologic symmetry (Table 1).

Domain Structure

Domains are one form of macroscopic defect in ferroelectric crystals; they can be revealed in lithium and barium-sodium niobate crystals by etching. The morphologic symmetry
of the domains is given for various directions and various growth conditions in Table 1 and
Fig. 6. A lithium niobate crystal grown on the *a* axis (Table 1) has domain symmetry described
by incorporation of antisymmetry elements, the overall symmetry being very much dependent
on the growth conditions.

For crystals grown on [000$\bar{1}$] [4], one can describe the domain structures in sections
perpendicular to the c axis via groups ∞ : m and 2 : m (Fig. 7).*

We examined the question via the solution forms of the crystals; cylindrical specimens
cut from crystals with one or more domains after growth on [00$\bar{1}$] were worked to class 12
finish and then were etched in a mixture of Hf and HNO_3; the solution forms were examined
with a GM3-52 goniometric microscope. The faces formed by dissolution were slightly convex,
and the cross sections were those of duplicated figures with three vertices.

The mode of etching varied most greatly in the directions with highest etching rate; a
small face on a figure with three vertices had convex cellular relief (Fig. 8a); as the edge of
the prism was approached, the orientation of the cells altered, and they became elongated.
The prism edge was covered by striations. The large face had relatively smooth relief; on it
were small hummocks or ridges perpendicular to the prism edge (Fig. 8b). Near the edge and
on it there were elongated cells closely overlapping one with another. The direction of these
is shown by the arrow in the figure.

---

*The groups of structural symmetry used here are given in the symbols used in the Interna-
tional Tables, while the antisymmetry groups are given in Shubnikov's symbols.

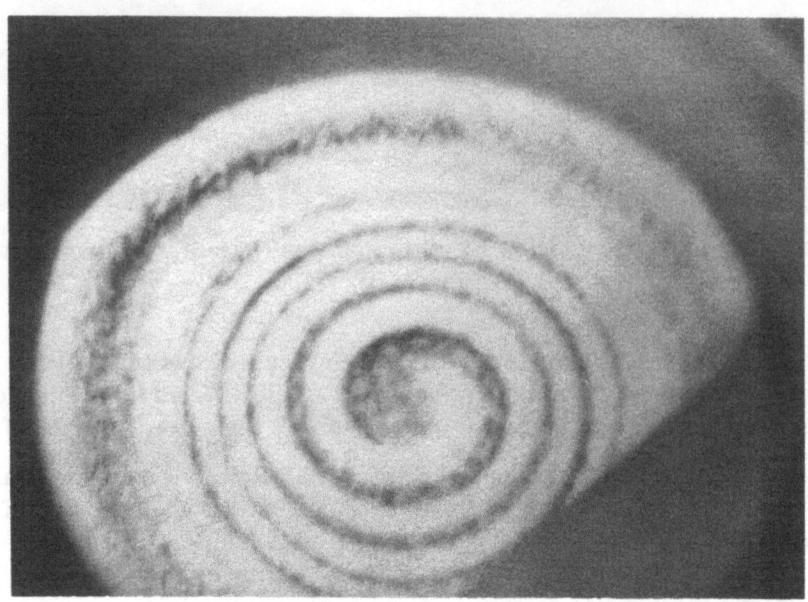

Fig. 7. Spiral domains in Li metaniobate crystal grown
on [000$\bar{1}$], ×5.

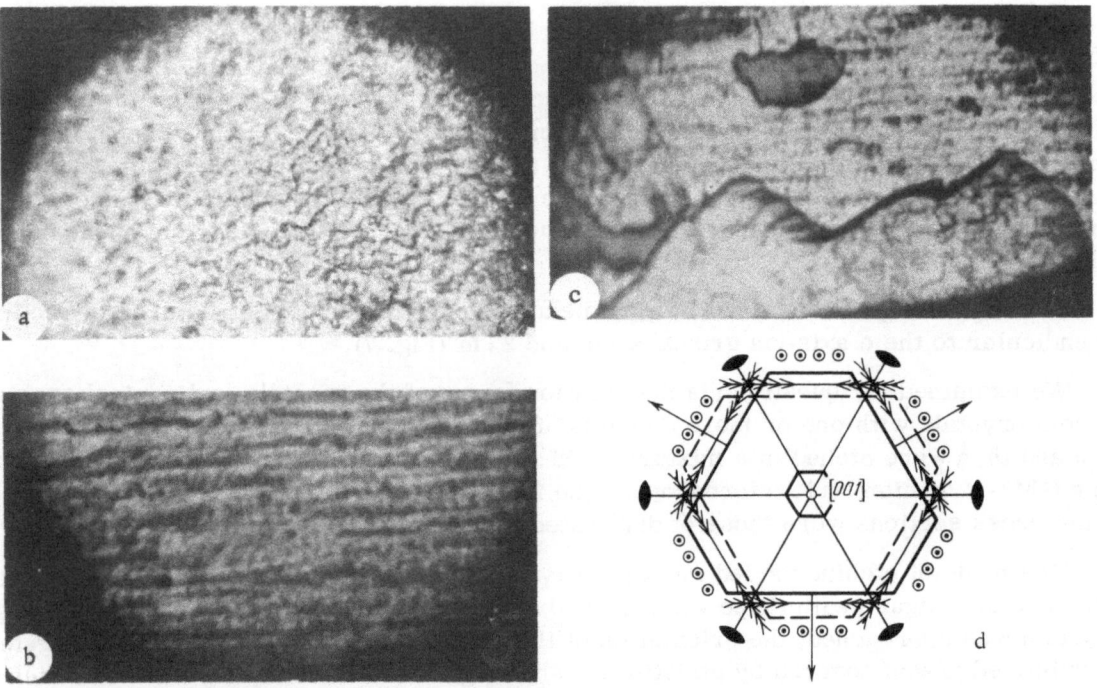

Fig. 8. Solution forms and etch areas on a polydomain Li metaniobate crystal: a) Small ditrigon face, ×530; b) large ditrigon face, ×530; c) difference
in domain etching, ×530; d) cross section of solution shape.

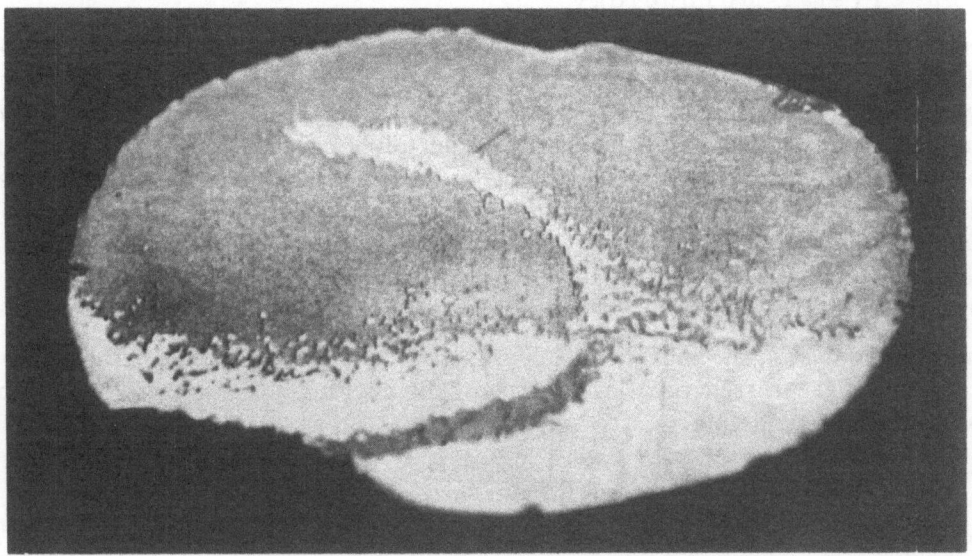

Fig. 9. Domain structure of Li metaniobate crystal grown on [10$\bar{1}$0]
in a medium with m.1:m symmetry, × 5.

The antiparallel domains take the form of rings at the surface of the prism; the surface of a heavily etched large face is ridged, while that of a weakly etched face is cellular (Fig. 8c). The change in the degree of etching of the domains is due to 180° rotation of the spontaneous-polarization vector, which affects the solution forms. Figure 8d shows schematically the cross section of the solution forms for a polydomain specimen; this is indicated by the full line, while a domain with the opposite direction for the vector is shown by a broken line. The cells on the large faces are denoted by small arrows, while convex relief on the small faces is denoted by a circle with a point at the center.

Group 3m describes the solution form of a monodomain crystal of lithium niobate grown on [000$\bar{1}$], while that for a polydomain one is $\bar{3}$/m, which corresponds to the structural symmetry of the crystal in the high-temperature state for the growth direction used. The textures differ in ease of etching, and they are identified with domains in the directions of the a and c axis, which indicates that classical models for the domain structures of barium titanate and Rochelle salt are not applicable to domains observed in these crystals; the model of [5] appears suitable.

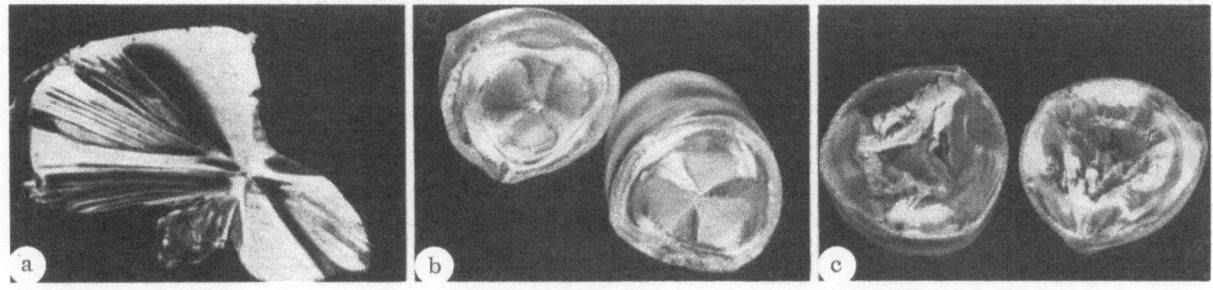

Fig. 10. Cleaved surfaces, natural size: a) [001] La—Na molybdate; b, c) [000$\bar{1}$] Li metaniobate.

Effects on Crystal Morphology

Crystal morphology and defect symmetry are affected by changes in growth conditions: growth rate, growth without rotation, introduction of additional elements, production of temperature gradients, application of electric fields, etc. When YAG crystals are grown on [001] in a system with a longitudinal temperature gradient (symmetry m in the medium), one gets a reduction in the morphologic symmetry of the crystal and also that of the bulk defect (4mm → mm2). Lithium metaniobate crystals grown on [10$\bar{1}$0] in a device with additional cooling and current passing through the melt are represented by a medium with symmetry m.1 : m, while the symmetry of the domain structure falls to 1.m. Usually, such a structure has 2.m morphologic symmetry in the growth direction (Fig. 9).

Then in these cases, the morphologic symmetry arises by superposition of the symmetry of the medium and the structural symmetry of the crystal in the growth direction, in accordance with Curie's principle.

Thermal Cracks

Growing crystals tend to crack when the temperatures change or the impurity composition of the melt is altered; the morphology of the cracks sometimes reflects some of the elements of the structural symmetry. Figure 10 shows cracks in a sodium lanthanum molybdate crystal and in a lithium niobate crystal grown on fourfold and threefold axes, respectively. The elements of the morphologic symmetry in these correspond to elements of the structural symmetry of the crystals in the growth directions; this crack configuration arises from the onset of defect production.

Conclusions

1. Elements from the structural symmetry of the crystal in the growth direction determine the morphologic symmetry and the symmetry of the macroscopic defects for crystals of lithium metaniobate, yttrium-aluminum garnet, barium-sodium niobate, and sodium-lanthanum molybdate.

2. One finds a superposition of the medium symmetry and the structural symmetry of the crystal, in accordance with Curie's principle, when one examines the morphologic symmetry of crystals of yttrium-aluminum garnet and lithium metaniobate grown by Czochralski's method, as well as the macroscopic defects in these crystals.

We are indebted to N. N. Sheftal for valuable comments on the results.

Literature Cited

1. K. Nassau, H. I. Levinstein, and G. M. Loiacano, J. Phys. Chem. Solids, 27:1966 (1963).
2. N. Nuzeki, T. Yamada, and H. Toyoda, Japan. J. Appl. Phys., 6:318 (1967).
3. W. Honigmann, Crystal Growth and Form [Russian translation], IL (1961).
4. V. P. Klyuev, R. M. Tolchinskaya, V. L. Farshtendiker, and S. A. Fedulov, Kristallografiya, 13:531 (1968).
5. K. Aidzu, J. Phys. Soc. Japan, 23:794 (1967).

SOME PHYSICOCHEMICAL GROWTH CONDITIONS
FOR SINGLE CRYSTALS IN VERNEUIL'S METHOD

É. B. Zeligman, B. K. Kazurov,
and S. D. Krasnenkova

There are three distinct kinds of condition for crystal growth in Verneuil's method: one dependent on the physicochemical properties of the materials, mechanical features in the growth zone, and thermophysical features there.

Here we consider the thermophysical conditions in the growth zone: the temperature distributions in the crystal and apparatus together with the heat transfer between the crystal and the medium.

The temperature distribution in a Verneuil oven is very uneven; this is due to the unilateral placing of the heat source (burner) in relation to the heated object, as well as to the large heat losses and the complex structure of a plane. One result of this thermal nonuniformity is a considerable temperature gradient in the growing crystal.

Many studies have shown that the perfection of a single crystal is dependent on the temperature gradients in the crystal and in the melt, especially in the phase-transition zone. To obtain perfect crystals one needs strictly fixed conditions during the crystallization period throughout the growth surface [1]. There is a relation between the microscopic growth rate and the degree of deviation from equilibrium [2]. The dislocation density and distribution are related to the temperature gradients, and in some cases it is stated that radial gradients dominate the picture [3, 4], while in others the main role is assigned to vertical ones [5]. The concentration nonuniformity is [6] related to the temperature distribution, this nonuniformity arising from uneven evaporation as the mixture falls through the flame when the dope concentration is low.

Bazhenova and Shorin [7] in 1969 derived the first quantitative characteristics relating the temperature distribution in the growing crystal to the temperature distribution in the oven and flame; they showed that the gradients in the crystal are determined mainly by the rate of radiated heat transfer from the crystal, primarily to the surface of the ovens. Their calculations [7] showed that the wall temperature substantially influences the value and difference between the axial and peripheral gradients (Fig. 1).

Heat transfer from the flame also plays an important part in the heat balance of the crystal; the flame structure in the burner is determined by diffusion combustion, and the burning occurs at the contact surface between the reacted gases, not within a volume, the latter being the case in kinetic burning. The rate of diffusion burning is governed by that of the slowest stage, i.e., molecular diffusion of the components [8]. Then the reaction zone, i.e., the true flame zone, is a thin layer whose temperature is the maximum theoretical temperature for

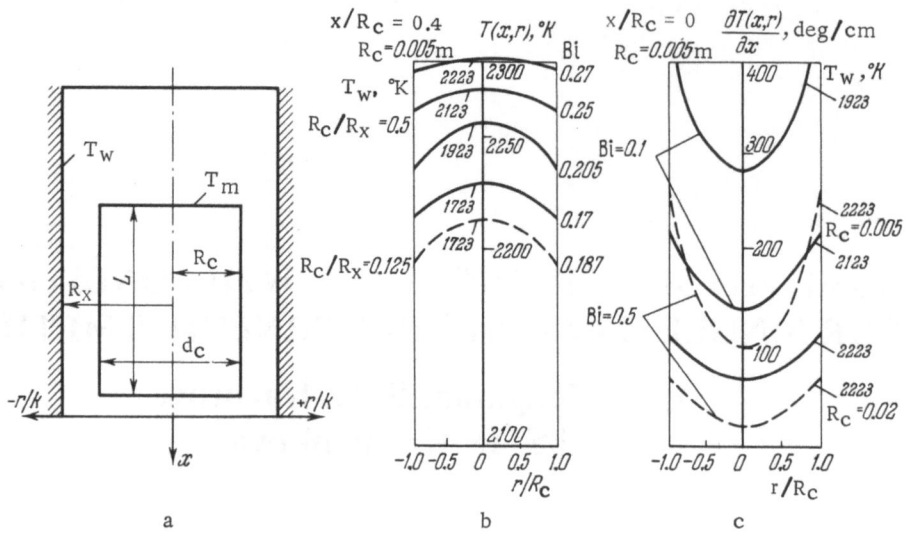

Fig. 1. Temperature distributions in a crystal [7]. a) Calculation
scheme; b) temperature distributions; c) vertical gradients.

combustion of a stoichiometric mixture [8, 9]. That temperature is 2810°C for $H_2 + O_2$ [10].
An important characteristic of a diffusion flame is the length, called the flame length L [8-10].

This information allows one to construct a qualitative model for a burner flame; Fig. 2
shows the flame structure when there is excess hydrogen (the usual condition for corundum).
The flame zone ends at the flow axis, and within it there is a flow of oxygen with a relatively
low temperature, while outside it and below there is a mixture of hydrogen and combustion products.

The heat transfer between the melt and the gases is controlled by the zone in contact with
the melt; bench tests show that the ratio of L to H (the distance from the burner to the surface
of the crystal) has to be adjusted to suit the growth of a crystal of a given diameter, and it
varies in relation to the temperature of the inside wall of the muffle and the heat loading in the
combustion volume in cal-cm³/sec. The last is dependent on the design of the gas composition,
and the relation between the gas flow rates.

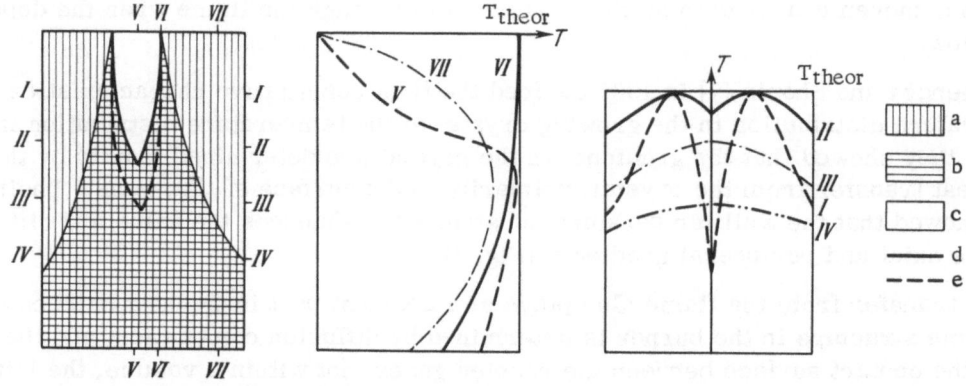

Fig. 2. Flame structure and temperature distribution. a) Hydrogen flow;
b) water vapor; c) oxygen flow; d) zone of $2H_2 + O_2 \rightleftharpoons 2H_2O$ equilibrium; e)
zone of equilibrium shift.

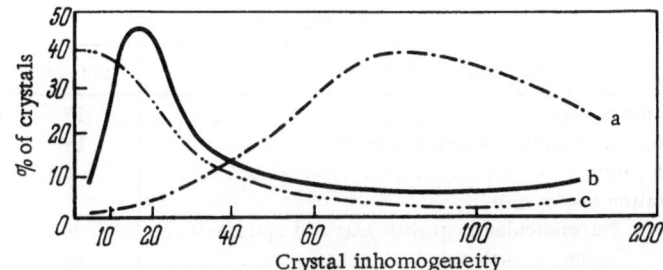

Fig. 3. Optical inhomogeneity of crystals grown at various $T_{cr}-T_{muf}$: a) $T_{cr}-T_{muf} = 200-250°$; b) $T_{cr}-T_{muf} = 55-100°$; c) $T_{cr}-T_{muf} = 20-50°$.

The higher the heat output in the combustion zone, and also the higher the muffle temperature, the less the L/H corresponding to a given crystal diameter; this relationship is readily explained via the heat balance of the crystal: the higher the muffle temperature, the less the radiative heat loss, and consequently the heating can be provided by a less powerful and shorter flame. The following is L/H for various growth conditions for a corundum single crystal of a given diameter under conditions of: ordinary A and intensified B combustion:

$T_{cr}-T_{muf}$, °C	A	B
300—400	1.7—1.8	1.4—1.5
200—250	1.1—1.2	0.95—1.0
20—50	0.9—1.1	0.6—0.8

Here L/H determines the flame zone in contact with the melt; if L is much larger than H, the center of the melt is in contact with gas colder than that at the periphery, and vice versa, which means that small L/H imply that heat transfer by radiation and with the gases facilitates temperature elevation at the center of the melt relative to the periphery. If the flame is long, one can get the reverse effect on the temperature distribution from the heat transfer with the gases.

The conditions of heat transfer from the melt to the gases readily explain the radial distribution of added materials in the crystal; the chromium distribution in corundum usually found (concentration at the center much less than that at the edge) arises when radiative heat transfer causes a higher temperature to be attained at the center of the melt; one gets a relatively even diametrical distribution of the chromium when the edge of the melt is strongly heated, and then the temperature gradients at the surface of the melt are less.

There are also smaller radial gradients in the dope concentration when the muffle temperature is high, i.e., the heat transfer is reduced.

These results indicate that temperature gradients at the surface of the melt are the cause of the radial nonuniformity in dope distribution; the evaporation from this surface is comparable with the evaporation on passing through the flame; in fact, if evaporation occurred only in the flame, the chromium concentration at the center would be higher than that at the edge, because the temperature is lowest at the axis throughout the path from the burner to the crystal.

The flame size and structure explain also the results of tests on passing the material through a flame [6]. There is elevated evaporation of chromium from a mixture entering along the axis of the oven, and this occurs lower down the flame, in the gas medium where the maximum temperature occurs at the axis of the flow.

TABLE 1

Crystal imperfection	250-300	20-50
Growth stresses, kg/mm^2	17	9
Radial gradient in dope concentration, %	18	7
Block size, mm.	1.1	2
Disorientation angle, min		
Perpendicular to growth axis and optic axis. . . .	30	7
Around growth axis.	20	10
Around optic axis.	8	5
Dislocation density, cm^{-2}.	$5 \cdot 10^6 - 1 \cdot 10^7$	$5 \cdot 10^5 - 5 \cdot 10^6$

The thermophysical growth conditions provide the basis for designing apparatus and techniques for producing crystals of higher quality; now, some heating systems enable one to have muffle temperatures of about 2000° (instead of the 1600° found in old systems), and then the temperature difference between the corundum crystal and the muffle at the growth level is only 20-40°.

Work on improved combustion conditions is directed mainly to development of short-flame states; the L/H show that short-flame combustion can be effective only when the muffle temperature is high, since then there are changes in the composition and temperature of the medium around the melt. The use of such combustion in high-temperature furnaces stabilizes the growth, evidently on account of reduced effect variations in gas flow rate and also from flame oscillations.

We made optical and x-ray measurements on corundum single crystals grown under various conditions; Fujiwara's x-ray method gives very accurately the block disorientation angles for three mutually perpendicular directions, as well as the block size and the general distribution of the blocks over the area of the specimen [11]. This method may be combined with optical studies and examination of the dislocations by etching, which gives the most complete objective picture for the structural imperfection.

Table 1 gives the mean quality characteristics of crystals grown with various differences in temperature between crystal and muffle ($T_{cr} - T_{muf}$, °C).

The x-ray studies showed that crystals grown at high muffle temperatures had parts free from blocks; also some of the interference spots had black-and-white lines, which indicate slight deformation of the crystallographic planes.

Conclusions

1. A qualitative model is given for the effects of temperature distributions on crystal uniformity.

2. Heat transfer between melt and flame is dependent on the flame characteristics; the flame size is related to other thermophysical characteristics such as muffle temperature and mode of combustion.

3. Evaporation of the dope from the surface of the melt plays the main part in producing the radial distribution.

4. The results show that the crystal perfection is dependent on the muffle temperature.

We are indebted to I. S. Rez, S. A. Borodin, S. N. Shorin, and O. N. Ermolaev for interest and valuable advice.

Literature Cited

1. N. N. Sheftal', in: Growth of Crystals, Vol. 5A, Consultants Bureau, New York (1968), p. 25.
2. B. Ya. Lyubov, in: Growth of Crystals, Vol. 5A, Consultants Bureau, New York (1968), p. 80.
3. B. M. Turovskii, Kristallografiya, 8:778 (1963).
4. G. P. Ivantsov, in: Growth of Crystals, Vol. 1, Consultants Bureau, New York (1959), p. 76; Vol. 3, Consultants Bureau, New York (1962), p. 53.
5. V. L. Indenbom, Kristallografiya, 9:74 (1964).
6. N. P. Tikhonova, F. K. Volynets, I. V. Tunimanova, and N. P. Mironova, Opt.-Mekh. Prom., No. 7, 45 (1967).
7. I. L. Bazhenova and S. N. Shorin, Papers from the 30th Conference [in Russian], Trudy Mosk. Inst. Khim. Mash., Vol. 1, Izd. MIKhM, Moscow (1969), p. 15.
8. L. N. Khitrin, Physics of Combustion and Explosion [in Russian], Izd. MGU (1957).
9. O. N. Ermolaev, "Combustion and Radiation of a Diffusion Flame," Thesis, Moscow Chemical Plant Design Institute (1959).
10. A. H. Gaydon and H. G. Wolfhard, Structure, Emission, and Temperature of Flames [Russian translation], Metallurgizdat, Moscow (1959).
11. E. P. Kostyukova, B. K. Kazurov, V. G. Lyuttsau, and V. I. Sidenko, in: Growth of Crystals, Vol. 7, Consultants Bureau, New York (1969), p. 124.

SUPERCOOLING AND THE GROWTH FORMS
OF CYCLOHEXANOL CRYSTALS

D. E. Ovsienko, G. A. Alfintsev,
and A. V. Mokhort

The growth form has a substantial effect on the impurity distribution and also on the perfection and properties of a crystal. Recent theoretical studies have shown that the form can be used to judge the structure of the phase boundary and the growth mechanism, and this gives great interest to research on the relation of shape to various factors that influence the growth.

There are numerous papers describing crystals, but only a few of them deal with the reasons of production of any particular form in relation to environmental conditions; particularly little is known about the forms in the practically important case of crystallization of molten metals, which is due mainly to difficulties arising from the high melting points and lack of transparency in metals.

We have examined the growth of crystals of cyclohexanol, which is a transparent organic material with a melting point of 25.46°C [1]. This resembles a metal in that it has a low entropy of melting and a face-centered cubic lattice, and it also resembles camphor and other substances [2-4] in crystallizing in somewhat the same way as do metals.

We have previously shown that small supercoolings (0.005-0.1°) produce cyclohexanol crystals that do not have clear-cut faces, while the growth rate is linearly dependent on the supercooling; the kinetic coefficients for the latter are $8.0 \cdot 10^{-5}$ cm/sec-deg for the $\langle 110 \rangle$ directions and $8.9 \cdot 10^{-5}$ cm/sec-deg for $\langle 100 \rangle$. It has been concluded from this evidence, in conjunction with the absence of characteristic signs of layer growth, that cyclohexanol crystals grow by the normal mechanism.

Here we examine the change in the growth form in relation to supercooling; the material was used in a sealed cell with plane-parallel thin glass windows that was 0.15 mm thick. The cell on all sides was flushed by water from a thermostat, which provided control and measurement of the temperature to ±0.002°. Cyclohexanol is extremely hygroscopic, so it contained 0.01% water in spite of careful vacuum distillation.

As a rule, the growth forms were examined on individual regularly oriented crystals, with the (100) plane parallel to the plane of the cell windows. Such crystals were obtained by slow melting and subsequent cooling of the whole preparation. After a short delay, which gave a crystal of 50-70 μm in size with a supercooling of 0.003-0.1°, the cell was rapidly transferred to another thermostat with the set temperature.

The growth forms were recorded photographically at appropriate film speeds.

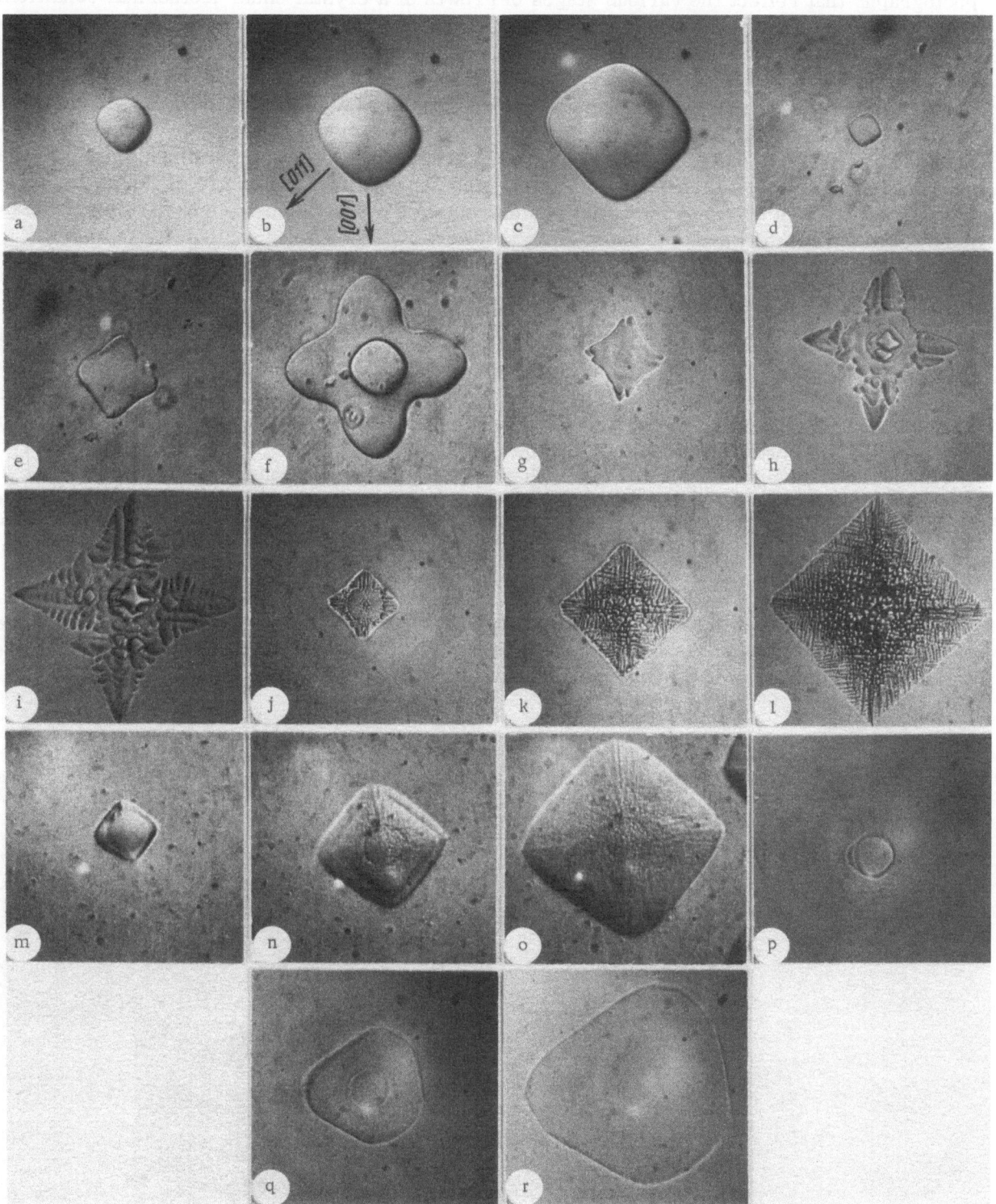

Fig. 1. Growth forms of cyclohexanol crystals at various supercoolings: a–c) 0.005°;
d–f) 0.3°; g–i) 0.45°; j–l) 2.35°; m–O) 3.55°; p–r) 10.25°.

Figure 1 shows growth forms for various supercoolings; for each crystal we give three photographs that reflect the various stages of growth of a crystal under isothermal conditions. When the supercooling was $\Delta T_B = 0.005°$ (Fig. 1a–c), the crystal was rounded and grew without change of shape. This continued until the size was 3–4 times the original size. For $\Delta T_B = 0.3°$ (Fig. 1d–f) the crystal initially grew without change of shape, but then the shape became unstable, and along the $\langle 110 \rangle$ directions there were produced dips, which increased in size during the isothermal period. This was due to the uneven distribution of impurities around the periphery of the crystal, and perhaps also to uneven heat loss. Naturally, the impurities tended to accumulate most at points of least curvature, since here diffusion is more difficult, and this unevenness was not balanced by a slight anisotropy in the kinetic coefficient [5, 6], which has been found for these crystals growing from the melt.

The less the supercooling, the larger the size to which the crystal shape remains stable; this explains the more even impurity distribution around the periphery at low supercoolings.

When $\Delta T_B = 0.45°$ (Fig. 1g–i), the kinetic coefficient is largest along $\langle 001 \rangle$, and one gets dendrites; in dendritic growth, much of the impurity is found in the space between the dendrites, and the distribution of it in the melt in front of a dendritic crystal is different from that for a flat front, as Figs. 2 and 3 show, where we see change in the growth rate and shape when two crystals meet. Parts a and b of Fig. 2 show two rounded crystals growing together for $\Delta T_B = 0.1°$; even at a distance of about 100 μm, there is an appreciable change in crystal shape; the crystals do not intergrow on reducing the temperature (Fig. 2c). The growth rate is much reduced when the crystals meet (Fig. 3a). These features of the growth are related to the more marked accumulation of impurity between the nondendritic crystals, which has an appreciable influence out to a substantial distance.

In dendritic growth, the rate is constant until the crystals meet (Fig. 3b); there is also no appreciable change in the shape of the dendrite (Fig. 2d–f), which indicates only slight impurity accumulation, at least at large distances from the dendrite.

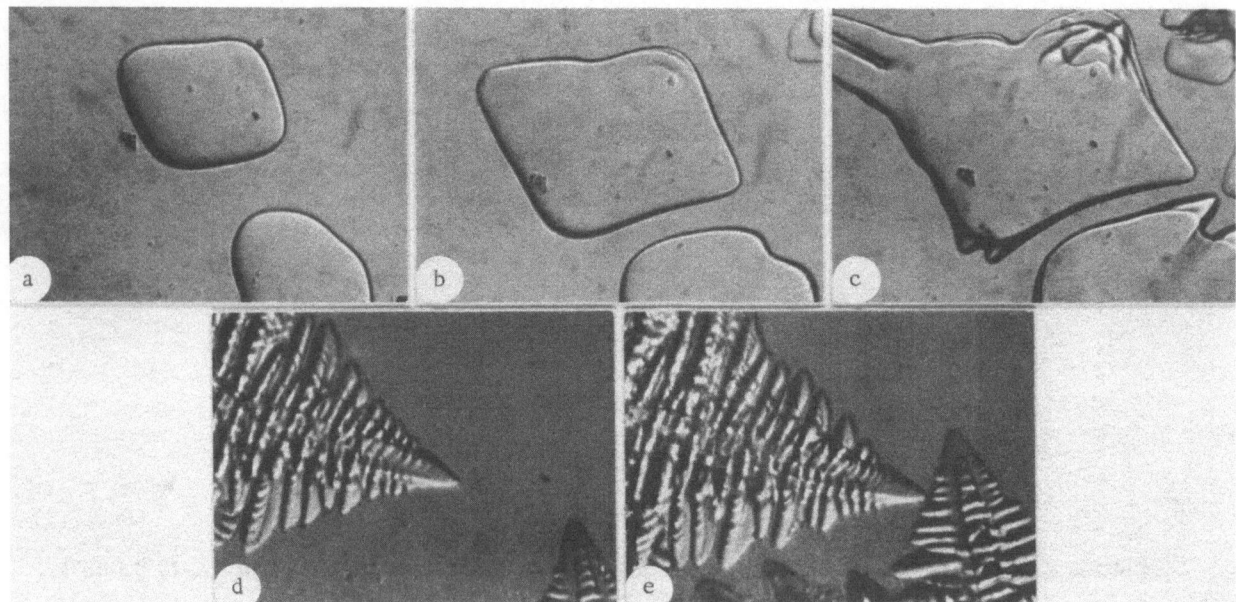

Fig. 2. Growth forms of two crystals growing close together. a–c) Rounded crystals of altered shape; d, e) dendritic crystals unaltered in shape.

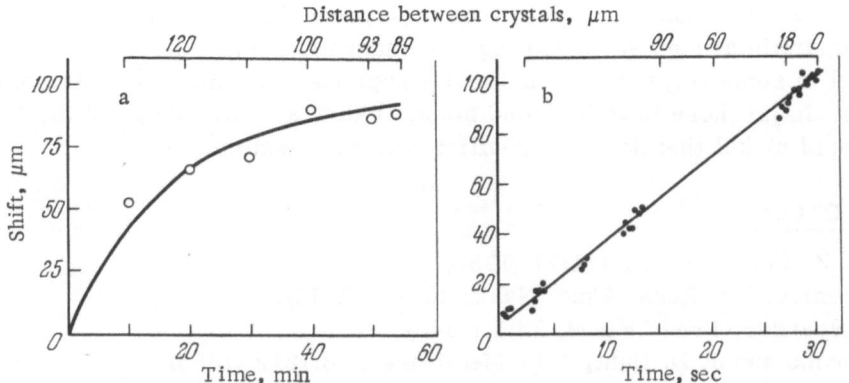

Fig. 3. Growth rate as a function of distance between crystals.
a) Rounded crystals; b) dendrites.

Figure 4 shows a crystal that initially grew at a supercooling of 0.02°, and then at ΔT_B = 0.2°, and finally at an even larger supercooling. Dendrites were produced only along $\langle 001 \rangle$ directions external to the crystal, while in internal analogous directions there were none such on account of the considerable accumulation of impurity and the worse conditions for heat loss.

Further temperature reduction (Fig. 1j–l) caused the dendritic structure to become finer, while the distances between the branches were reduced. The shape of the crystal is nearer then to cubic than at small supersaturations. When one grows certain metallic crystals, the shape is sometimes determined by decantation, and the adhesion of liquid to the crystal [7, 8] sometimes gives a false impression of the crystal habit and the production of layer growth.

For ΔT_B = 3.55° (Fig. 1m–o), the dendritic structure vanished and the crystal again became rounded, as at low supercoolings (Fig. 1a–c). It was difficult to determine whether the crystal was truly smooth or was divided into thin needles, but dendrites were clearly not present. The structure observed on the surface can arise on account of poor heat transfer through the crystal.

At even larger supercoolings, the crystallization front became microscopically smooth and the growth form was stable, as parts p–r of Fig. 1 show for the growth of a crystal with irregular orientation. Here no great amount of impurity accumulates around the periphery of the crystal because otherwise it would be extremely unevenly distributed in the presence of the high growth rate and the growth form would be unstable. One assumes that most of the impurity is taken up by the crystal.

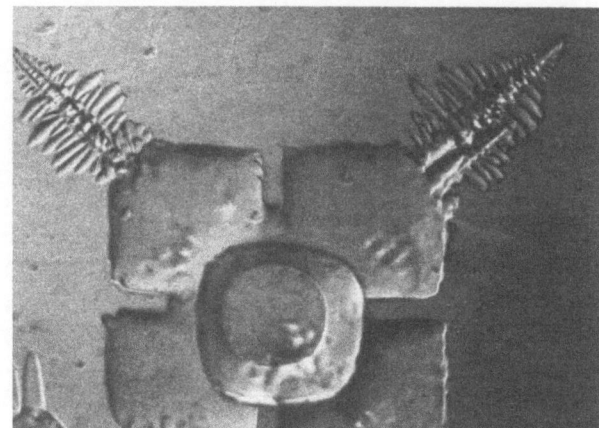

Fig. 4. Change in crystal shape with stepwise
cooling.

These results show that the growth from a supercooled melt varies with the supercooling; it may not be dendritic at high supercoolings, at least for some substances. Cyclohexanol resembles a metal in some ways, so one naturally supposes that metal crystals can grow in a similar fashion. In [9] there is to be found some evidence on the shape of the interphase boundary in supercooled nickel that does not conflict with this assumption.

Literature Cited

1.   J. Lange, Z. Phys. Chem., 161:77 (1952).
2.   N. N. Efremov, Izv. Ross. Akad. Nauk, 23:539 (1915).
3.   I. N. Fridlyander, Trudy VIAM, 95:1 (1949).
4.   K. A. Jackson and J. D. Hunt, Acta Metallurg., 13:1212 (1965).
5.   A. A. Chernov, Kristallografiya, 8:87, 499 (1963).
6.   A. A. Chernov and B. Ya. Lyubov, in: Growth of Crystals, Vol. 5A, Consultants Bureau, New York (1968), p. 7.
7.   F. Weinberg, Trans. Metallurg. Soc. AIME, 224:628 (1962).
8.   J. Barthel and R. Scharfenberg, Canad. J. Phys., 42:1411 (1964).
9.   G. A. Colligan and B. S. Bayles, Acta Metallurg., 10:895 (1962).

THE EQUILIBRIUM PARTITION COEFFIENTS FOR IMPURITY CATIONS IN ALKALI HALIDE CRYSTALS

O. L. Kreinin, K. M. Rozin, and M. P. Shaskol'skaya

The equilibrium partition coefficient K_0 serves to describe the entry of impurities into a growing single crystal; thermodynamic calculations [1] give

$$\ln K_0 = \frac{\Delta H^f - \Delta H^S}{RT} + \frac{\sigma - \Delta S^f}{R} + \ln \gamma \ldots, \tag{1}$$

where ΔH^f and ΔS^f are the heat and entropy of fusion of the substance, ΔH^S is the partial heat of mixing for the impurity, σ is the change in the vibrational entropy of the system, and γ is the activity coefficient of the impurity in the melt.

So far as we are aware, K_0 has been calculated via (1) only for some impurities in germanium and silicon [2].

The object of this study has been to develop a method of calculating the equilibrium partition coefficients for cation impurities in alkali halides KCl and NaCl, and also to compare the calculated K_0 with the experimental ones.

We used the general relation of K_0 to the thermodynamic functions expressed via (1), which shows that one needs to consider ΔH^S, σ, and γ, and to choose these via appropriate calculated expressions. All the other quantities in the formula are tabulated ones.

To determine ΔH^S we assumed that the impurities formed some substitutional solid solutions with the host lattice; the problem then is to determine the heat released on cation substitution. Consider the energy of this process when a divalent impurity enters a single crystal of KCl or NaCl. The method will be considered in relation to the incorporation of Pb^{2+} into KCl. Here we have to remember that this entry causes an additional cation vacancy.

Let g_i be the free energy needed to remove two separate but arbitrary K^+ ions from a pure KCl crystal in a state of rest to infinity and let the Pb^{2+} ion previous to this be in a state of rest at infinity and enter to fill one of the vacancies. We assume that G_1 and G_2 are the Gibbs free energies of the pure crystals of KCl and $PbCl_2$, respectively, relative to ions at rest at an infinite distance apart. Then

$$PbCl_2 = Pb\,(\text{in } KCl) - (g_i - 2G_1 + G_2). \tag{2}$$

Then G_1 and G_2 are tabulated values of the lattice energies of KCl and $PbCl_2$, and the problem amounts to determination of g_i. To do this, we consider the following energy cycle: we remove the K^+ ions from two regular lattice nodes and we place the Pb^{2+} ion in one of these released cation vacancies. Then

$$g_i = W_{Pb^{2+}(KCl)} - 2W_{CV^-(KCl)} \cdot \tag{3}$$

189

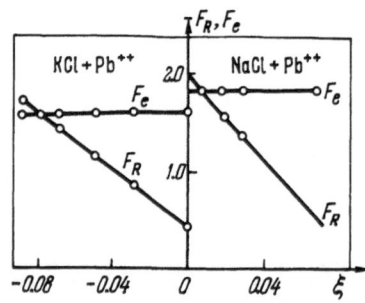

Fig. 1. Graphical solution of (3).

Then we have to determine the work required to form a point defect (a foreign ion or a vacancy) in the lattice of KCl; such calculations [3] are based on determining the work needed to remove an ion from the lattice, which is the mean of the potential energies of the ion in the field together with the displacements and polarizations of the ion in the initial position and in the displaced one.

To calculate ion energies in alkali halides, we can neglect the van der Waals energy and the zero-point vibration energy, and need to consider only two forms of interaction between the ions: the electrostatic and the repulsion. These quantities can be calculated with various degrees of accuracy corresponding to the number of ions around the defect that is considered. We have carried out the calculation to a first approximation with allowance for the first coordination sphere of a defect, while the effects of more remote neighbors were considered in the continuum approximation.

Methods have been given for calculating the energy of formation of cation vacancies and the energies needed to introduce ions into KCl and NaCl [4-6]. These calculations have not been done before for Pb in KCl and NaCl. The main difficulty arises over the determination of ξ, the displacements of the ions closest to the defect. This can be found from the condition for a minimum sum of the electrostatic force F_e and the repulsion force F_R at an ion in the first coordination sphere of the defect. Figure 1 shows graphical determination of ξ for Pb in KCl and NaCl. The other quantities needed to calculate ΔH^S are deduced from the result for ξ and are given in the table.

Consider the terms in σ and $\ln \gamma$ in (1); σ in (1) indicates the change in the vibrational entropy of the system on account of entry of the impurity into a lattice, the melting point of the impurity being T_A, while that of the host crystal is T_B. If we assume that the vibrational

TABLE 1. Data for Equilibrium Partition Coefficients of Pb^{2+} in NaCl and KCl [6-12]

Parameter	Crystal		Parameter	Crystal	
	NaCl	KCl		NaCl	KCl
$Me^+ - CV^-$	0.096	0.092 [6] *	G_2, ev	21.6	21.6 [9]
Shift ξ for Pb^{2+} in place of Me^+	0.0075	—0.08	$(2G_1 - G_2)$, ev	5.84	7.24
			ΔH^S, ev	—0.13	0.05
$W_{CV^- (KCl)}$, ev	4.65	4.35	$\ln \gamma$	—0.5	—1.6 [11]
$W_{Pb^{2+} (KCl)}$, ev	15.27	15.89	σ, cal/mole-deg	—3.9	—2.4
$[2W_{CV^- (KCl)} - W_{Pb^{2+} (KCl)}]$, ev	—5.97	—7.19	K_0	0.15	0.016
G_1, ev	7.88	7.18			

*Literature citations in square brackets.

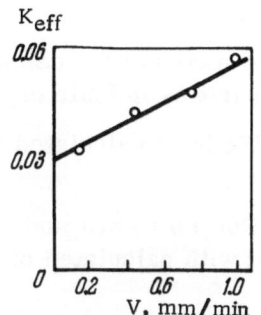

Fig. 2. Kinetic dependence of K_{eff} for Pb in KCl.

spectrum of the host crystal is largely unaltered by the addition of several impurity atoms, then

$$\sigma = 3k \ln \frac{T_B \theta_A}{T_A \theta_B}. \tag{4}$$

Here k is Boltzmann's constant, θ is the Debye temperature, and the subscripts A and B refer, respectively, to atoms of the impurity and the host crystal. We took the values of γ from published experimental results on the emfs of cells for the corresponding salt mixtures [11].

It was thus possible to calculate the equilibrium distribution coefficients for impurities in alkali halide crystals; Table 1 gives the calculated values for the basic quantities needed in (1).

The calculation was checked by measuring the K_0 for lead in KCl and NaCl crystals grown with $C_0 = 0.2$ wt.% by the Kyropoulos method at rates from 0.04–1.0 mm/min. The Pb contents of the crystals were determined polarographically to ±5%. The effective partition coefficients were calculated for the melting point. Figures 2 and 3 show K_{eff} for Pb as a function of crystallization rate for KCl and NaCl single crystals; extrapolation of the straight lines to zero rate gives the equilibrium results for Pb^{2+} in NaCl and KCl:

Crystal Results	NaCl	KCl
calculated	0.15	0.016
experimental	0.12±0.01	0.03±0.003

The error in measuring the equilibrium coefficients is due to the errors of measurement for the concentration and growth rate. We used least-squares fitting in constructing the kinetic curves and found that the error in determining K_0 is ±10%.

We found good agreement between the calculated and observed K_0.

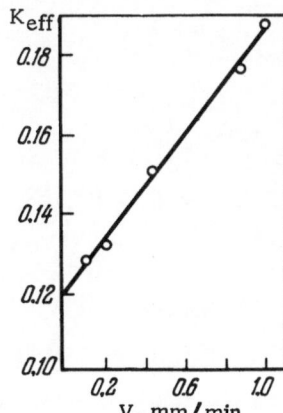

Fig. 3. Kinetic dependence of K_{eff} for Pb in NaCl.

Conclusions

1. A method has been proposed for calculating the equilibrium partition coefficients for cation impurities in ionic crystals.

2. We have calculated equilibrium partition coefficients for single crystals of KCl and NaCl.

3. For Pb in KCl and NaCl crystals we have found values of K_0 that are in reasonable agreement with calculated ones.

Literature Cited

1. C. D. Thurmond and S. D. Struthers, J. Phys. Chem., 5:831 (1953).
2. K. Weiser, J. Phys. Chem. Solids, 7:118 (1958).
3. A. Lidiard, Ionic Conduction in Crystals [Russian translation], IIL, Moscow (1962).
4. N. E. Mott and M. J. Littlton, Trans. Faraday Soc., 34:485 (1938).
5. F. Bassani and F. G. Fumi, Nuovo cimento, 11:274 (1954).
6. P. Brauer, Z. Naturforsch., 7a:372 (1952).
7. M. Born and J. Mayer, Z. Phys., 75:1 (1932).
8. C. Kittel, Introduction to Solid-State Physics [Russian translation], IL, Moscow (1962).
9. V. D. Kuznetsov, Solid-State Physics [in Russian], Izd. Tomsk. Univ. (1949).
10. J. Lumsden, Thermodynamics of Alloys [Russian translation], Metallurgizdat, Moscow (1959).
11. M. F. Lantratov and A. F. Alabyshev, Zh. Prikl. Khim., 26:263 (1953).
12. J. Lumsden, Thermodynamics of Molten Salt Mixtures, New York (1966).

MORPHOLOGY AND NEODYMIUM DISTRIBUTION
IN YTTRIUM ORTHOVANADATE CRYSTALS

I. N. Guseva, V. S. Krylov, I. L. Poluéktova, and V. I. Popov

Yttrium orthovanadate crystals doped with neodymium are no worse than yttrium-aluminum garnet ones as regards chemical stability, melting point (1850°C) [1], and lasing features [2]. We have made [3] crystals by Verneuil's and Czochralski's methods, as well as by crystallization from high-temperature solutions. The most perfect crystals are those grown by Czochralski's method, and a laser has been based on these [2, 4]. Here we consider the crystals.

We tried three methods of making the initial mixture: sintering $Y_2O_3 + V_2O_5 = 2YVO_4$ and two methods of crystallization from aqueous solution: $Y(NO_3)_3 + K_3VO_4 = YVO_4 + 3KNO_3$; $Y(NO_3)_3 + NH_4VO_3 + H_2O = YVO_4 + NH_4NO_3 + 2HNO_3$.

In our subsequent work we used the last method, which gave the best-quality crystals with isomorphous neodymium. The crystals were grown on single-crystal seeds oriented parallel to the optic axis. The initial material was placed in an iridium crucible in an argon atmosphere. The crystals were identified by chemical and x-ray methods [1]. Spectral analysis revealed the following elements: Fe < 0.005%, Cr < 0.001%, Co < 0.0005%, Mn < 0.0005%, Si < 0.02%, Cu < 0.0006%.

The crystals had four well-developed (110) prism faces elongated along the specimen; these coincide with the direction of perfect cleavage. The surface had thin small spirals or hummocks of various sizes arranged along the crystal in the form of steps, whose pitch was dependent on the relation between the pulling rate and the frequency of the temperature fluctuations. We used a polarizing microscope, a slit ultramicroscope [5], a pinhole method [6], an x-ray microanalyzer (JEOL, type JXA-3), and also spectrochemical analysis with crystals with neodymium contents of 1.9, 3.5, 5.4, and 8.4 wt.% of neodymium. Most of the crystals were uniform in refractive index, as the pinhole method revealed; only some small parts of certain of them showed signs of growth zones as flat or slightly curved surfaces reflecting the shape of the crystallization isotherm. The same isotherm surfaces were observed on rapidly detaching the growing crystal from the melt. The partition coefficient for neodymium was determined at the start of growth by x-ray and spectral methods, which gave the values 0.75 and 0.73, respectively. The melt becomes enriched in neodymium as the crystal pulling proceeds, and there is a monotonic increase in neodymium concentration from the start of crystallization toward the end. In what follows we give typical results for a crystal where the initial mixture contained 8.4 wt.% neodymium.

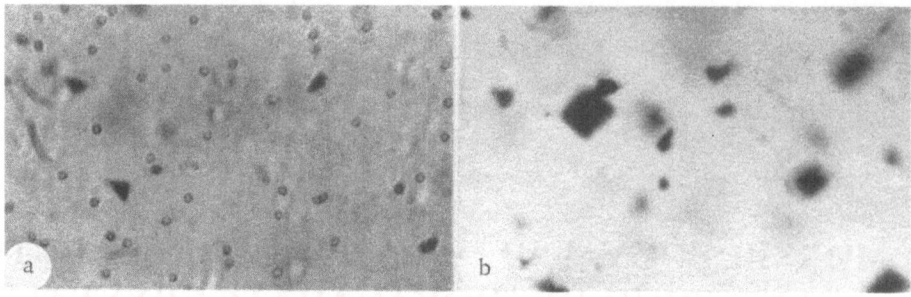

Fig. 1. Inclusions in yttrium orthovanadate crystals. a) Triangular; b) rectangular.

Crystal length from seed, mm	0	1	2	3	4	5	6	7
Neodymium concentration,* wt.%	6.35	6.44	6.50	6.57	6.64	6.71	6.77	6.88
Crystal length from seed, mm	8	9	10	11	12	13	14	
Neodymium concentration,* wt.%	6.91	7.00	7.00	7.11	7.19	7.28	7.30	

*Nd assayed to ± 0.02 wt.%.

The crystals contained inclusions, some of which had sharp triangular or hexagonal outlines
(Fig. 1a) with sharp or truncated corners (sizes of 10-40 μm), while others were rectangular
(Fig. 1b) and 40-60 μm in size. The x-ray microanalyzer established that the triangular in-
clusions (Fig. 2a) contain calcium. We recorded photographs of parts with such inclusions in

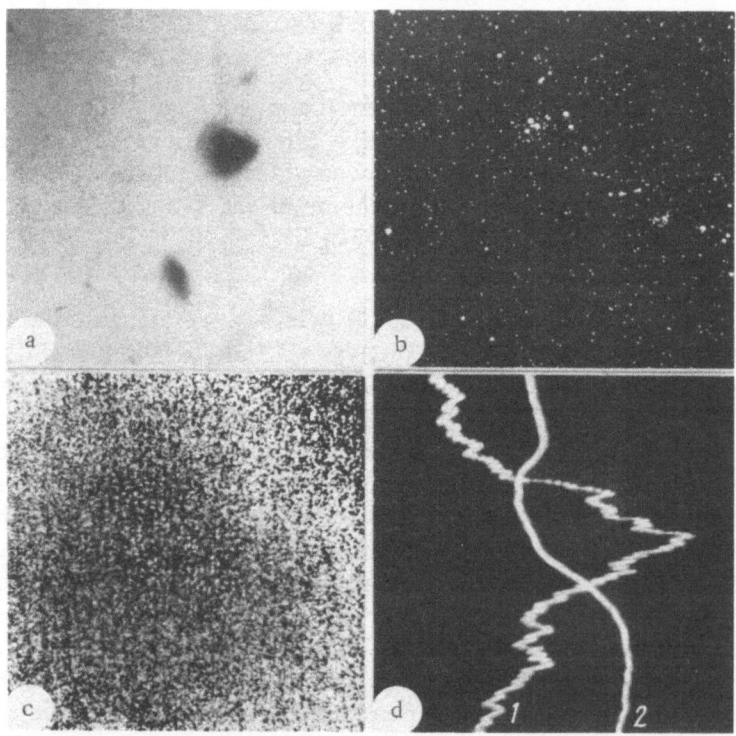

Fig. 2. Part of yttrium orthovanadate crystal with
a triangular inclusion. a) In field of view of microscope attached to x-ray analyzer, ×400; b) in Ca x-
ray, ×600; c) in V x-rays, ×600; d) distributions of
Ca and V along the Ca K_{α_1} (1) and V K_{α_1} (2) lines.

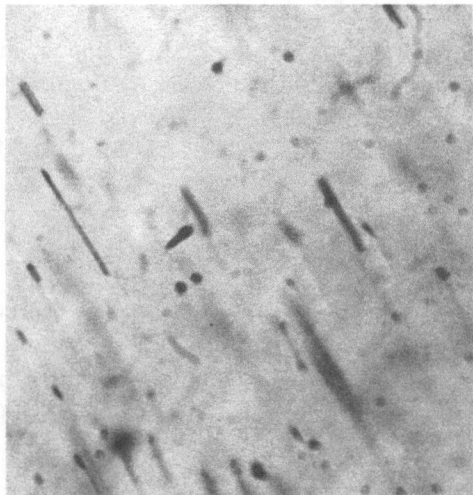

Fig. 3. Microcracks in yttrium orthovanadate crystals.

the characteristic x-rays of calcium (Fig. 2b) and vanadium (Fig. 2c). The lighter parts correspond to higher contents of the element. Along the spectral line (Fig. 2d) there was a peak on the curve for calcium and a minimum on the vanadium curve, which corresponds to the position of a triangular inclusion. Analogous studies for the rectangular inclusions indicated the presence of iron and traces of calcium and silicon entering from the initial mixture; other elements were not detected.

Some crystals had microcracks up to 100 μm long (Fig. 3), which were oriented on the prism. These arose from rapid cooling of the crystal.

All of these defects scatter light and were observable with the slit ultramicroscope (Fig. 4).

These studies of the material enable us to characterize the real structure of neodymium-doped yttrium orthovanadate crystals.

We are indebted to Kh. S. Bagdasarov for a discussion and valuable advice.

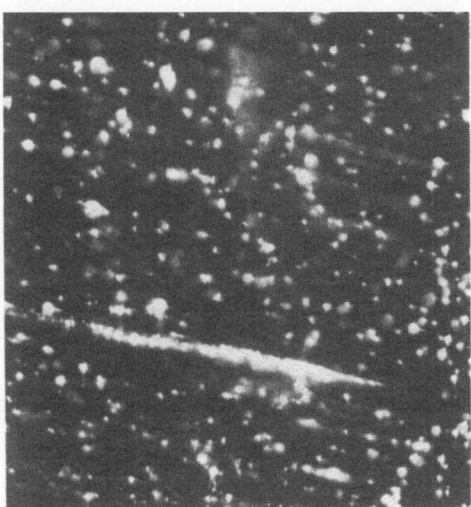

Fig. 4. Scattering at inclusions in yttrium orthovanadate crystals.

Literature Cited

1.   V. A. Timofeeva, V. I. Popov, and T. S. Kon'kova, Kristallografiya, 14:150 (1969).
2.   A. A. Kaminskii, G. A. Bogomolova, and L. Li, Izv. Akad. Nauk SSSR, ser. neorg. mat., 5:673 (1969).
3.   Kh. S. Bagdasarov, I. L. Poluéktova, V. I. Popov, and V. S. Krylov, Kristallografiya, 5:617 (1969).
4.   Kh. S. Bagdasarov, G. A. Bogomolova, A. A. Kaminskii, and V. I. Popov, Dokl. Akad. Nauk SSSR, 180:1347 (1968).
5.   E. V. Antonov, I. N. Guseva, N. M. Melankholin, and G. D. Shnyra, in: Methods and Instruments for Monitoring the Quality of Ruby Crystals [in Russian], Nauka, Moscow (1968), p. 54.
6.   I. N. Guseva, N. M. Melankholin, and G. D. Shnyra, in: Methods and Instruments for Monitoring the Quality of Ruby Crystals [in Russian], Nauka, Moscow (1968), p. 59.

PURIFICATION OF METALS BY ZONE MELTING

P. S. Vadilo

Introduction

It has been stated [1] that metal purification by zone melting is dependent only on the ratio of the solubilities for the impurities in the solid and liquid phases as well as concentration ratio in the solid and in the adjacent liquid layer. However, we have found [2] that the impurity uptake by a growing crystal is determined also by the number of subindividuals present; the impurity uptake increases with this number for nonisomorphous substances, and none of the latter may be taken up if there are no subindividuals. The latter tend to displace impurities from one towards another, and so the impurities accumulate between the subindividuals in relatively larger amounts. We have shown [3] that subindividual formation is dependent on the supercooling differences between parts of the surface, which arise from concentration or convection currents. If the differences are substantial, a part in contact with the more supercooled melt starts a new layer before the previous one has spread entirely over the face, and so subindividuals can arise there. The name critical is given to the supercooling difference needed to initiate subindividual formation; it is less for fast-growing faces.

The probability of a critical difference is dependent on the supercooling of the melt as a whole. If subindividuals are already present on the growing faces, they vanish when the supercooling is small, since the crystallization occurs primarily in the reentrant spaces between them. The number of subindividuals increases with the general supercooling, but the sizes are reduced.

Here we give experimental evidence on the effects of subindividuals arising mainly on fast-growing faces as regards impurities in the crystal.

Experimental

Single crystals of KCl of weight up to 800 g were grown by the Kyropoulos method from KCl of reagent grade; H_2O and starch failed to reveal iodine, but this reaction on cleavage planes of the KCl blocks did reveal iodine between the subindividuals produced from highly supercooled melts.

In another series, a potash alum solution (10 liters) of medium supersaturation (20 g/liter) was used with homogeneous crystals of this alum; the (111) faces were the most extensive, but (100) and (110) ones were also present. Within a day, subindividuals had formed on the latter (Fig. 1), while the (111) faces were flat. Subindividuals were produced on all faces (Figs. 2 and 3) when the temperature was reduced by 15°C.

In a third series, a drop of NH_4Cl solution saturated at room temperature and containing $5 \cdot 10^{-4}$ wt.% $CoCl_2 \cdot 6H_2O$ was placed between two glass plates held apart by a few sand grains

P. S. VADILO

Fig. 1. Potash alum crystals grown at a supersaturation of 20 g/liter, natural size. Subindividuals occur only on the (100) and (110) faces, with none on the large (111) faces.

Fig. 2. Potash alum crystals. 1) Grown at a supersaturation of 35 g/liter; 2) without subindividuals, grown at 10 g/liter supersaturation.

Fig. 3. Potash alum crystal composed of subindividuals after growth from a highly supersaturated solution at 20°, natural size.

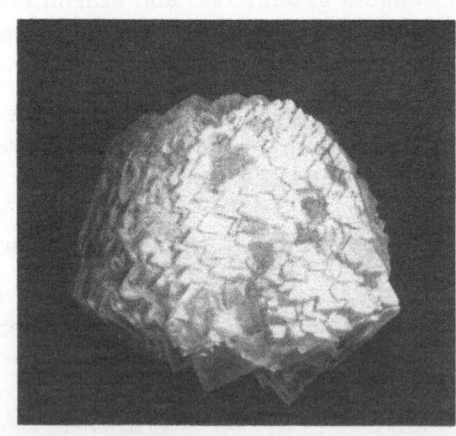

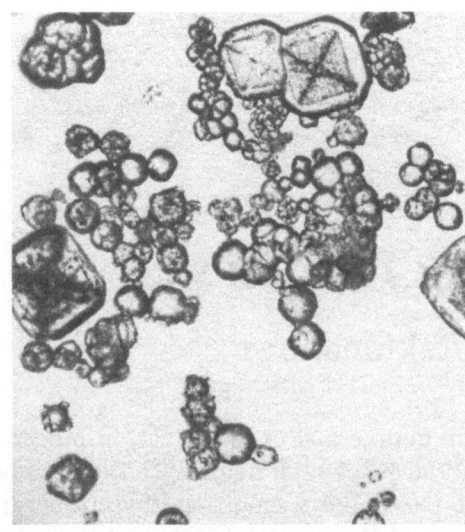

Fig. 4. Large NH_4Cl crystals grown at high supersaturation with a trace of $CoCl_2 \cdot 6H_2O$, which is taken up only by the central parts, $\times 50$.

1 mm in diameter. The drop dried up in the course of a few days and produced several isometric NH_4Cl crystals having (100) faces in contact with both glass surfaces. At the sides there were prominent (100) faces and slight development of (111). On (111) there were large subindividuals, with $CoCl_2 \cdot 6H_2O$ between them in the form of a cross. Then the glass plates were turned one relative to the other; small particles were produced, which grew into crystals and thereby reduced the supersaturation. No subindividuals arose on the small crystals, while those on the large ones vanished. The large crystals continued to grow without subindividuals, and the peripheral parts did not contain impurities (Fig. 4).

Discussion

Iodine was found between subindividuals of KCl grown from a melt in which iodine was undetectable, which shows that impurities accumulate between subindividuals.

The alum experiments (Fig. 1) show that the subindividuals arise on rapidly growing faces at supersaturations lower than those needed for slowly growing faces, because the critical supersaturation differences are larger in the second case.

The NH_4Cl tests show that nonisomorphous components collect in the spaces between subindividuals; these vanish when the supersaturation is reduced considerably, and the crystal grows free from impurity.

Conclusions

Uptake of nonisomorphous impurities is very much dependent on the presence of subindividuals. Allowance must be made for the critical supercooling in discussing the impurity distribution along a crystal purified by zone melting.

Literature Cited

1.   J. Wernik, in: Ultrapure Metals [Russian translation], Metallurgiya (1966), p. 50.
2.   P. S. Vadilo, Zh. Prikl. Khim., 36:2666 (1963).
3.   P. S. Vadilo, Zh. Éksp. Teor. Fiz., 8:1374 (1938).

MORPHOLOGY OF EUTECTIC COLONIES

V. V. Nikonova and É. P. Rakhmanova

There has recently been increased interest in the structure and crystallization mechanisms of eutectics; the first theoretical study appeared in 1958 [1], which dealt with the steady-state growth of eutectic colonies of platy structure, which predicted a functional relationship between the pitch of the eutectic structure, the supercooling at the growth front, and the growth rate. Subsequently a revised theory was published [2], and then a reconsideration of the whole subject [3].

Interest in eutectic crystallization has also increased because of the observation of special properties in eutectics of fibrous structure [4-7].

It was found that fibrous eutectic colonies have elevated strength because the fibers in them represent defect-free whiskers. It was also found that the size and orientation of these can be controlled under conditions of directed growth, which gave great scope for making materials reinforced by whiskers of one of the phases, which in some instances provided strength increases by more than a factor of 3.

All these studies were concerned largely with the steady-state growth of platy and fibrous structures, but these forms do not exhaust the variety of eutectic structures. Also, steady-state growth is a rare and artificial case, whose advantage consists in the simplicity of the theoretical description, which allows of a mathematical treatment.

A complete understanding of eutectic crystallization is impossible without discussing nucleation and free growth of eutectic colonies of various types, together with elucidation of the physical reasons leading to all forms of structure. The complexity of the problem can be seen from a survey of the numerous classifications based on various parameters: Shiele's classification via the presence or absence of eutectic colonies (normal and anomalous eutectics), Tiller's from the form of the particles constituting a colony (globular, rod-shaped, platy, like Chinese characters, etc.), and Jackson's via the entropy of melting of the phases, etc. Here we consider the structures of eutectic colonies freely growing from a liquid of eutectic composition. The temperature and composition of the liquid far from a colony are identical in all directions. The concentration and temperature nonuniformities near the colony are set up by the features of the growth; such conditions are idealized and occur in colony growth when the distances between neighboring crystals are sufficiently large and neither thermal nor diffusion fields overlap, otherwise the shapes of the crystals would be distorted.

There are numerous eutectic structures in metal and organic systems, which indicate that the shapes of the eutectic colonies are determined by the crystallographic features of one of the phases, which in future we call the principal phase.* Nonuniform concentration distribution at the surface of the principal phase leads to nonequilibrium relief of the crystal surface and strange growth forms; the crystals of the second phase fill the recesses in this skeleton and indicate the resulting structure while facilitating its further development.

At present there is no properly developed theory of the stability loss in various crystal shapes that incorporates the thermal and diffusion processes at crystal surfaces [8-13]. If there is a positive temperature gradient ahead of a growing crystal, the stability limit may be correlated roughly with the conditions for occurrence of diffusion supercooling, which are dependent on the state parameters, or rather on

$$\psi = \frac{mvC\,(1-k)}{D}\,, \tag{1}$$

where C is the concentration at the crystallization front in the melt, k is the equilibrium partition coefficient, m is the slope of the liquidus line, D is the diffusion coefficient of the second component in the melt, and v is the growth rate.

There is [13] a discussion of the stability of a spherical crystal growing from a supersaturated (supercooled) melt, and it has been shown that stability loss sets in when the crystal has reached a certain definite size. The limiting radius for a stable crystal is calculated as exceeding by at least a factor of 7 the critical nucleation radius, and it can be shown that this limiting radius is dependent on the ψ of (1).

In a liquid of eutectic composition, there is a difference in value between the ψ for the α and β phases, which consequently are not under equivalent conditions; the phase with the larger ψ is the first to lose stability and initiates the production of a eutectic colony; therefore, one can predict from the phase diagram which of the phases will be the first to lose stability and give rise to the skeleton of the eutectic colony.

If the growth is free, i.e., if the heat is lost through the liquid phase, the structure of a eutectic colony is determined by the growth forms of the principal phase; if that phase grows as rounded crystals without prominent faces, it is likely that the splitup will occur by the radial growth of fibers, and the eutectic colony will take the form of a spherulite (Fig. 1). Structures of this type may appear as globular when the colony is cut through peripherally, because only the end faces of the fibers are seen in section. Such colonies are readily oriented by the heat transfer, because the fibers grow perpendicular to the crystallization front, and they form oriented reinforced structures.

If the growth rate has pronounced anisotropy, the splitting begins at different times on the various faces; faces of largest growth rate have the largest value of ψ and they are the first to split, so the anisotropy of the principal phase determines the type of eutectic colony.

---

* The term principal phase is often used in a different sense, which must be considered here to avoid misunderstanding. When the eutectic colony grows, often one of the phases runs somewhat ahead of the other throughout the whole process; up till now there has been no clear treatment of the reasons for this, and therefore it is difficult to say whether the phase that is principal as regards nucleation is also so as regards growth, but general considerations indicate that the phase leading to formation of the eutectic colony may lose its principal role, which will then result in a change in the structure of the colony. There are many published discussions of the conditions for passage from platy structures to rod ones, and from these to globular ones during steady-state growth of eutectic, but no one has shown that the change in particle shape is due to replacement of the principal phase during growth.

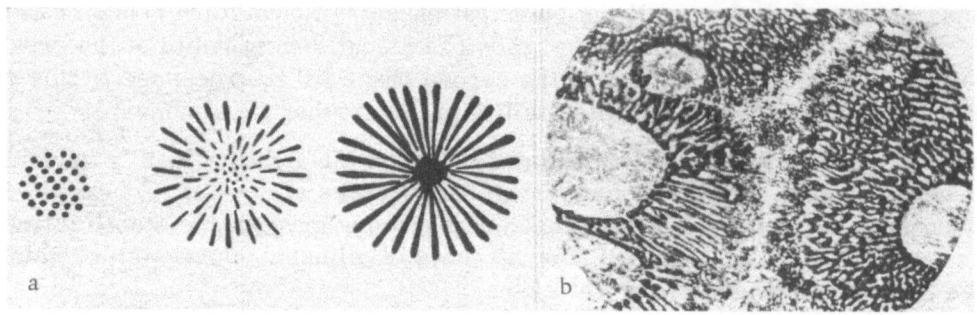

Fig. 1. a) Cross sections of eutectic colonies; b) structure of a
radial fibrous eutectic colony in the Sn−Pb system, ×150.

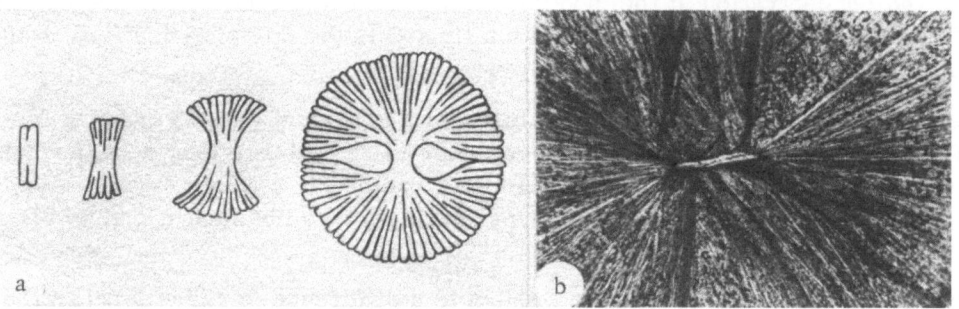

Fig. 2. a) Formation of a Shubnikov two-bladed form; b) eutectic
colony in a piperonal crystal (azobenzene−piperonal system), ×30.

If the crystal growth rate along one axis is much greater than that on another, one will
get splitting along that direction, which gives rise to a Shubnikov flower pattern (Fig. 2). Structures of this form are particularly common in organic crystals; under real conditions, the
thermal and diffusion fluxes in adjacent crystals often distort the colonies and prevent the formation of closed two-bladed flower forms, so various intermediate stages are produced.

If the principal phase grows as equiaxial crystals with good facets, the structure of the
colonies in detail often reproduces the shape of the primary crystal; such colonies are pseudo-
crystals of the principal phase containing inclusions of the second phase as fibers perpendicular

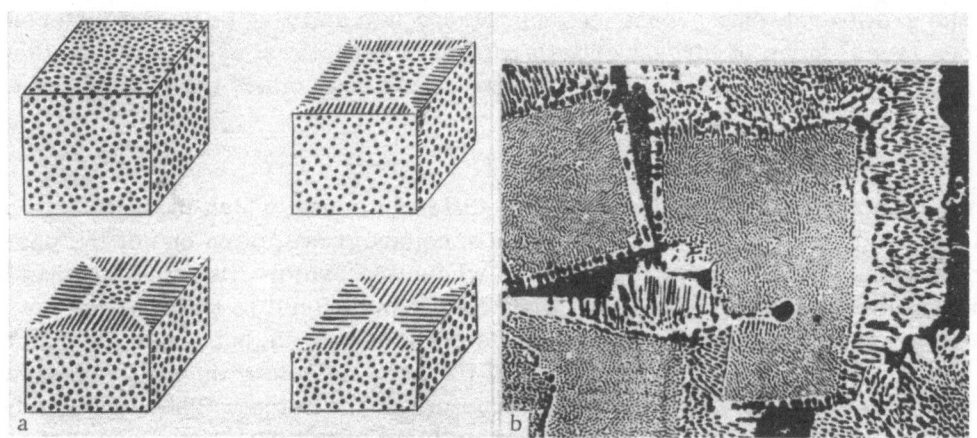

Fig. 3. a) Structure of a eutectic colony of pseudocrystal type; b)
eutectic colonies in the Sn−Bi system, ×50.

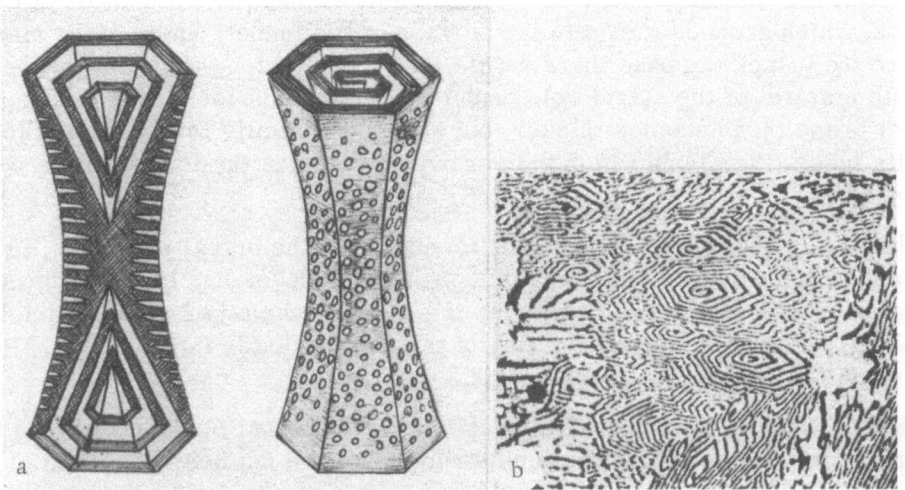

Fig. 4. a) Eutectic colony of spiral type; b) eutectic structure
in the Zn−MgZn$_2$ system, ×300.

to the faces of the eutectic colony (Fig. 3). Such structures are common in metallic and semi-
metallic systems. In metallography they are often seen as globular or platy in accordance with
the orientation of the section. Figure 3 shows schematically various sections of a colony
parallel to the faces of the pseudo-crystal. Some morphologic details of this type of colony
have been described previously [14].

The skeletal growth forms of anisotrpic crystals of the principal phase can produce
spiral eutectic structures (Fig. 4). The only known example of this type occurs in the system
MgZn$_2$−Zn, where the spiral eutectic is formed in accordance with the metastable phase diagram (the MgZn$_2$ phase has a hexagonal structure). The colonies are preceded by skeletal
growth of the MgZn$_2$ phase, which forms a growth funnel on the pinacoid plane. The elevated

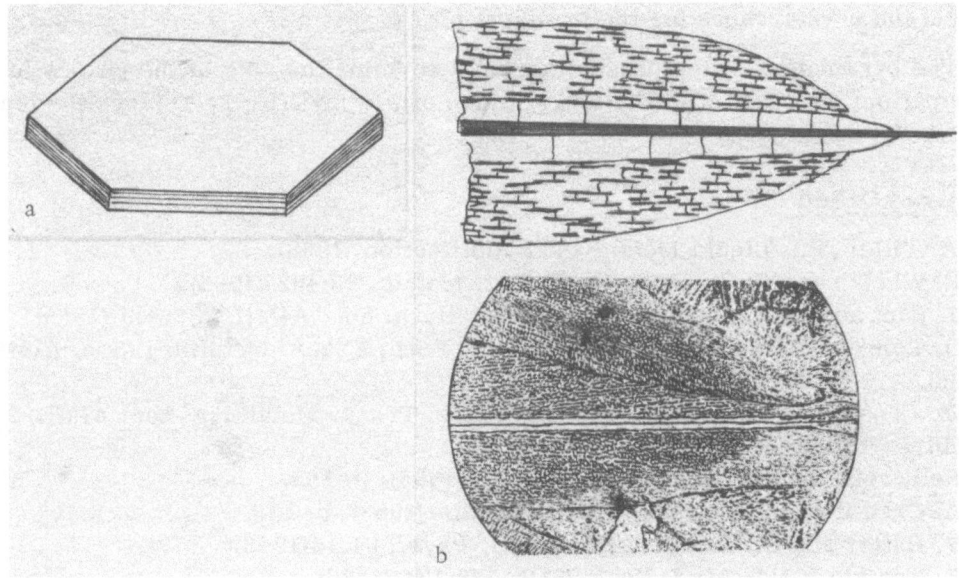

Fig. 5. a) Structure of a eutectic colony arising from a platy crystal; b) eutectic colony in the Sn−Zn system, ×30.

concentration of the second component within the funnel leads to production of nuclei of the second phase, which grow outwards to the surface of the funnel, which gives rise to conditions for growth of the principal phase there. This joint growth along the walls of the funnel leads to filling with a spiral of the spiral role of the two eutectic phases. The hexagonal prism of the principal phase then becomes thicker, but at a substantially lower rate; colonies are formed at its faces (Fig. 3), in which the second phase takes the form of rods perpendicular to the prism faces.

A scheme has been proposed [15] for formation of the spiral structure, in which the colonies start to arise when a pair of plates of the two phases has been produced. The plate with the higher growth rate tends to overlap the edges of the other plate, which leads to a change in the growth direction and twisting of the pair of plates into a spiral. However, this scheme is unable to explain the morphological details.

Skeletal funnels on all faces of a crystal of the principal phase could lead to spiral eutectic formations on each face, but such structures have not been observed.

When facetted needle crystals grow, one often gets combined forms of colonies; the end faces of the needles split into funnels (Fig. 2), and the thickening of the needles occurs by the production of pseudocrystal colonies (Fig. 3). Such structures have been observed in the Fe_2B-Fe system.

Platy crystals of the principal phase can result in eutectic colonies in flat form; the most common example of this type occurs in the $Sn-Zn$ system (Fig. 5), where the hexagonal crystals are highly anisotropic, since the surface energy of the pinacoid plane is minimal, while the growth rate is very small. The growth rate of the prism faces is larger by several orders of magnitude. The small radius of curvature of the edges of the plate enables these to grow to a considerable size without splitting. Usually, splitting starts at very late stages in the growth. Thickening of the plate occurs via the formation of a foliated dendrite as described in detail by Saratovkin [16].

Conclusions

1. The structure of a freely growing eutectic colony is determined by the growth forms and relief of the crystal faces for the principal phase.

2. The parameters of the phase diagram determine the role of the phases in eutectic colony formation; the phase with the larger value of $\psi = mvC(1-k)/D$ plays the leading part in producing the colonies.

Literature Cited

1. W. A. Tiller, in: Liquid Metals and Solidification (1958).
2. E. P. Whelan and C. W. Haworth, J. Inst. Metals, 93:402 (1965).
3. J. D. Hunt and K. A. Jackson, Trans. Metallurg. Soc. AIME, 236:843 (1955).
4. F. D. Lemkey, R. W. Hertzberg, and J. A. Ford, Trans. Metallurg. Soc. AIME, 233:334 (1965).
5. P. W. Crossman, A. S. Jue, and A. E. Vidoz, Trans. Metallurg. Soc. AIME, 245:397 (1969).
6. A. Hellawell, in: Solidification of Metals (1968), p. 155.
7. G. A. Chadwick, in: Solidification of Metals (1968), p. 138.
8. J. W. Rutter and B. Chalmers, Canad. J. Phys., 31:15 (1953).
9. D. E. Temkin, Dokl. Akad. Nauk SSSR, 138:174 (1960).
10. W. A. Tiller, K. A. Jackson, J. W. Rutter, and B. Chalmers, Acta Metallurg., 1:428 (1953).

11. S. R. Coriell and R. L. Parker, J. Appl. Phys., Vol. 37 (1966).
12. S. R. Coriell and R. L. Parker, J. Phys. Chem. Solids, Suppl., 1:703 (1967).
13. W. W. Mullins and R. F. Sekerka, J. Appl. Phys., 34:323 (1963).
14. Yu. N. Taran, E. K. Pirogova, and I. M. Galushko, Dop. Akad. Nauk Ukr. RSR, ser. A, No. 1, 79 (1967).
15. R. L. Fullman and D. L. Wood, Acta Metallurg., 2:188 (1954).
16. D. D. Saratovkin, Dendritic Crystallization [in Russian], Metallurgizdat, Moscow (1957).

DENDRITE GROWTH AND STRUCTURE FOR SOME
BINARY SEMICONDUCTOR COMPOUNDS WITH
THE SPHALERITE STRUCTURE

M. Ya. Dashevskii, A. N. Poterukhin,
A. V. Zakharova, and L. I. Kaikova

There are several papers [1-9] on the growth of dendrites with the structures of diamond and sphalerite. It is considered by some [2-4] that the active centers in dendrites are reentrant angles formed by {111} faces on twin planes, while others [6-8] consider that the centers are the convexities formed by two twinned tetrahedra on twin planes. The evidence [9] shows that both approaches are inadequate to represent fully the dendrite growth mechanism (Fig. 1).

We have examined the structure of the dendrites of doped gallium arsenide to determine whether there is a concentration dependence for dope segregation.

The Te-doped dendrites were grown along $\langle 211 \rangle$ from a melt covered with a layer of B_2O_3 flux by the method of [10]; the Te content varied from 0.01 to 1.5 at.%. The Te segregation was revealed by etching on {211} cross sections and on ones at 3, 5, and 12° to the B(111) face. After grinding and mechanical polishing, the etching was done in 0.02 M $K_2Cr_2O_7$ + 0.3 M H_2SO_4 [11].

Figure 2 shows a photomicrograph of a $(2\bar{1}\bar{1})$ section for 0.01 at.% Te in the melt, where there are two twin planes. Figure 3 shows an inclined section at 5° to B$(\bar{1}\bar{1}\bar{1})$, with indication of the twin-plane region (about 100 μm between planes in the inclined section). There are lines perpendicular to the twin planes and lying between them (shown by arrow).

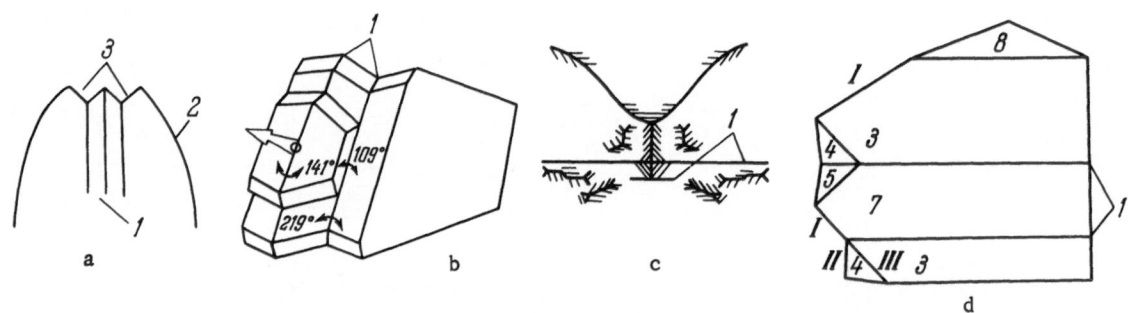

Fig. 1. Supposed dendrite structures. a) In [4]: 1) three twin planes, 2) tip of dendrite, 3) reentrant angles; b) in [3]: 1) two twin planes; c) in [6], with traces of impurity segregation in a dendrite cross section; the active center is a pair of twinned {111} tetrahedra (convex corner): 1) two twin planes; d) [9]: 1) two twin planes, I) tetrahedral edge, II, III) octahedral edges, 3, 4, 5, 7, 8) A{111} and B{1$\bar{1}$1} faces.

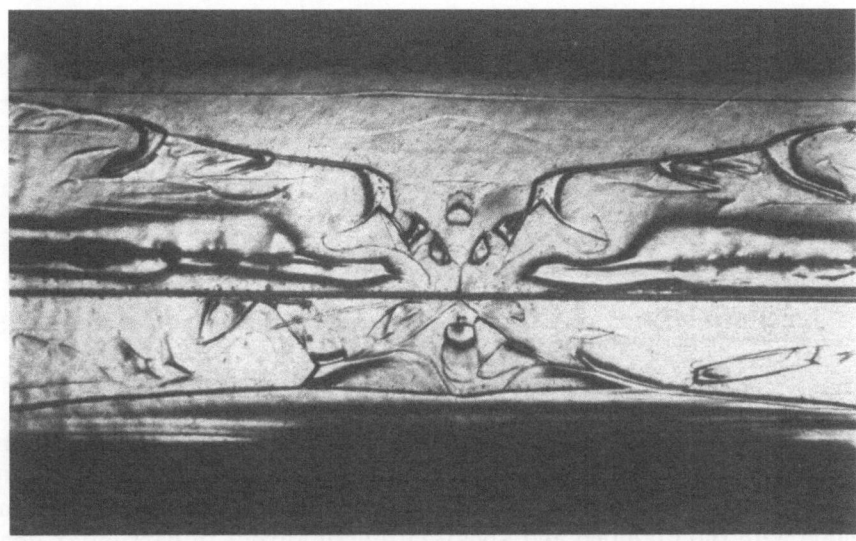

Fig. 2. Cross section of a GaAs dendrite, ×70.

Figure 4 shows an inclined section of a dendrite (about 1.5 at.% Te in melt) at 3° to B($\overline{1}\overline{1}\overline{1}$); there are traces of segregation between twin planes (shown by arrow) as in the lightly doped dendrite (Fig. 3).

Our results on Te segregation in ⟨112⟩ gallium arsenide dendrites are as follows.

1. Two regions can be distinguished as regards segregation: I) between twin planes, II) away from twin planes. In I the signs of Te segregation do not coincide with {111} traces, while in II they largely do so.

2. The segregation patterns are [4, 6] assigned to dendrite growth patterns, so the results can be explained as follows: a thin platelet (active growth center) is formed between twin planes, but the structure of the crystallization front (points of attachment of the nuclei) is not clear. It may be that the crystallization front (if considered on {110}) is curved and that nuclei are initially attached in the region of the twin planes. Perhaps a reentrant angle is formed where the plate contacts a twin plane. Te concentration change in the melt (supersaturation or supercooling) affects the plate size (thickness, extent, etc.) but does not essentially affect the structure of the active growth center, which persists provided that there are not less than two planes.

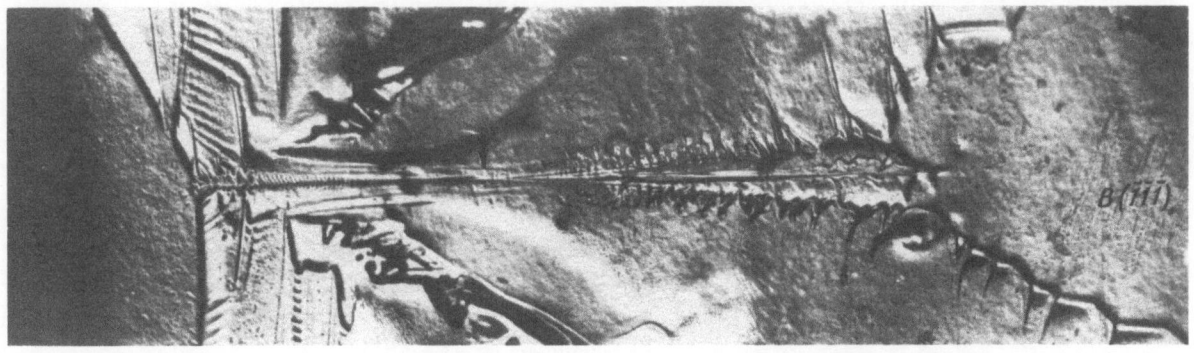

Fig. 3. Inclined section of a GaAs dendrite. 1) Twin-plane region, ×40.

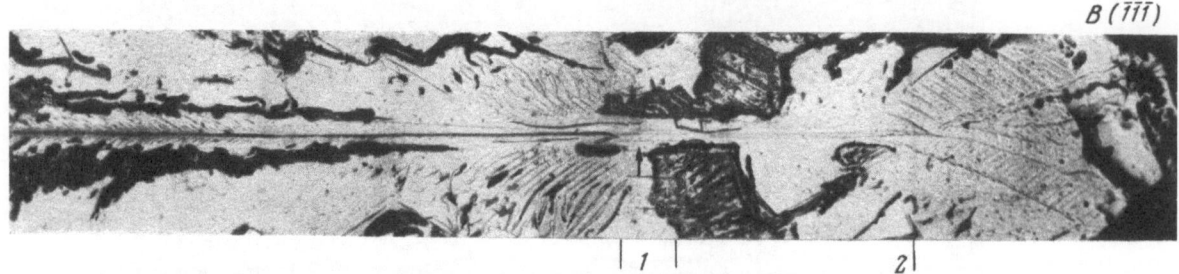
$B(\bar{1}\bar{1}\bar{1})$

Fig. 4. Inclined section of a GaAs dendrite with three twin plates, ×25. 1) Two planes; 2) one plane.

These results cast doubt on the convex-angle concept [6-8], in which it is supposed that the active center of a $\langle 112 \rangle$ dendrite is formed by two twinned tetrahedra from $\{111\}$. The Te segregation in region II in GaAs dendrites closely resembles dope segregation in dendrites of other substances with the diamond and sphalerite structures, in particular dendrites of Ge [6, 7] and InSb [9]. The structure of the crystallization front in region II is probably related to growth on $\{111\}$ planes and is dependent on the dope concentration [6, 9].

Literature Cited

1. E. Billig, Proc. Roy. Soc., A 229:346 (1955).
2. R. S. Wagner, Acta Metallurg., 8:57 (1960).
3. D. R. Hamilton and R. G. Seidensticker, J. Appl. Phys., 34:3113 (1963).
4. G. E. Bolling and W. A. Tiller, in: Metallurgy of Elemental and Compound Semiconductors, New York (1961), p. 101.
5. M. Yu. Dashevskii, G. V. Kukuladze, V. B. Lazarev, and M. S. Mirgalovskaya, Dokl. Akad. Nauk SSSR, 172:403 (1967).
6. A. A. Bukhanova and D. A. Petrov, Izv. Akad. Nauk SSSR, ser. neorg. mat., 4:1439 (1968).
7. A. A. Bukhanova, "Structural Studies on the Dendritic Growth of Germanium," Thesis, Moscow Aviation Engineering Institute (1966).
8. D. A. Petrov, Dokl. Akad. Nauk SSSR, 177:1075 (1967).
9. M. Yu. Dashevskii and A. N. Poterukhin, Izv. Akad. Nauk SSSR, ser. neorg. mat., 4:1478 (1968).
10. M. Yu. Dashevskii and A. V. Zakharova, Izv. Akad. Nauk SSSR, ser. neorg. mat., 4:1471 (1968).
11. L. N. Vozmilova and É. V. Buts, Izv. Akad. Nauk SSSR, ser. neorg. mat., 4:1340 (1968).

DENDRITE SHAPE AND GROWTH RATE IN
A SUPERCOOLED METAL

V. T. Borisov and A. I. Dukhin

We have examined the growth kinetics of mercury needles, the metal being purified by filtration through cloth, passage through 10% nitric acid, and double distillation in vacuum. Figure 1 shows the apparatus. The mercury drop 6 was placed between a mica plate 30 μm thick and the glass plate 3; the mica was attached by a slip of copper foil 0.1 mm thick, which was cemented to a lucite base. To the center of the foil underneath was soldered a copper-constantan thermocouple 9. The entire assembly was placed at the bottom of the massive copper vessel 2, which was in the closed chamber 7 made of lucite, which was cooled by the vapor from boiling liquid nitrogen, which was fed in through a hole 10 and escaped via a hole 8. The nitrogen flow rate governed the temperature and cooling rate of the specimen, and it itself was adjusted via an oven placed in the vessel containing the liquid nitrogen. The glass 3 flattened out the drop of mercury and from above bore a ring 4 that vibrated with a frequency of 100 Hz in response to the electromagnet 5. The drop was observed and photographed via an MBS-2 microscope. Mercury is readily supercooled, and crystallization was initiated by a seed of solid mercury placed on the small rod 1.

Photographs (Fig. 2) show the structure of the surface of the drop; at supercoolings up to 1°C there were single needles, which in time extended throughout the drop away from the seed (Fig. 2a). Usually, the top of a needle was rounded, but sometimes there were needles with sharp ends. There were also secondary branches. The number of needles in the specimen increased with the supercooling, as did the number of branches of higher order, and the branches became shorter and thinner. The structure became dendritic, as is clear from parts b and c of Fig. 2. Supercoolings of 30°C or more made the structure so fine-grained that the individual branches could not be distinguished.

The angles between the first- and second-order dendrites were 96°46' ± 2°05'; mercury is [1] rhombohedral with α = 98°14'. This value agrees satisfactorily with the previous, which indicates that the main and secondary axes of the dendrites grow along the edges of $\langle 100 \rangle$ rhombohedron.

This agrees with the rule [2] that the axis of a dendrite does not coincide with the closest-packed axis but is perpendicular to a grid of second or third order as regards reticular density. The angles between directions of closest packing in the lattice of mercury are 70°32'. In four cases out of thirty, we observed angles close to 70° between the branches of dendrites.

We examined the needle growth also by cinematography; the magnification in this case was 3.86, and the filming rate was 64 frames per second. Figure 3 shows a series of frames illustrating the crystallization at a supercooling of 5°. The needles grow under the surface of the mercury. The frames give the growth rate of the main and side branches of the dendrite

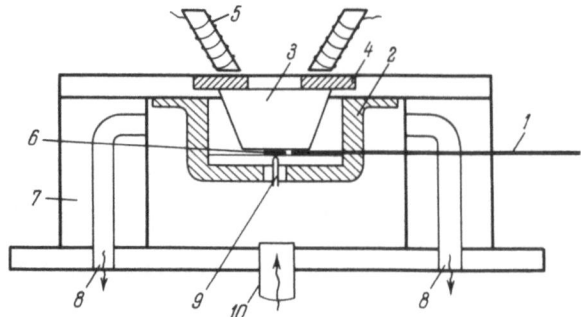

Fig. 1. Apparatus for crystallization of mercury.

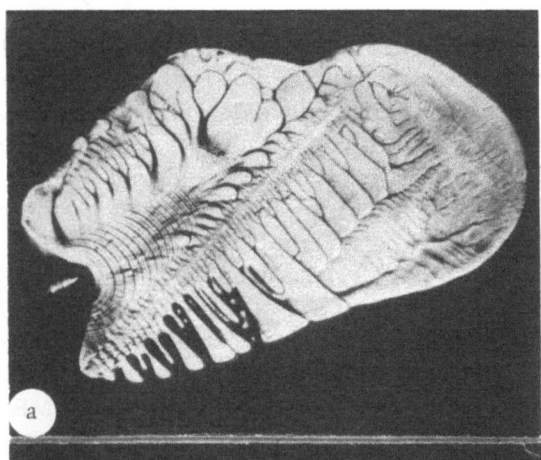

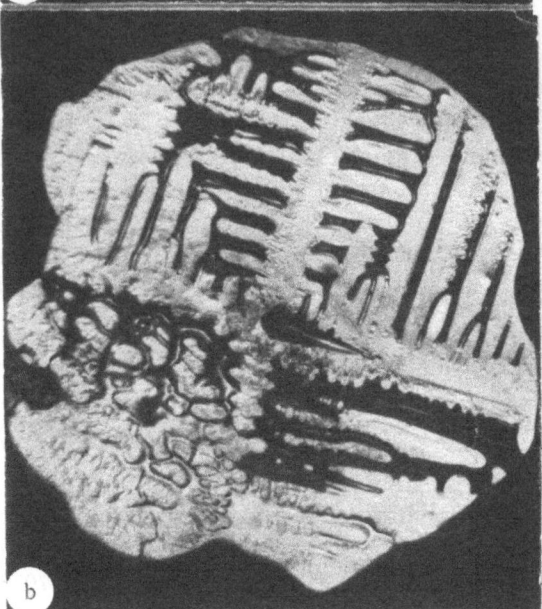

Fig. 2. Surfaces of mercury drops crystallized at super-coolings ΔT of (a) 3.5°C, (b) 9.2°C, and (c) 16.5°C.

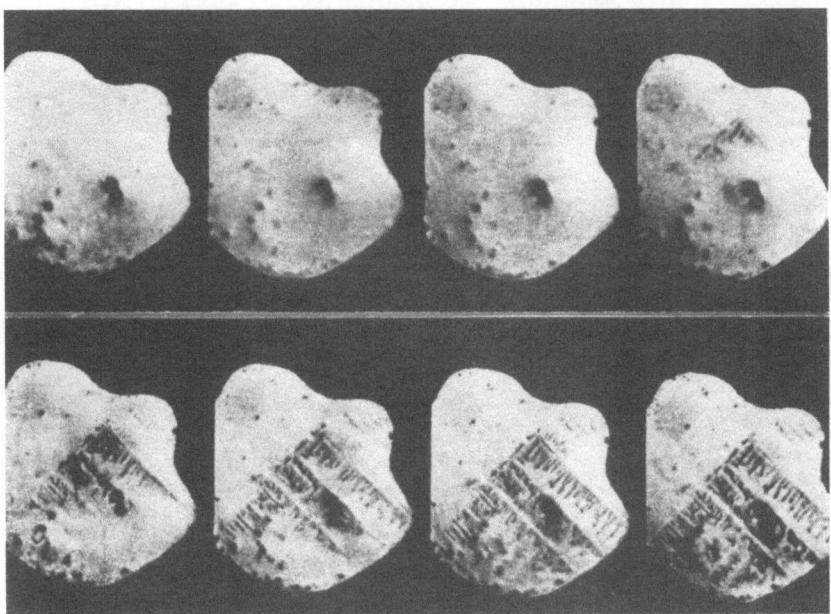

Fig. 3. Dendrite growth in a supercooled melt.

(Table 1); at supercoolings greater than 7°, the growth rate was so large that it was not possible to follow the details of the process; at $\Delta T = 7.5°C$ we merely estimated a lower limit of 10.2 cm/sec. Table 1 shows that the needles grow at very high speeds in the crystallization of mercury.

The measurements also showed that the needle temperature rises very rapidly after formation; Fig. 4 shows the result given by two thin constantan wires 30 μm in diameter immersed in supercooled mercury; the marked step in the emf occurs at the instant when a crystal is produced at one of the wires, where one has a constantan—mercury contact. The limiting temperature was attained within 0.01 sec (the period of the wave on the oscillogram is 0.01 sec).

We also observed the temperature change when needles grew in a supercooled drop; Fig. 5 shows an oscillogram typical of this case. In part A-B the drop temperature slowly falls by external cooling, and the metal becomes somewhat supercooled (in this case by 1.3°C). At the time corresponding to B, a crystal arises on the thermocouple or close to it, and the

TABLE 1. Growth Rates of Dendrite Main and Side
Branches for $\Delta T = 5°C$

Frame	Dendrite axis A_0		Branch A		Branch B		Branch C	
	l, cm	v, cm/sec	l, cm	v, cm/sec	l, cm	v, cm/sec	l, cm	v, cm/sec
0	0		0		0		0	
1	0.060	3.83	0.053	3.37	0.046	2.94	0.06	3.83
2	0.113	3.4	0.104	3.29	0.074	1.75	0.106	2.94
3	0.135	1.4	0.120	1.01	0.113	2.48	0.119	0.85
4	—	—	—	—	0.123	0.66	—	—

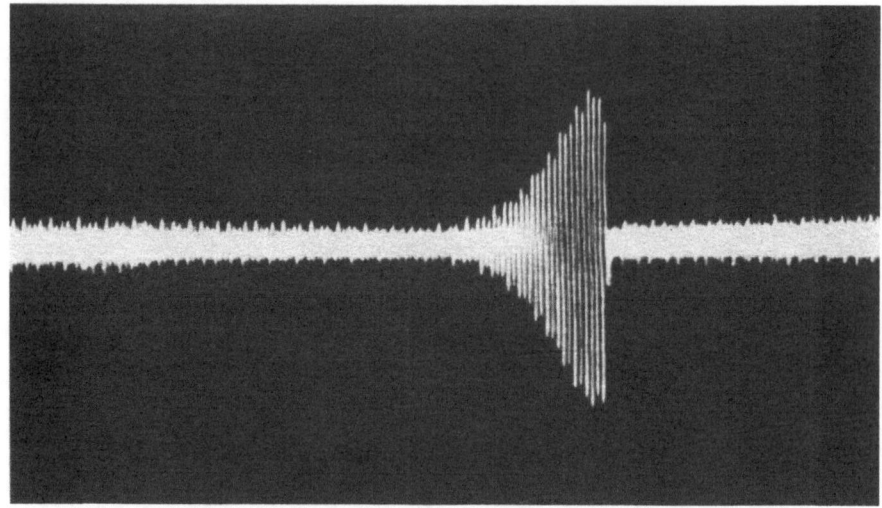

Fig. 4. Temperature rise in crystal nucleation (0.09°C per mm).

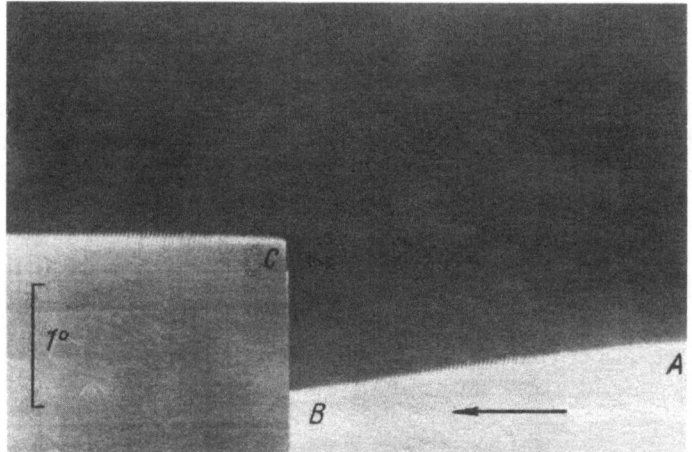

Fig. 5. Temperature variation in crystal growing a supercooled drop (0.01 sec between adjacent peaks).

TABLE 2. Crystal Sizes Obtained with Various Initial
Supercoolings

$\Delta T°$, C	t, sec	Y_0	$\Delta T°$, C	t, sec	Y_0
	Hg at $t_0 = 1 \cdot 10^{-5}$			к at $t_0 = 2 \cdot 10^{-5}$	
18.1	0.02	3.2	1.3	0.02	2.0
18.1	0.03	3.2	2.2	0.03	3.4
17.1	0.03		2.0	0.02	3.1
14.6	0.02		1.9	0.02	2.9
14.6	0.50	2.6	1.9	0.07	2.9
14.6	0.02		1.7	0.05	2.6
			1.7	0.02	2.6
			1.6	From 0.10 to 0.02	2.5

temperature very rapidly rises to point C, which corresponds to the melting point, and it remains at this level for about 10 sec until the whole drop has crystallized.

Table 2 gives results from other similar experiments; within the time t, the initial supercooling ΔT is eliminated to the extent of ~95%. Table 2 gives estimates of the theoretically expected time t_0, the calculations being based on the observed kinetic coefficient K = 10 cm/sec-deg. If the supercooling is small, with $(T_k - T_0)/\theta = \varepsilon \ll 1$, the temperature $T(t)$ at the surface of a crystal growing in accordance with $y' = K[T_k - T(t)]$ is given by the following relationship:

$$\frac{T(t) - T_0}{T_k - T_0} = 1 - e^\tau \operatorname{erf} \sqrt{\tau} \approx 1 - \frac{1}{\sqrt{\pi \tau}}, \qquad \tau = \frac{(K\theta)^2}{a} t. \tag{1}$$

where T_k is the crystallization temperature, T_0 is the temperature of the liquid, $\theta = q/c$, q is the latent heat of fusion, a is the thermal diffusivity, and c is the specific heat. Then the size of the crystal in the one-dimensional case is

$$Y(t) = \frac{4}{\sqrt{\pi}} \varepsilon \sqrt{at} - \frac{2\varepsilon a}{K\theta}(1 - e^\tau \operatorname{erf} \sqrt{\tau}) \simeq \frac{4}{\sqrt{\pi}} \frac{T_k - T_0}{\theta} \sqrt{at}. \tag{2}$$

The direct measurements indicate that t is about 0.01 sec, whereas the actual loss of the supercooling should occur in an appreciably shorter time t_0; clearly $t > t_0$ on account of errors in the methods. These very high rates of rise of temperature make the lag in the recording instruments appreciable, especially when a loop oscillograph is used. Also, a crystal does not always arise directly at the active point in the thermocouple, and in any case its size is initially less than that of the active area, which also reduces the rate of rise in temperarue.

Table 2 gives values for the size Y_0 calculated from (2) for the crystals that is attained at time t_0. The values are not very large. Also, it is unlikely that a large number of nuclei will arise together, because the formation of the first very much reduces the probability of production of others even when the heating is slight, so one assumes that the main body of the supercooled liquid crystallizes at a temperature very close to the equilibrium temperature.

Literature Cited

1.   B. F. Ormont, Structures of Inorganic Substances [in Russian], GITTL (1960).
2.   J. Schlipf, Z. Krist., 107:35 (1956).

THE HABIT OF DISLOCATION-FREE
SILICON SINGLE CRYSTALS

É. S. Fal'kevich, N. I. Bletskan,
K. N. Neimark, and M. I. Osovskii

It has been found over many years of production of dislocation-free silicon single crystals that the shape differs from that of single crystals containing dislocations that have been grown under the same thermal conditions. Here we consider the external features characteristic of dislocation-free single crystals with a view to explaining their nature and the conditions facilitating their occurrence. We find that these distinctive features occur more or less prominently in all cases without reference to the growth method or crystal diameter.

We grew dislocation-free single crystals under vacuum by crucible-free zone melting and under vacuum or in helium by Czochralski's method; in the second case, we used quartz crucibles of various sizes. We varied within wide limits the weight of material, the pulling rate, the rotation rates of crucible and crystal, and the screening and heating systems. The doping employed various elements: phosphorus, boron, and antimony.

When we grew dislocation-free single crystals by the zone melting method we employed one- and two-turn coils fed at 5.28 MHz, while varying the rate of displacement and speeds of rotation of rod and seed. The seeds were of square cross section 3×3 mm and oriented on [111].

The dislocation-free crystals began to show differences right at the start; it was clear, especially in the zone melting, that there was curvature of the crystal at the conical end and in the region where the diameter was being established,* and this curvature increased with the nonuniformity in the temperature distribution. A distinctive hump was formed (Fig. 1) in these parts. The humps were absent from single crystals made with dislocations under these conditions.

The expansion section and the part of constant diameter in a dislocation-free crystal had very prominently a characteristic luster due to unseen faces. When Czochralski's method was used, clearly seen faces persist over the entire length of the single crystal and are most prominent when the diameter is reduced and at the end of the crystal (Fig. 2). Any expansion of the crystal caused them to appear most prominently when the crystallization front was concave; a slightly convex or a flat front caused the faces to become narrow. The shape of a dislocation-free crystal was substantially different from that of one with dislocations when the steady diameter had been established.

If the crystal had been grown by zone melting, there was a characteristic edge that ran from the above hump along the entire length of the crystal (Fig. 1); the number of these edges

---

*That is, at the transition to a constant diameter.

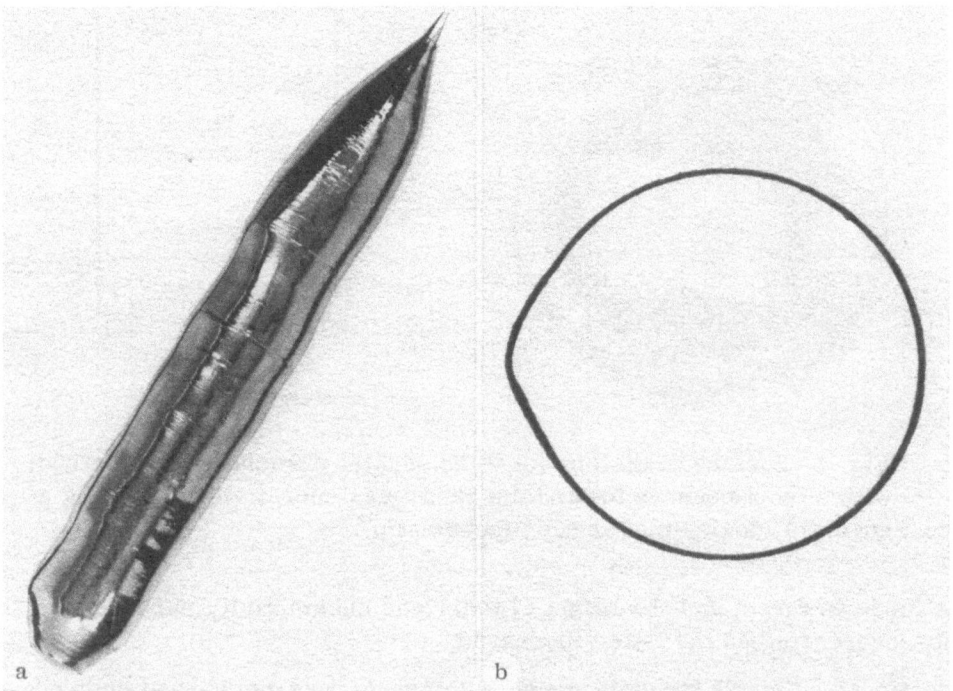

Fig. 1. Dislocation-free single crystal from zone melting. a) Photograph; b) scheme of cross section through hump.

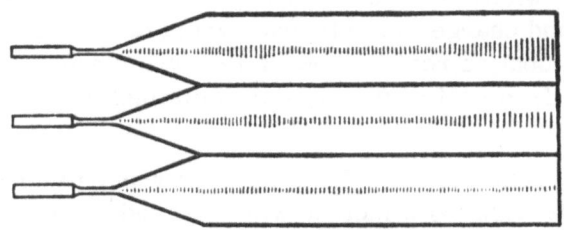

Fig. 2. Scheme for a dislocation-free crystal grown by Czochralski's method (surface unfolded), with width variation in the faces.

Fig. 3. a) Photograph; b) scheme for a single crystal grown by Czochralski's method, with transition from a dislocation-free upper part to a lower part with dislocations. The arrow in b indicates the transition point.

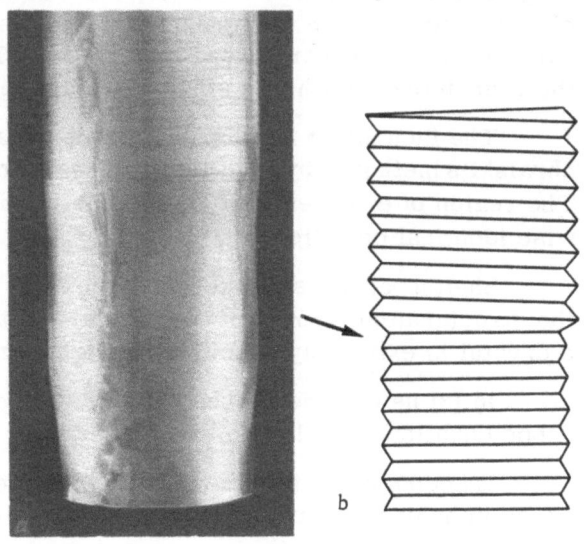

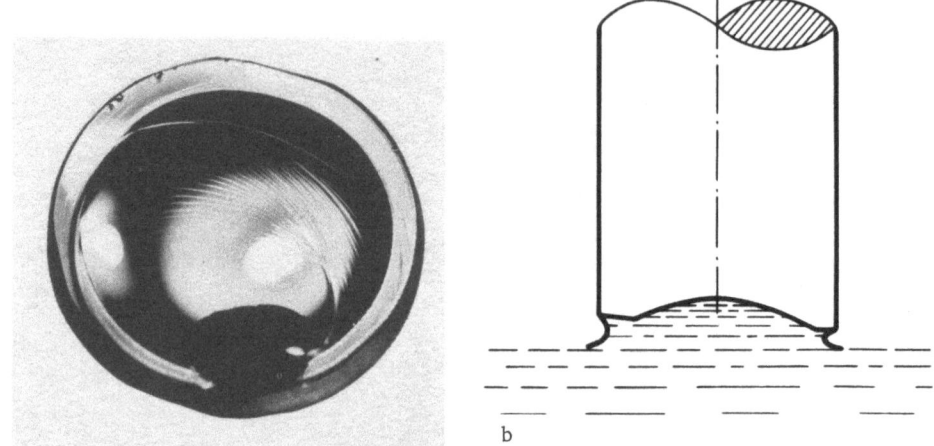

Fig. 4. Surface of dislocation-free crystal (Czochralski's method)
with a front concave toward the seed. a) General view (flat ring at
edge); b) longitudinal section (schematic).

might be as large as three, and the height of them was substantially dependent on the crystal
rotation rate, decreasing as the latter increased.

Dislocation-free single crystals made by Czochralski's method had such edges much less
commonly and less prominently. The surfaces showed growth bands in the form of a screw
thread, and the thread depth and pitch were dependent on the pulling rate, the rotation rate of
the crucible, the crystal diameter, the symmetry of the temperature distribution, any fluctuations in the growth conditions, etc. This screw thread depth was much greater in growing
dislocation-free crystals. Figure 3 shows a place where a dislocation-free crystal acquires
dislocations; after a distinct change in diameter and change in the thread, which is characteristic of the transition point, the surface in the dislocation region shows a less prominent
thread appearance. The thread in the dislocation-free part of the crystal is distinctly rectilinear on parts where there are distinct faces.

Also, the surface formed on removing the crystal suddenly from the melt was quite different as between the two types of crystal when Czochralski's method was used under conditions producing a concave growth front. The interface in the dislocation-free case had a flat
ring of variable width at the edge, and a concave part at the center (Fig. 4). The flat ring was
clearer and its width was greater for the dislocation-free case, and the sides increased in
parts where there were broad clear faces. The narrow part of the ring always runs ahead of
the rest in the growth of a single crystal, which leads to inclination of the interface.

The face effect is very prominent for a dislocation-free single crystal grown by Czochralski's method or by zone melting; it is not nearly so prominent for a crystal with dislocations.
The region of emergence of (111) faces on the interface was considerably larger under otherwise identical conditions.

Sheftal' appears to have been the first to direct attention to the close connection between
the structural uniformity of the crystal and the shape [1, 2]; in [2] he stated that the shape of
a crystal is extremely sensitive to slight variations in the lattice.

In [3] he extended Gibbs principle for equilibrium forms of crystal and introduced into
Gibbs equation an additional term related to the bulk free energy.

It seems clear that the bulk free energy of the crystal is responsible for these differences in habit of the two types of crystal; but the mechanism of atomic uptake during growth

that affects the shape still remains to be explained. As regards the curvature of a dislocation-free crystal, the features of the growth responsible for this may be as follows.

We used industrial-type equipments to obtain single crystals of silicon, and the thermal axis in this as a rule does not coincide with the growth axis, which leads to an unsymmetrical temperature distribution at the crystallization front; in general, if crystals are grown in such an apparatus without rotation, the axis is curved, and the crystal deviates towards the cold side. To eliminate this effect, the crystal or the crucible (or the two together) may be rotated. Stresses are set up during rotation in a crystal with dislocations, and these are sufficient for plastic deformation, which causes the axis to straighten out. In the case of a dislocation-free single crystal, which has a higher yield point, the stresses are insufficient for plastic deformation, and the crystal grows with a curved axis.

Literature Cited

1. N. N. Sheftal', Dokl. Akad. Nauk SSSR, 31:33 (1941).
2. N. N. Sheftal', in: Growth of Crystals, Vol. 1, Consultants Bureau, New York (1959), p. 5.
3. N. N. Sheftal' and I. V. Gavrilova, in: Growth of Crystals, Vol. 4, Consultants Bureau, New York (1966), p. 24.

PART V

CRYSTAL GROWTH FROM THE VAPOR

EPITAXIAL GROWTH UNDER ANOMALOUS CONDITIONS AS A METHOD OF RESEARCH ON VAPOR CRYSTALLIZATION MECHANISMS

V. F. Dorfman, M. S. Belokon', B. N. Pypkin, and I. D. Khan

There are numerous series and parallel stages involved in crystallization from the vapor state, especially via transport reactions, and the rates of these stages are usually nonlinearly dependent on the conditions. This explains why it is difficult to give a theoretical analysis of experimental results. In order to reveal some one of the microscopic processes, one has to use anomalous conditions, which may lead to extensive changes in the working processes and consequent growth in stability. Such studies are therefore usually avoided in experimental researches, especially technological ones, but anomalous conditions provide a useful means of physicochemical research. This is illustrated here by reference to the iodide method of crystallization.

It is relatively easy to determine the role of mass transfer in the vapor state. Even the elimination of convection gives rise to smoother layers with lower dislocation densities [1], and the layers near the surface have an even greater extent. Any lack of flatness in the crystal leads to disturbance in the temperature and concentration gradients and hence in the diffusion flows. For instance, a spherical deviation from flatness causes the equations to imply a bunching of the flux by a factor 3 at the vertex and fall to zero at the edge of the hump [2]. These microscopic flows in turn determine the subsequent development of the relief, i.e., there is feedback, which is usually positive during growth and negative during etching.

These local microscopic fluxes could be suppressed, one would obtain smoother and structurally perfect films; these considerations have been tested in two series of experiments. In the first, the substrate (or source) was treated before the start or in the initial stage to produce projections and recesses of various heights, as well as patches of SiO of width 100–200 μm. In the second series, the reaction volume was compressed in the normal direction by screening the substrate with a passive or active screen at distances ranging from a few μm to 1 mm, and also in the tangential direction by crystallization in cells of diameter 0.05–3 mm. In the sandwich method, where the surface layer takes up all the reaction volume, the unevenness develops in a very much autocatalytic way, and it is comparatively easy to examine theoretically what occurs [3]. Experiment confirms the form of the calculated relationship (Fig. 1) and it is found that a recess is formed in the film near an actively growing projection. Passivating nonuniformities on the other hand repel the mass flows to the edges, where the growth rate increases considerably, so one gets vicinals and an increase in the dislocation density; one gets highly defective and even porous transition regions. The flux density increases so greatly at the vertex of a sharp-ended hummock that one gets bundles of whiskers. Substrate screening in the dynamic and static methods [1] reduces the height of the projections by 1–2 orders of magnitude, while the dislocation density falls to the level found in the sub-

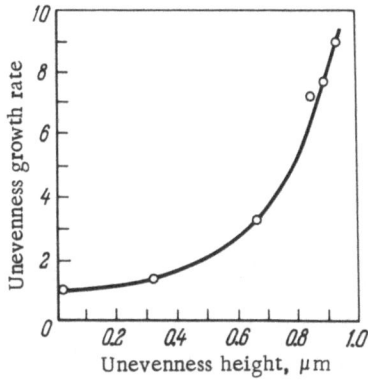

Fig. 1. Unevenness growth rate as a function of height for a substrate temperature of 640°C, gap 1.7 mm, film ground rate 36 μm/hr.

strate. The sandwich method has given good reproducible growth of dislocation-free films of Ge on appropriate substrates, for example. We merely vary the distance between the screen and the substrate to get all types of growth figures characteristic of the dynamic, static, and sandwich methods. A passive screen suppresses expansion of individual microcrystals and also enables one to produce polycrystalline layers with a thickness uniformity of ±10% at 550-600°C.

Tangential compression of the flux leads to production of smooth columns with isolated vicinals that occupy the entire area of a cell and form an angle of about 5" with the surface. The free part of the surface under the same conditions has large vicinals with inclinations of several degrees or else relief characteristic of the diffusion region.

While the evolution of the unevenness is substantially dependent on the microscopic fluxes, the initiation of them is related to the properties of the surface and the behavior of the heterogeneous processes. The mechanisms of the heterogeneous stages have been examined over a wide range of crystallization conditions in the systems GaAs−I, GaP−I, and Ge−Si−I. The need to reduce the role of mass transfer determines the choice of the sandwich method with a high temperature gradient; to eliminate convection, the hot zone was placed above the cold one, while the reaction volume did not exceed 10-15 cm³ for a cross section of 6 cm². The apparatus provided a temperature gradient of 500°C/cm and operation over a wide pressure range (up to 30 atm). Crystallization rates of 2500 μm/hr were obtained for GaAs and GaP, or 1000 μm/hr for InAs and InAs and G. Figure 2 shows some of the experimental re-

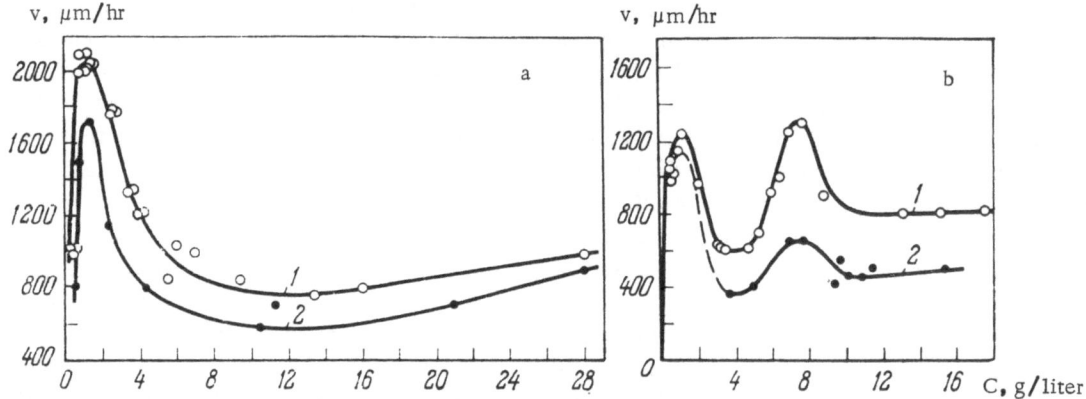

Fig. 2. Layer growth rate as a function of initial iodine concentration ($\Delta T/\Delta x =$ 500 deg/cm, $\Delta x = 0.035$ cm). a) Gallium arsenide ($T_{av} = 715$°C): 1) (111) A side, 2) (111)B; b) gallium phosphide, (III)A side: 1) $T_{av} = 770$°C, 2) $T_{av} = 750$°C.

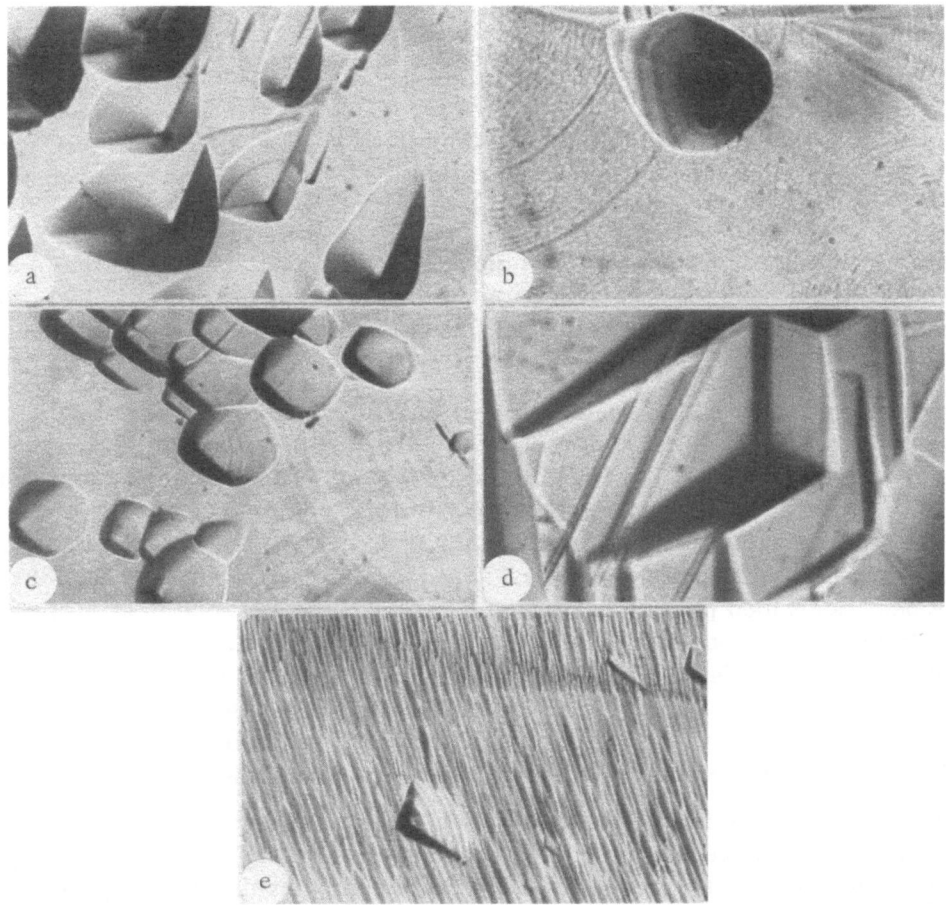

Fig. 3. Surfaces of (III)A gap films (×300) grown at initial iodine concentrations (g/liter) of (a) 1.3, (b) 3.2, (c) 5.4, (d) 7.5, and (e) 17.5.

sults. In the case of GaAs, concentrations above 5 g/liter gave rise to continuous epitaxial growth on the (111)A and (111)B faces, the second face giving more perfect and smoother layers with dislocation densities of 10^2-10^3 cm^{-2} or less. At concentrations below 5 g/liter, the (111)B side did not give rise to a continuous epitaxial layer [4]. We found that in this case there arise in the initial stages identically oriented twins on the (111)A side, on which the crystallization subsequently proceeds. As a result, one gets an open columnar structure (for this region, curve 2 of Fig. 2a characterizes the average growth rate as determined from the gain in weight of the substrate). A characteristic feature of these relationships is the increase in the growth rate at concentrations above 10 g/liter for GaAs and 6 g/liter for GaP, with production of a second peak (in the case of GaAs, this second peak appears at higher temperature gradients). There is also a change in the apparent activation energy. These effects are due to change in the mechanisms of the heterogeneous stages; they can be explained because the transport occurs as a result of the reaction $3GaI \rightleftharpoons GaI_3 + 2Ga$ followed by the synthesis $Ga + {}^1/_4 As_4 ({}^1/_2 P_2) \rightleftharpoons GaAs(GaP)$, and also as a result of the overall reaction that should be of higher order and consequently should predominate at higher pressures, at which the transport of gallium via the first reaction is considerably reduced. If the temperature is raised, or if the gradient is reduced, the kinetic obstacles are eliminated, the second maximum vanishes, and the shape of the curve throughout the pressure range agrees with the calculated one for the quasiequilibrium approximation. We may reasonably assume that these two mechanisms

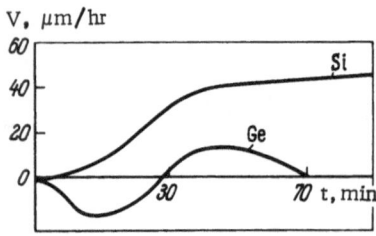

Fig. 4. Growth rate as a function of time in the
initial stages for the Ge−Si−I system.

show different selectivity for the A and B directions of the polar axis, which is responsible for the change in the mode of crystallization on the (111)B side of GaAs. The change in the mechanism and kinetics of the transport is also reflected in the microrelief of the finished layers (Fig. 3).

The measurements on the systems Ge−I, Si−I, and Ge−Si−I were made by static method in a pressure range on 0.003 to 30 atm and above. The activation energy in the Si−I system was higher than the value for Ge−I, while the iodides of Si are thermally more stable than those of Ge. Above 3 atm, Si can displace Ge from its iodide. The growth of Si on Ge results in transient initial stages (Fig. 4), in which the Ge and Si are transported in opposite directions, and there is consequently a porous transition region consisting of Ge−Si solid solutions, which has been described [5] as a cavitation layer. The occurrence of this is completely eliminated by introducing ready-made iodides. The difficulty of growing Ge on Si is due mainly to the stability of the oxide film on Si, but at pressures above 2 atm, we were able to make epitaxial films of Ge on Si at 600-700°C without preliminary vapor-phase etching, which is due to increased sorption of the vapor component, which plays an important part in epitaxy [6, 7], in conjunction with possible etching of the substrate by the Ge iodides.

One way of establishing the mechanisms of the heterogeneous stages is to examine the effects of impurities; increase in the dope concentration in the growth of Ge on (111) results in various sequential effects: reduction of the rate and poisoning of the growth figures (for instance, transition from trigonal vicinals via hexagonal ones to inversed trigonal ones), and then to local followed by continuous disruption of the epitaxial state, and finally to local formation of a liquid phase and transition to the VLS mechanism (as dopes we used Sb, Au, As, and Ga). Then whiskers grew either normal to the plane of the substrate or along all four {111} directions, the detailed conditions being the decisive feature; at 750-850°C, the growing layer also ceased to be continuous, and at 850-900°C there were conical figures some mm in diameter and height, while the rest of the substrate was etched.

The method of anomalous crystallization conditions enables one not only to observe some features of the epitaxial growth from the vapor but also to establish some features of practical significance.

Literature Cited

1.   V. F. Dorfman, K. A. Bol'shakov, and I. P. Kislyakov, Izv. Akad. Nauk SSSR, ser. neorg. mat., 1:37 (1965).
2.   L. D. Landau and E. M. Lifshits, Electrodynamics of Continua [in Russian], Fizmatgiz, Moscow (1958).
3.   V. F. Dorfman, Kristallografiya, 13:140 (1968).
4.   F. A. Pizzarello, J. Electrochem. Soc., 110:1059 (1963).
5.   R. C. Newman and J. Wakefield, J. Electrochem. Soc., 110:1068 (1963).
6.   V. F. Dorfman, I. P. Kislyakov, and K. A. Bol'shakov, Zh. Fiz. Khim., 39:996 (1965).
7.   V. F. Dorfman and M. S. Belokon', in: Growth of Crystals, Vol. 8, Consultants Bureau, New York (1969), p. 128.

EFFECTS OF CRYSTALLIZATION CONDITIONS ON THE MORPHOLOGY OF ZnSe CRYSTALS

A. A. Simanovskii

Chemical transport reactions have provided considerable advances in the production of single crystals of various substances [1]. Single crystals of compounds of type $A^{II}B^{VI}$ were first reported as made in this way by [2, 3].

Here we report the growth of zinc selenide crystals produced by transport reaction in the system $Zn-Se-I$ in a closed tube. Particular attention was given to polarity in the ZnSe crystals during growth and etching in order to relate the difference in morphology of the {111} polar faces to the conditions of crystallization.

It has been shown [4] that the system can be described essentially via the equations

$$ZnSe + I_2 \rightleftarrows ZnI_2 + \frac{1}{2}Se_2, \tag{1}$$

$$I_2 \rightleftarrows 2I \tag{2}$$

and one may assume to a first approximation that the vapor phase contains the four components ZnI_2, Se_2, I_2, and I. The partial pressures P of the individual components have been calculated for various conditions such as temperature, or initial iodine concentration, via equations for the equilibrium constants of (1) and (2) in conjunction with the conservation of the iodine and retention of stoichiometry in the ZnSe [4, 5].

The ZnSe crystals were grown in sealed quartz tubes 100-200 mm long and of internal diameter 14-20 mm; see [4] for the methods. The face morphology was examined with polarizing and metallographic microscopes in reflected light, and also by interference contrast. The faces were examined directly after production and also after etching and various chemical treatments. Thermochemical calculations indicated the best growth conditions as an initial iodine concentration of 4-6 mg/cm^3, crystallization temperature of 750-850°C, and a temperature difference of 15-25°C.

Most compounds of $A^{II}B^{VI}$ type, including ZnSe, show polymorphism [6]; when ZnSe is grown by chemical transport [4, 7], one always gets crystals of the cubic modification with a structure of sphalerite type. Two factors have to be taken into account in explaining this. (1) Growth by chemical transport occurs at relatively low temperatures, where the cubic modification is the more stable, although we lack exact values for the transition temperature between the two modifications. (2) A considerable part is played by the composition of the gas phase; thermochemical calculations show that the iodide method always has a comparatively high partial pressure of selenium under any conditions, while the zinc is present only in combined form as ZnI_2 [4, 5]. However, the literature shows that the presence of excess chalcogen in the gas phase facilitates formation of the cubic phase of $A^{II}B^{VI}$ [8, 9].

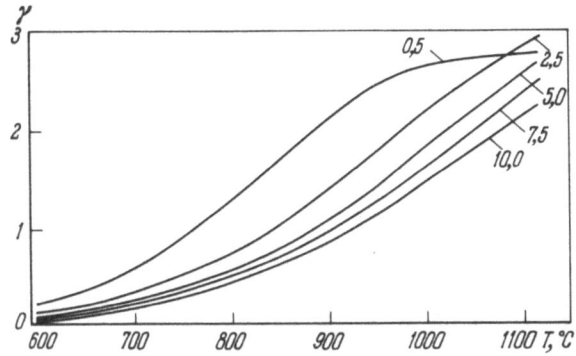

Fig. 1. Dilution factor γ as a function of temperature and initial iodine concentration.

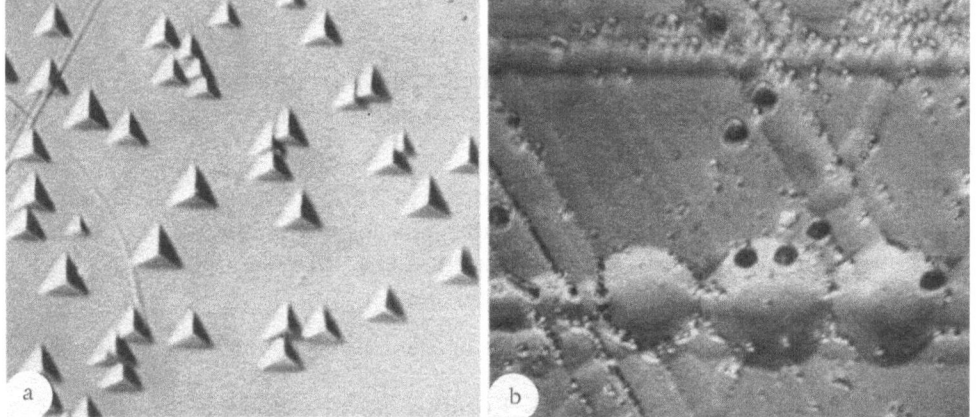

Fig. 2. Growth figures on (a) B$\{\overline{1}\overline{1}\overline{1}\}$ and (b) A$\{111\}$ of ZnSe; ×650.

Goniometry showed that the ZnSe crystals contain the $\{111\}$ and $\{100\}$ forms; very rarely, there were weak reflections from $\{110\}$ faces, which were observable only under the microscope in the form of narrow bands on the edges between $\{111\}$ and $\{\overline{1}\overline{1}\overline{1}\}$ faces.

Crystals of cubic zinc selenide belong to symmetry class $\overline{4}3m$; the lack of a center of symmetry causes the $\langle 111 \rangle$ directions in the sphalerite lattice to be polar, and the A$\{111\}$ faces differ in properties from the B$\{\overline{1}\overline{1}\overline{1}\}$ faces, which is quite appreciable on etching crystals of $A^{II}B^{VI}$ [10, 11]. There is an equally appreciable difference in the morphology of fresh $\{111\}$ faces of ZnSe crystals [7], and the degree of difference increases with the ratio of the number of ZnI_2 molecules to the number of iodine atoms in the vapor phase, which we call the dilution coefficient γ (Fig. 1).* It may be that a change in γ leads to altered kinetics in (1) and hence to altered face morphology.

The B$\{\overline{1}\overline{1}\overline{1}\}$ faces are more prominent, i.e., their growth rate is less than that of A$\{111\}$ faces, and this difference is more appreciable when γ is large; the growth figures on B$\{\overline{1}\overline{1}\overline{1}\}$ faces are either almost absent or else have a symmetry corresponding to a threefold axis (Fig. 2a). The A$\{111\}$ faces do not have regular growth figures (Fig. 2b), and these faces are usually matt at large γ. The surfaces of B$\{\overline{1}\overline{1}\overline{1}\}$ faces often have growth layers of various thicknesses. Twinning occurs easily in a structure of sphalerite type, so sometimes the layers can grow in a twin position, as is clear from the opposite orientation of the growth figures on two adjacent layers (Fig. 3). Growth spirals are usually not seen. The $\{100\}$ faces are usual-

*$\gamma = P_{ZnI_2}/(P_I + 2P_{I_2})$.

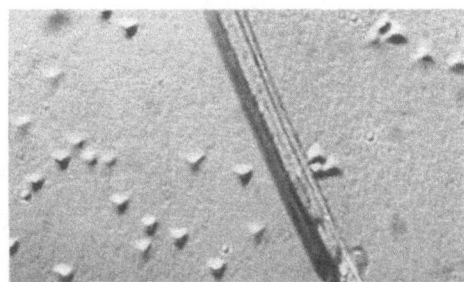

Fig. 3. Growth layers on a B$\{\bar{1}\bar{1}\bar{1}\}$ face in the twin position; ×250.

ly mirror-smooth; sometimes they have growth figures whose symmetry corresponds to that of the face (Fig. 4).

We did some preliminary tests on ZnSe crystals grown with excess of selenium or zinc (concentrations ~0.5 mg/cm³); excess selenium had little effect on the face morphology, because the partial pressure of the selenium is relatively high even when the initial composition is stoichiometric [5]. On the other hand, excess zinc even at low concentrations shifts the equilibrium in (1) considerably at a given temperature, and therefore substantially alters the crystallization conditions. The $\{111\}$ faces then show a series of characteristic defects (Fig. 5), which are still being studied.

We had to obtain a relation between the growth pictures and etching patterns of ZnSe crystals in relation to face symbol, but the literature carried no reliable evidence, so we made

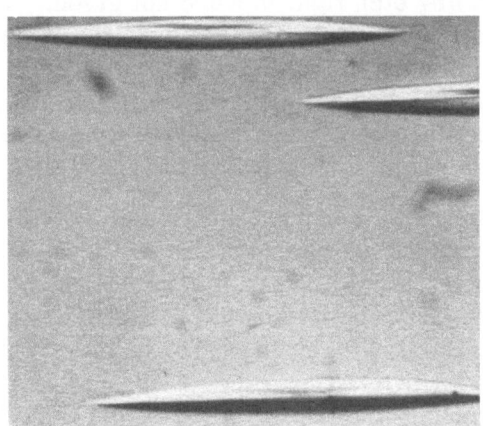

Fig. 4. Growth figures on a $\{\bar{1}00\}$ face of ZnSe; ×650.

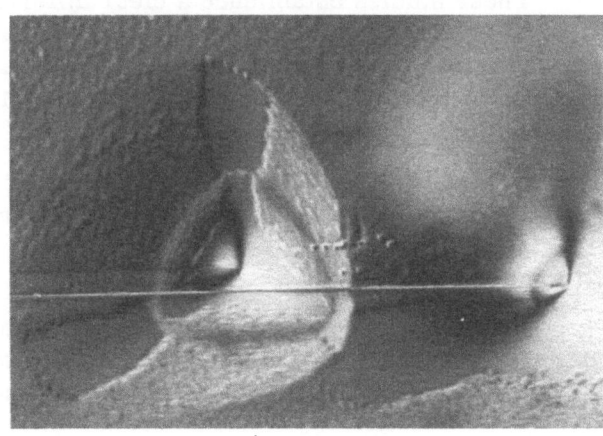

Fig. 5. Growth figures on a B$\{\bar{1}\bar{1}\bar{1}\}$ of a ZnSe crystal in the presence of excess zinc; ×250.

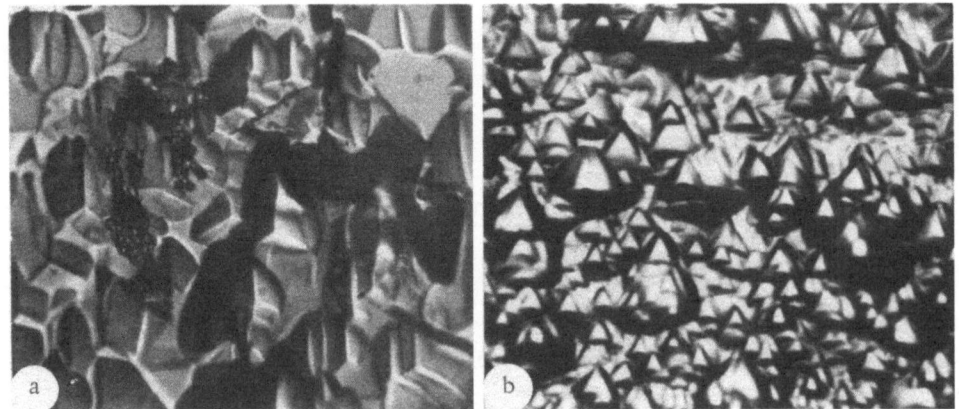

Fig. 6. Etch patterns ($\times 450$) for ZnSe: a) A$\{111\}$, b) B$\{\bar{1}\bar{1}\bar{1}\}$.

absolute identifications of the polar faces, which was done via the intensities of the x-ray reflections from these near the absorption edge of one of the components [12]. This method proved valueless, because we could not use a sufficiently powerful source of monochromatic radiation of appropriate wavelength (~ 1.28 Å). Also, the continuum method [13] gave a large error on account of the appreciable background, and we were unable to observe any real difference in the reflection intensities.

It has been stated in the literature that the polar faces of ZnSe crystals may be identified absolutely by an x-ray method [14]; but the corresponding etch figures were not given. We identified the polar faces via the sign of the piezoelectric effect, but in this case there were considerable difficulties because the piezo modulus is small in magnitude for cubic ZnSe ($\sim 10^{-12}$ Cu/N [15]). The measurements were facilitated in that we only needed to know the sign of the effect, which was determined statically. Silver electrodes were applied to opposite $\{111\}$ faces, while the charge sign indicator was an electrometer amplifier of type VI-2.

We found that compression of a ZnSe crystal along $\langle 111 \rangle$ caused the smooth faces with regular growth figures to be positively charged; our results may be compared with the sign of the piezoelectric effects for cubic ZnS and ZnTe crystals [16], which indicates that these faces are B$\{\bar{1}\bar{1}\bar{1}\}$, i.e., are formed by selenium atoms. This determination of the indices enabled us to determine the reason for the difference in the etching of the $\{111\}$ faces of ZnSe crystals; the best results were obtained with a mixture $HNO_3 : CrO_3 : H_2O = 1 : 4 : 3$. The etch patterns enabled us to determine simply and unambiguously the nature of the A and B$\{111\}$ faces (Fig. 6), and the dislocation pits are formed only on A$\{111\}$ faces.

These studies established a clear difference in the growth and etch patterns of the polar faces of ZnSe crystals; change in the crystallization conditions appreciably affected the face growth pattern. Detailed explanation of the features requires incorporation of the bond configurations in the surface atoms of the A and B$\{111\}$ faces, as well as differences in adsorption and reactivity.

I am indebted to Professor N. N. Sheftal' for direction in the work and discussion of the results, and to Yu. G. Kostyuk and M. A. Garkakle for assistance in the experiments.

Literature Cited

1.   H. Schäfer, Chemical Transport Reactions [Russian translation], Mir, Moscow (1964).
2.   R. Nitsche, H. U. Bölsterli, and M. Lichtensteiger, J. Phys. Chem. Solids, Vol. 21 (1961).
3.   R. Nitsche and D. Richman, Z. Elektrochem., 66:709 (1962).

4. A. A. Simanovskii, in: Growth of Crystals, Vol. 7, Consultants Bureau, New York (1969), p. 224.

5. A. A. Simanovskii and N. N. Sheftal', in: Semiconductor Physics [in Russian], Izd. SO AN SSSR, Novosibirsk (1968), p. 34.

6. A. S. Pashinkin, The Polytypes of Zinc and Cadmium Chalcogenides [in Russian], Nedra, Leningrad (1963).

7. A. A. Simanovskii, Kristallografiya, 14:1098 (1969).

8. K. V. Shalimova and N. K. Morgova, Kristallografiya, 9:559 (1964).

9. L. A. Sysoev and V. M. Koshkin, Crystallization Mechanisms and Kinetics [in Russian], Nauka i Tekhnika, Minsk (1968), p. 161.

10. E. P. Warekois, M. C. Lavine, A. N. Mariano, and H. C. Gatos, J. Appl. Phys., 33:690 (1962).

11. G. A. Wolff, J. J. Franley, and J. R. Hietanen, J. Electrochem. Soc., 111:22 (1964).

12. D. Coster, K. S. Knol, and J. A. Prins, Z. Phys., 63:345 (1930).

13. H. Cole and N. R. Stemple, J. Appl. Phys., 33:2227 (1962).

14. A. N. Mariano and G. A. Wolff, Z. Kristallogr., 126:244 (1968).

15. D. Berlincourt, H. Jaffe, and L. R. Shiozawa, Phys. Rev., 129:1009 (1963).

16. G. Arlt and P. Quadflieg, Phys. Status Solidi, 25:323 (1968).

THE ACTUAL STRUCTURE OF A CRYSTAL AS THE FACTOR GOVERNING NUCLEAR NUCLEATION AND GROWTH

G. I. Distler

Usually, crystallization (including epitaxy) is considered in isolation from other heterogeneous processes, but it is extremely revealing to represent crystallization as a particular case of an extensive class of chemical reactions generally and surface reactions in particular. The active centers in the substrate surface or seed are to be treated as components of the corresponding crystallization reaction. Our laboratory has recently obtained some new experimental evidence that confirms the decisive role of the real crystal structure in nucleation and growth [1-3].

The exceptional selectivity of nucleation is seen on decorating x-rayed alkali halide crystals [4, 5], when one gets color centers, which are point defects with a definite electronic structure. In fact, the density of the decorating particles at the surface of the colored specimens increases substantially for NaCl, which reflects the production of new point defects. An interesting experimental point is that there are groups of decorating particles in the form of squares, triangles, trapezia, etc., which in many cases are mutually parallel and are also oriented in a definite way relative to crystallographic directions. This grouping of decorating particles reflects the presence of complex active centers. It has been shown [6, 7] that color centers, in particular F centers, are active sites for many catalyzed reactions, including polymerization. It is quite likely that compound active centers behave in some heterogeneous reactions as single entities; they are simultaneously adsorption centers and provide the basis for the initial stages of heterogeneous processes.

We examined synthetic NaCl crystals grown doped with $PbCl_2$, and the cleaved surfaces showed anisometric inclusions of width 500-1000 Å and length 1-5 μm [8, 9]. Each such inclusion represents a new phase deposited in the decomposition of $NaCl-PbCl_2$ solid solutions, and these inclusions are surrounded by shells of decorating gold particles, with the width of the shells extending out to 2000 Å but decreasing as the temperature is increased. The density of the decorating particles in the shells is about 10^{12} cm^{-2}. In 1945, Frenkel' [10] theoretically considered the production of double electrical layers of charge vacancies at the surfaces of ionic crystals, and these layers were subsequently discussed by others [11, 12], and it was shown that these layers can arise also at internal boundaries of pores, inclusions, and other inhomogeneous areas. The shells we have observed around inclusions represent the first experimental proof that there are double electrical layers at the boundaries of two solid phases (Fig. 1). This decoration of a double electrical layer consisting mainly of charged impurities and vacancies shows clearly that the active centers in nucleation are electrically active (charged) point defects.

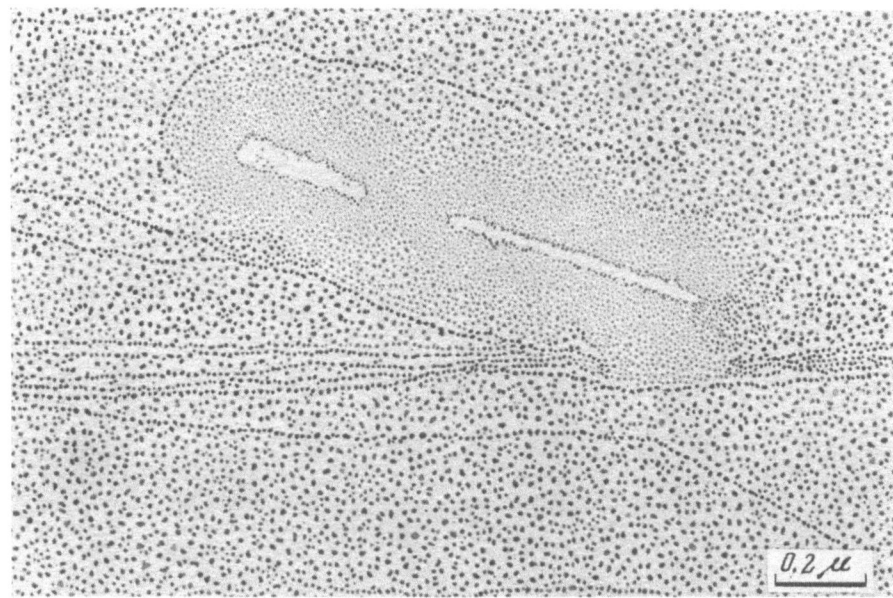

Fig. 1. Double electrical layers revealed by gold decoration
of a cleaved face of NaCl doped with $PbCl_2$.

The nuclei of silver chloride decorating triglycine sulfate and LiF are not oriented relative to the latter [13, 14], i.e., the primary nuclei in epitaxy are not obliged to be oriented. This is an important experimental result, since in many studies [15, 16] it has been assumed (but not proved) that orientation of the primary nuclei determines the character and degree of orientation of the films during growth.

Of course, point defects or active centers cannot be considered without reference to the basic structure, because the structure of such centers is inevitably linked with the chemical composition, symmetry, and structure of the matrix; therefore, the principles of geometrical and crystallochemical correspondence [17, 18] are still relevant, but they must be applied to local active centers that reflect the symmetry and properties of the host.

Oriented coalescence is the next important stage after nucleation; this stage is governed by the real structure of the crystal substrate, namely by the electrical properties (the sign and magnitude of the charge and potential) of the parts where the coalescence occurs. In the early stages of deposition of silver chloride on triglycine sulfate, the electrically active point defects give rise to AgCl nuclei, which are disposed at random [14, 19]. As the thickness of the AgCl layer increases, oriented coalescence of nuclei and larger particles begins only at the surfaces of the negative domains; the difference in rate of coalescence between domains differing in size is so great that the negative domains give rise to a single-crystal film of AgCl when the layer thickness is 90 Å, whereas coalescence is only just beginning on the positive domain. This oriented coalescence of AgCl occurs in different ways not only on the surfaces of the domains but also on different local parts of surfaces of other ionic and semiconducting compounds. For instance, Fig. 2 shows the crystallization pattern of AgCl on a cleavage surface of LiF [13]. The surface has localized areas of size up to some thousands of Å where oriented coalescence of nuclei and particles occurs at substantially different rates; these regions, by analogy with the coalescence of AgCl, may be considered as positively and negatively charged. Locally charged areas differing in sign have also been detected on decoration of NaCl crystals [20]. There are many local charge regions within an alkali halide crystal, as has been shown [21] from electrical measurements. The rate of oriented coalescence is dependent on the electrical relief of the substrate crystal for metals (Ag, Au), semiconduc-

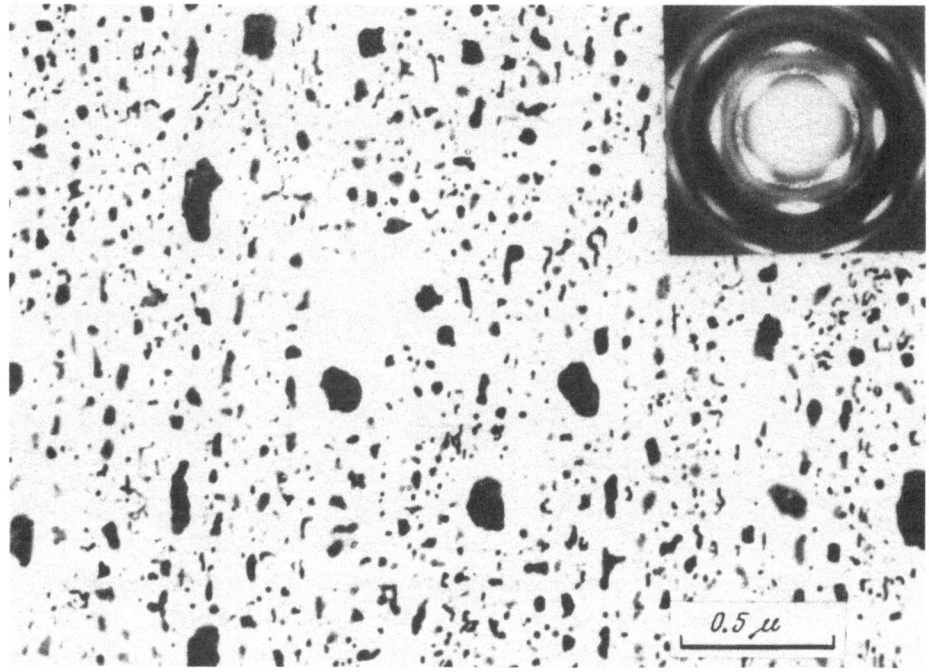

Fig. 2. Oriented coalescence of AgCl on negatively charged parts of a cleaved surface of LiF and initiation of coalescence on positive parts.

Fig. 3. Selective crystallization on a cleaved surface of triglycine sulfate: a) AgCl; b) NaCl; c) AgI; d) AgBr.

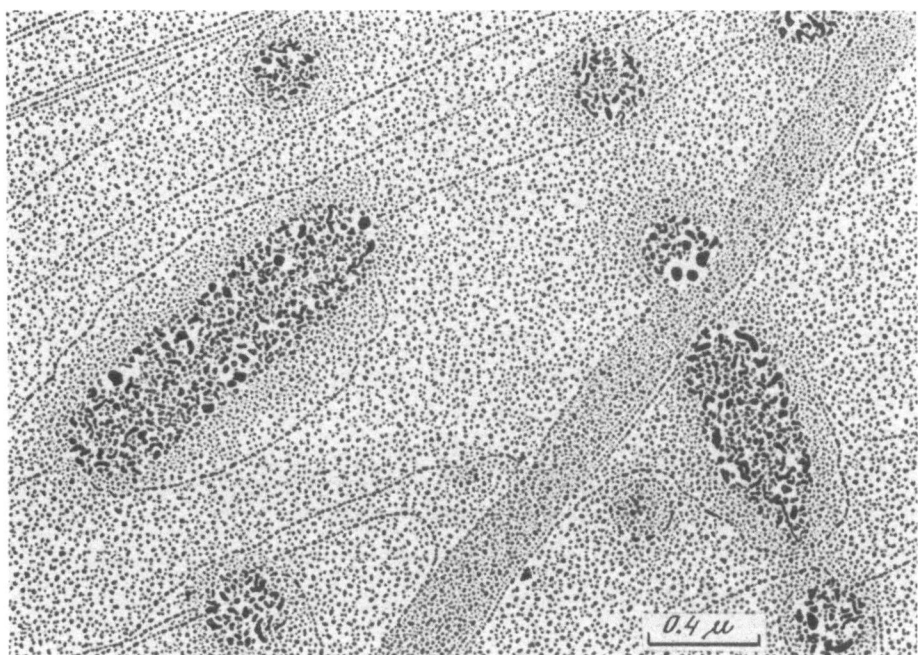

Fig. 4. Complex structure of the double electrical layer surrounding a PbCl$_2$ inclusion as revealed by silver decoration.

tors (PbTe, CdS, SnSe), and insulators (AgCl, AgI, NaCl) [22] (Fig. 3). Electrical nonuniformity makes itself very strongly felt for NaCl crystals doped with PbCl$_2$ [9]; when silver is deposited thermally on cleaved surfaces of these crystals (Fig. 4), one gets oriented coalescence
of silver on the internal parts of the double electrical layers, the rate being maximal here, as
for negative domains in triglycine sulfate [23]. On the outer part of the double layers, and also
on the rest of the surface, there are local areas where one gets oriented coalescence with
appreciable rates and also parts where coalescence is practically absent. The first areas are
clearly charged negative and the second positive. The negative parts are more active on account of physical adsorption of thin layers of water, which act as a lubricant [24].

Growth and cleavage steps play a particular part in nucleation and growth; but in decoration the AgCl deposited on triglycine sulfate [19, 25] accumulates more at cleavage depths
intersecting domains of different size preferentially relative to smooth parts of the domains.
When silver chloride is deposited on cleaved surfaces of NaCl, one finds that some cleavage
depths either have no AgCl at all or else an uneven distribution (Fig. 5a and 5b). This difference in crystallization rate between steps or between parts of a given step is due to the most
rapid deposition occurring at those parts of the surface with the highest negative charge and
potential; it is not the step itself as an element of geometrical relief that initiates the crystallization but the electrical properties found there. When silver decorates the surface of NaCl
crystals doped with PbCl$_2$ [9] (Fig. 6), the silver particles are formed selectively only at the
surfaces of double electrical layers around PbCl$_2$ inclusions and at cleavage steps. This
selective crystallization occurs because these steps have higher negative potentials than the
adjacent smooth parts of the surface; but the internal part of the double layer has an even
larger negative potential, which facilitates oriented coalescence, and the orientation of the
silver crystals on these parts of the surface is more perfect.

The electrical properties of the surface also control the long-range crystallization
mechanism in which structural information is transmitted through boundary layers previously

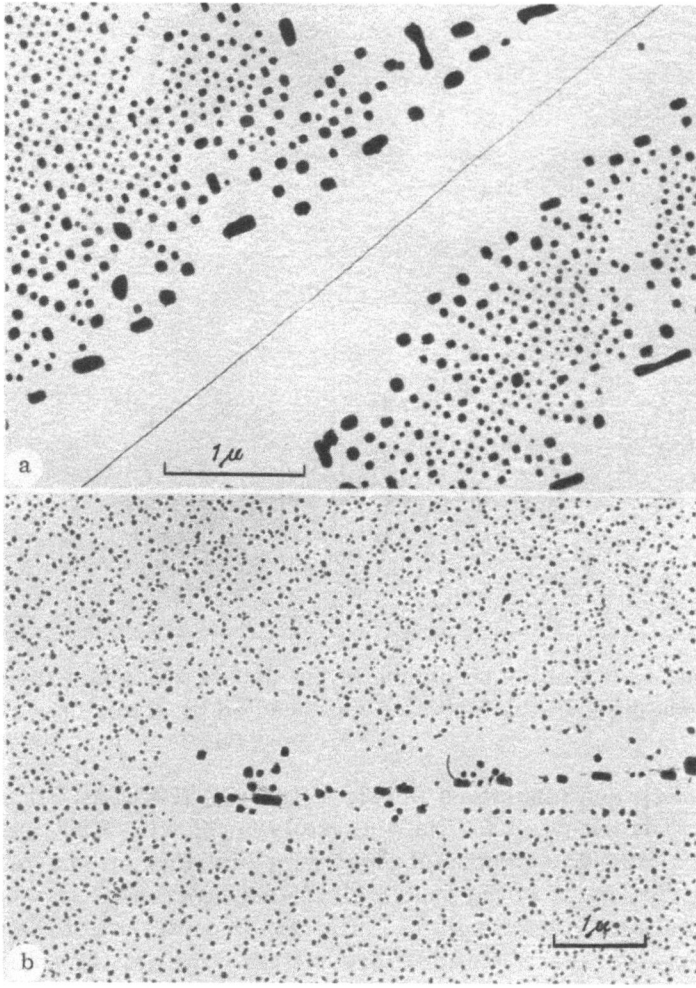

Fig. 5. Decoration of a cleaved NaCl surface with
AgCl. a) No crystallization; b) uneven crystallization.

prepared on the surface [1, 26-28]. Electrically active elements in the surface induce in the
boundary layers polarization structures of electret type, which reflect these elements.

The electret mechanism may be demonstrated by using films of polyvinyl chloride as
boundary layers, as these have thermoelectric properties [29, 30]. It has been found that
the contact sides of these films after detachment from the NaCl crystals reproduce the electrical relief of the underlying surface. The copying occurs at the nucleation stage and during
the coalescence of nuclei (crystallization of AgCl) [31], and also during the crystal growth
(crystallization of anthraquinone) [29, 30]. The external electric fields needed to produce the
electret state in the films are provided by the microfields of point defects and groups of these,
whose field strengths at distances of atomic order may be even thousands of kV/cm.

The photoelectric mechanism of structural information transition was demonstrated by
making layers of amorphous selenium, which has photoelectret properties [32, 33]. When
anthraquinone is sublimed onto NaCl crystals, the amorphous selenium layer about 200 Å
thick gives rise to a biaxial texture in the anthraquinone crystals, which persists after the
selenium layers have been separated from the NaCl (Fig. 7a), which shows the memory effect
in the layers, which can be suppressed by additional illumination.

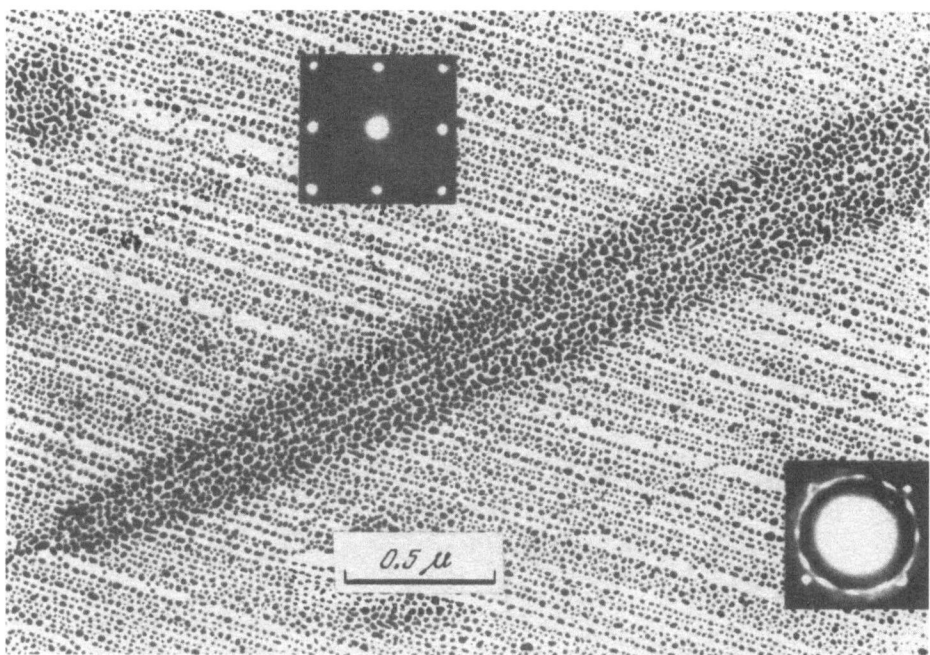

Fig. 6. Selective crystallization of silver on cleavage steps on NaCl and on the surface of an inclusion surrounded by a double electrical layer.

Single-crystal structural information can be transmitted by polycrystalline boundary layers [34-36], which is of importance for the theory of crystal growth; sublimation of anthraquinone onto the outside of polycrystalline ZnO layers up to 250 Å thick on NaCl causes the anthraquinone to crystallize with a biaxial texture (Fig. 7b). The ZnO crystals in the boundary layer are randomly disposed relative to the NaCl surface, so the induced polarization structures providing the information exist without reference to the crystallographic directions of the microscrystals in the boundary layer. Polycrystalline boundary layers of ZnO not only transmit information but also store it, since their contact side still allows anthraquinone to crystallize in an oriented fashion. The memory in ZnO layers can be erased by light, as in the case of selenium.

It has been found that metal boundary layers made by evaporation at about 10^{-5} mm Hg have information-storage properties [37]. On the outside of films of silver and gold, it has proved possible to produce biaxial textures of anthraquinone that reflect the electrical relief of the surfaces of NaCl crystals. If the thickness of the silver boundary layer is more than 100 Å, the texture axes of the anthraquinone crystals are rotated through 45° from ⟨110⟩ to ⟨100⟩ (Fig. 7c). After separation from the NaCl crystals, the contact sides of the silver films store the structural information, whereas gold films have no memory. The storage features of metallic films may be ascribed to local semiconductor and insulating microstructures, which are formed when the layers are prepared as the result of interaction of the atoms or ions of the metal with residual gases near electrically active centers in the substrate. This means that metal films made under real conditions have microscopic properties that determine at the elementary level heterogeneous processes such as crystallization; in this respect they are similar to dielectric and semiconductor films.

This evidence on the electrical relief in relation to crystallization requires us to reconsider the Kossel–Stranski theory and the dislocation growth theory. In fact, experiment does not confirm the postulate in these theories that the surface of the substrate crystal or

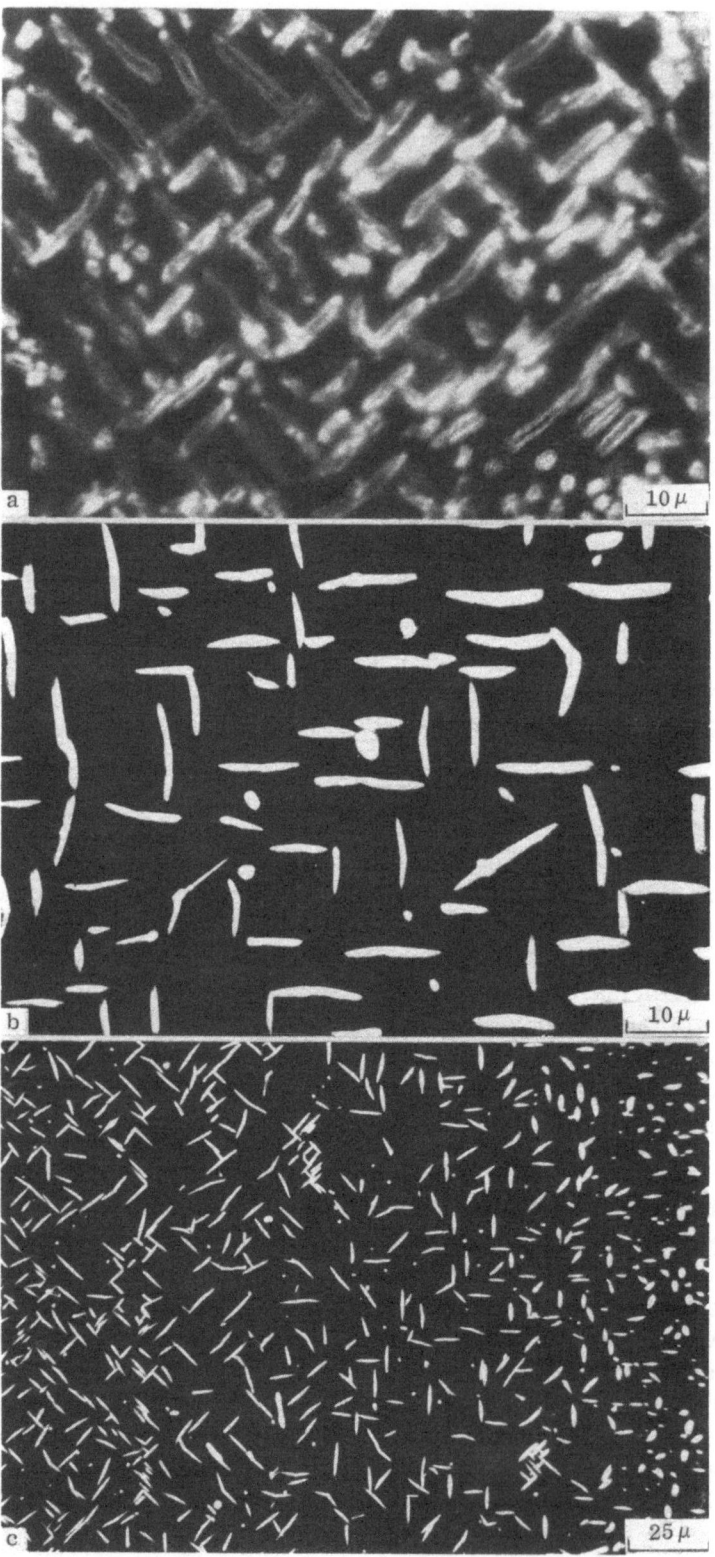

Fig. 7. Oriented crystallization of anthraquinone (a) on the contact side of an amorphous selenium film separated from a NaCl crystal, (b) on the outer surface of a polycrystalline film of zinc oxide about 200 Å thick, and (c) on the outside of a polycrystalline silver film about 100 Å thick.

seed is inactive otherwise than via its geometrical relief. In fact, it has been reliably established that the surface activity of a crystal is determined by the defect structure, which is represented primarily by electrically active point defects and groups of these. The activity of the geometrical relief (whether linear growth steps or the points of emergence of screw dislocations) makes itself felt via the electrical properties, as many experiments show; therefore, only when the elements of the geometrical relief are most suitable for crystallization will the crystal growth occur by the Kossel–Stranski mechanism or via the mechanism involved in the dislocation theory. These theories are therefore applicable to particular cases of crystal growth.

In our view, crystallization should be considered as a matrix replication process that reflects the electrical structure of the crystal substrate or seed.*

Literature Cited

1. G. I. Distler, Izv. Akad. Nauk SSSR, ser. fiz., 32:104 (1968).
2. G. I. Distler, J. Cryst. Growth, 3(4):175 (1968).
3. G. I. Distler, Kristall und Technik, 5:73 (1970).
4. G. I. Distler, V. N. Lebedeva, and V. V. Moskvin, Fiz. Tverd. Tela, 10:3489 (1968).
5. G. I. Distler, V. N. Lebedeva, and V. V. Moskvin, Kristallografiya, 14:664 (1969).
6. J. H. Lunsford and I. P. Jayne, J. Phys. Chem., 69:2182 (1965).
7. V. A. Kargin, N. A. Platé, I. A. Litvinov, V. P. Shibaev, and E. G. Lur'ye, Vysokomol. Soed., 3:1091 (1961).
8. G. I. Distler, V. N. Lebedeva, V. V. Moskvin, and E. I. Kortukova, Fiz. Tverd. Tela, 11:239 (1969).
9. G. I. Distler, V. N. Lebedeva, V. V. Moskvin, and E. I. Kortukova, Fiz. Tverd. Tela, 12:1149 (1970).
10. Yu. I. Frenkel', Kinetic Theory of Liquids [in Russian], Izd. AN SSSR, Moscow (1945).
11. K. Lehovec, J. Chem. Phys., 21:1123 (1953).
12. I. M. Lifshits and Yu. E. Geguzin, Fiz. Tverd. Tela, 7:62 (1965).
13. G. I. Distler and V. P. Vlasov, Fiz. Tverd. Tela, 11:2226 (1969).
14. V. P. Vlasov and G. I. Distler, Kristallografiya, 16:663 (1971).
15. Single-Crystal Films, edited by Francombe and Sato [Russian translation], Mir, Moscow (1966).
16. Anderson (ed.), The Use of Thin Films in Physical Investigations, Academic Press, London (1966).
17. L. Royer, Bull. Franc. Mineral., 51:7 (1928).
18. P. D. Dankov, Zh. Fiz. Khim., 20:853 (1946).
19. G. I. Distler and V. P. Vlasov, Kristallografiya, 14:872 (1969).
20. G. I. Distler and É. G. Sarovskii, Fiz. Tverd. Tela, 11:547 (1969).
21. M. I. Kornfel'd, Fiz. Tverd. Tela, 10:2422 (1968).
22. V. P. Vlasov, Yu. M. Gerasimov, and G. I. Distler, Kristallografiya, 15:346 (1970).
23. G. I. Distler, V. P. Konstantinova, Y. M. Gerasimov, and G. A. Tolmacheva, Nature, 218:762 (1968).
24. G. I. Distler and V. V. Moskvin, Abstr. Internat. Conf. Crystal Growth, Marseille (1971), p. 44a.
25. G. I. Distler and V. P. Vlasov, Thin Solid Films, 3:333 (1969).
26. G. I. Distler, in: Growth of Crystals, Vol. 8, Consultants Bureau, New York (1969), p. 91.
27. G. I. Distler and S. A. Kobzareva, Dokl. Akad. Nauk SSSR, 172:1069 (1967).

----

* Editor's note: These generalizations, which extend over numerous studies and papers by the author, are to be considered as somewhat controversial.

28. Yu. M. Gerasimov and G. I. Distler, Kristallografiya, 14:1101 (1969).
29. G. I. Distler and S. A. Kobzareva, Naturwissenschaften, 56:325 (1969).
30. G. I. Distler and S. A. Kobzareva, Dokl. Akad. Nauk SSSR, 188:811 (1969).
31. G. I. Distler and E. I. Tokmakova, Thin Solid Films, 6:203 (1970).
32. G. I. Distler and V. G. Obronov, Nature, 224:261 (1969).
33. G. I. Distler and V. G. Obronov, Dokl. Akad. Nauk SSSR, 191:584 (1970).
34. G. I. Distler and V. G. Obronov, Dokl. Akad. Nauk SSSR, 197:819 (1971).
35. G. I. Distler, J. Cryst. Growth, 9:76 (1971).
36. G. I. Distler and V. G. Obronov, Nature, 229:242 (1971).
37. G. I. Distler, Yu. M. Gerasimov, and V. G. Obronov, Kristallografiya, Vol. 16, No. 6 (1971).

GROWTH KINETICS OF EPITAXIAL SILICON FILMS WITH RADIATIVE HEATING

B. V. Deryagin, O. V. Spirin, B. V. Spitsyn, and D. V. Fedoseev

Heating with focused radiation enables one to examine the properties of almost any material at high temperatures and with a controlled gas atmosphere [1]; such heating is of considerable interest as research method for crystallization of extremely pure materials.

There are many papers on the growth of silicon by reduction of $SiCl_4$ by hydrogen [2-4]. The seed crystals are heated from below by high-resistance silicon or pure graphite fed with induction currents.

It is of interest to examine the growth of silicon films during radiative heating of single-crystal plates; we used a radiative heating apparatus described in detail in [5]. The apparatus consisted of a system for purifying the hydrogen and saturating it with $SiCl_4$, a quartz reaction vessel, and the radiation heating system. The source of pure hydrogen and the saturation system were similar to those used in [4]. At the outlet from the silicon chloride evaporator there was a glass valve with a Teflon seal. The growth of the silicon occurs in accordance with the overall reaction

$$SiCl_4 + 2H_2 = Si + 4HCl$$

and it was examined in the range 850–1390°C with silicon chloride concentrations from 0 to 32 mol.%. In our tests were used $SiCl_4$ of grade 04 for semiconductor purposes.

The silicon specimens before use were washed in heated benzene for 10 minutes, were flushed with alcohol, and then were inserted in a holder for emplacement in the center of the spherical reacting vessel, which was made of fused quartz, and which lay at the focus of the radiation heating system. The flask diameter was 70 mm. The specimens were laid out in the holder as a circle of 12 and were brought in turn to the focus by rotating a joint. Each specimen was a plate $3 \times 15 \times 0.2$ mm on which had been cut a narrow area $1 \times 1 \times 0.2$ mm by etching, which separated a part 3×3 mm from the rest of the plate, which served to mount the plate in the holder. The epitaxial films were grown on surfaces of (111) orientation that had been polished with AM-1 diamond powder. We used p-type silicon plates of specific resistance 0.03 ohm-cm.

The temperature of the specimens was measured by a modulation method via a quartz plane-parallel window by means of an OPBIR-09 optical pyrometer graduated by reference to a black body up to 700°C; the effective degree of blackness was determined from the brightness of the specimen at the melting point, and the true temperature was calculated on the basis of the known blackness of silicon at 0.65 μm [6].

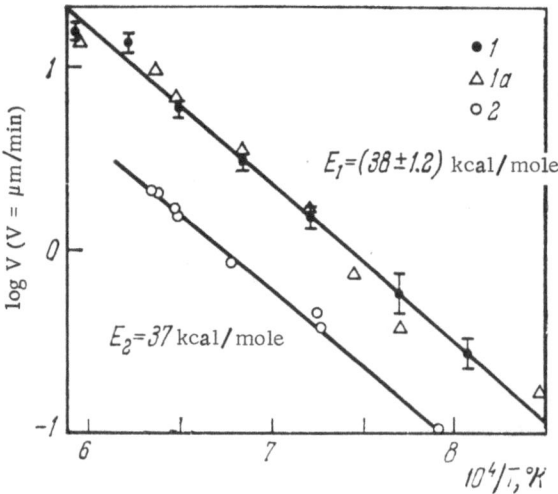

Fig. 1. Linear growth rate of a silicon film as a function of 1/T. 1) Under UV irradiation; 1a) without UV; 2) from [2].

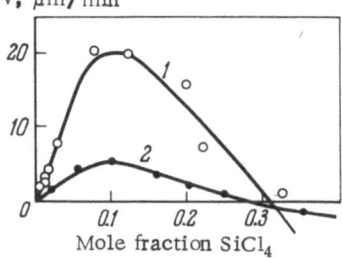

Fig. 2. Linear growth rate of a silicon film as a function of the mole fraction of $SiCl_4$ vapor. 1) Our results; 2) from [2].

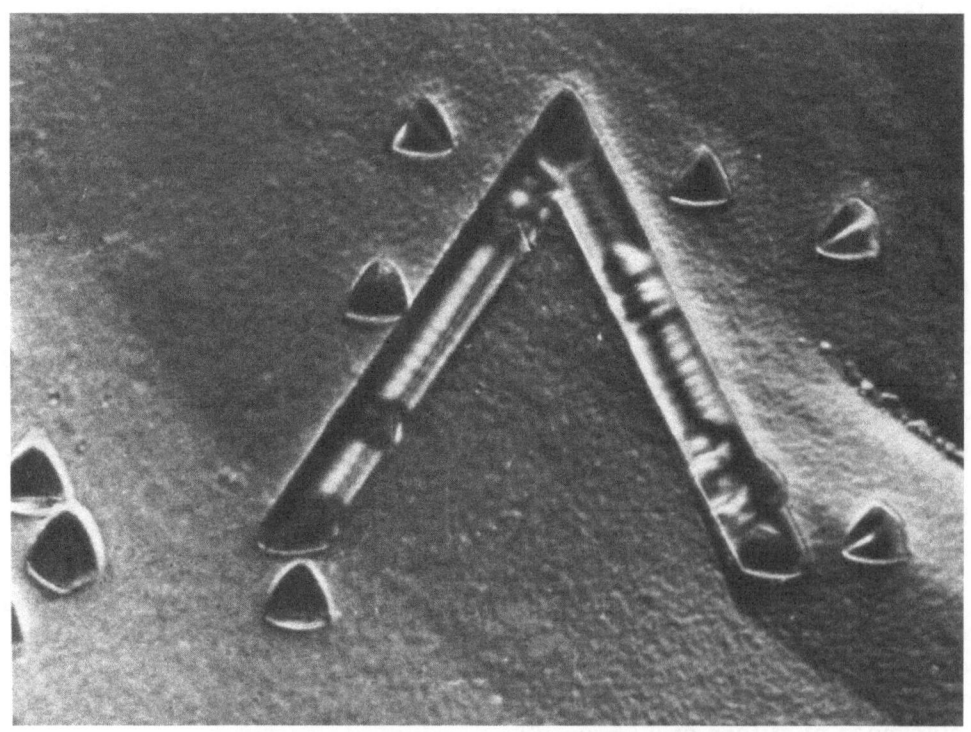

Fig. 3. Etched surface of film, ×250.

Traces of oxides were removed by heating the specimens in a flow of pure hydrogen (1 liter/min) at 1250°C for 15 min, after which the temperature was set at the working point and the hydrogen mixture of set composition was passed through the reaction vessel. A constant composition was maintained by means of a condenser in the inverted position in which the excess chloride vapor was condensed after production in the evaporator at a higher temperature (the temperature was 0-27°C in the condenser). The growth time ranged from 5 to 30 min, but the usual working time was 10 min. The thickness of the finished film was determined to ±0.02 mg by weighing. The nonuniformity in film thickness was determined essentially by the temperature nonuniformity (±10°C) over the surface.

Figure 1 shows the growth rate of silicon films as a function of temperature for a chloride concentration of 2 mol.% and a gas flow rate of 1 liter/min. The observed reaction rate was three times the rate recorded in [2]; over a wide temperature range, the rate was in accordance with a fixed activation energy of 38 ± 1.2 kcal/mole. Therefore, the growth occurred in the kinetic region and was governed by processes at the surface of the silicon. Above 1300°C, there was a slight diffusion retardation of the reaction: the film growth rate at which the retardation became appreciable was greater than that described in [3], which was due to natural convection along the surface of the vertical specimen. There was no bulky support as required in high-frequency heating, which accentuated the effect even more. Another undoubted advantage of the method is that one heats only the growing specimen and a very small volume of the gas immediately adjoining it.

It has been reported [7] that ultraviolet radiation affects the growth rate of silicon epitaxial layers; we used a xenon arc lamp to heat the specimen, and 10% of its emission falls in the ultraviolet region [1], so in continuing a previous study [8] we examined the kinetics of the film growth in the absence of the hard ultraviolet radiation, which was cut off below 0.28 μm by a filter made of molybdenum glass.

Successive exclusion of longer wavelengths in the spectrum forms the subject of ongoing studies on radiative heating; there is some evidence [7] for acceleration of the growth and reduction of the epitaxy temperature on exposure to UV with radiation densities up to 10 W/cm^2. In our experiments, the density of the UV was up to 100 W/cm^2.

Figure 2 shows the growth rate as a function of SiCl$_4$ concentration and also the results of [2]; both curves were recorded at 1270°C. A high growth rate was observed at all concentrations, probably because of the effects of the high-density light on the chemical and physicochemical processes.

The resulting films were of single-crystal type and had perfect epitaxial overgrowth on the substrate; etching in a dislocation-revealing solution showed that the film did have some dislocations (Fig. 3). The density of these was of the same order as in the initial silicon crystal and was $3 \cdot 10^4$ cm^{-2}.

Oriented overgrowth of silicon was observed down to 900°C; at 850 and 800°C, the silicon film was not continuous, and instead we obtained islands of crystallization of effective thickness less than 2 μm. The temperature for onset of epitaxy of silicon on silicon is low because the radiative heating produces local warming of the seed crystal, and the process occurs under very clean conditions.

We are indebted to V. V. Smirnov and A. V. Smol'yaninov for assistance in the experiments.

Literature Cited

1.   G. G. Lopatina, V. P. Sasorov, B. V. Spitsyn, and D. V. Fedoseev, Optical Furnaces [in Russian], Izd. Metallurgiya (1969).

2.   H. C. Theuerer, J. Electrochem. Soc., 108:649 (1961).

3.   F. G. Bylander, J. Electrochem. Soc., 109:1171 (1962).

4.   E. I. Givargizov, Thesis, Institute of Crystallography, Academy of Sciences of the USSR (1965).

5.   B. V. Deryagin, G. G. Lopatina, B. V. Spitsyn, and D. V. Fedoseev, Zh. Fiz. Khim., 42:2360 (1968).

6.   D. Ya. Svet, Thermal Emission of Metals and Certain Substances [in Russian], Izd. Metallurgiya (1964).

7.   M. Kumagawa et al., Japan J. Appl. Phys., 7:1332 (1968).

8.   B. V. Deryagin, B. V. Spitsyn, D. V. Fedoseev, and O. V. Spirin, in: Physicochemical Crystallization [in Russian], Izd. Kazakh. Univ., Alma Ata (1969), p. 100.

EFFECTS OF PCl$_3$ ON THE GROWTH MECHANISM OF AUTOEPITAXIAL GERMANIUM FILMS

A. N. Stepanova and N. N. Sheftal'

Growth mechanisms are basic to research on crystallization processes; two basic growth mechanisms may be distinguished: layer growth, when the crystal grows by tangential propagation of steps, and normal growth, when parts of the surface are displaced as a whole on account of independent addition of atoms [1-7]. The two mechanisms may, of course, occur together. In some cases [7] it is possible to demonstrate the preferential action of one of them via kinetic evidence or crystal morphology.

We have used morphologic features to determine the growth mechanisms of faces of germanium in the chloride process [8, 9] and we here discuss how PCl$_3$ in the vapor phase influences the results.

Methods

The germanium films were grown by reducing GeCl$_4$ in a flow of hydrogen; the reaction tube was vertical. The PCl$_3$ was added to the vapor from a separate source containing either PCl$_3$ or a mixture of this with GeCl$_4$. The molar concentration of PCl$_3$ in relation to the GeCl$_4$ varied from $1.5 \cdot 10^{-5}$ to 0.5. The measurements were made at substrate temperatures of 680 and 850°C. In all cases the molar ratio of GeCl$_4$ to hydrogen was $2.6 \cdot 10^{-3}$, while the gas flow speed in the cold part of the reaction tube was 1.4 cm/sec. The germanium substrates were oriented on (111), (110), and (100).

The morphology of the film surfaces was examined by optical microscopy (interference method) and via light figures; the specific resistance was measured with a four-probe technique (the substrates were plates of high-resistance germanium). The undoped films were p type and had ρ of several ohm-centimeters.

Results

The layer growth mechanism is observed via the growth steps visible in the optical microscope, and also the smooth parts of the faces visible on the surface after a certain period of growth. An additional feature is the formation of humps around the points of emergence of defects of microtwin type, which act as growth step sources. However, this feature is not unambiguous, because such a source can arise around an enclosed impurity, as is observed on (111) surfaces at high PCl$_3$ concentrations.

The PCl$_3$ substantially alters the concentration of the structural defects; for instance, the three-fold pyramids on the (111) films are lost, and another effect is a change in the form of the figures related to structure defects and surface relief. We do not intend to describe or

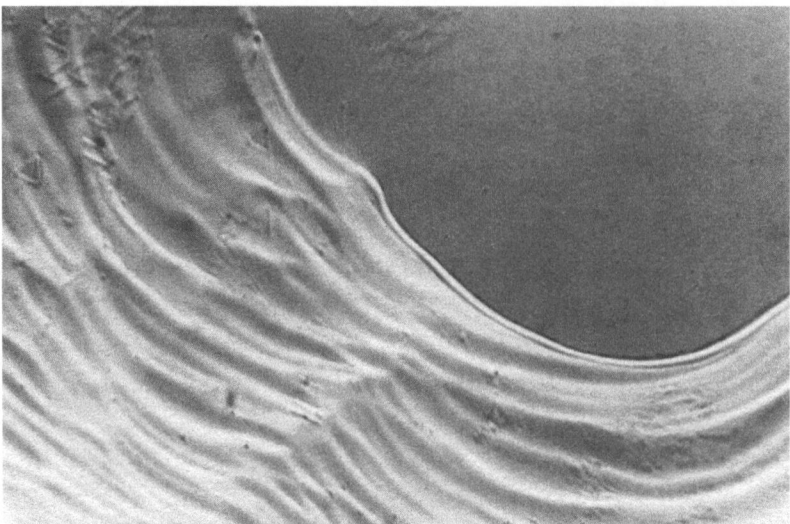

Fig. 1. Surface of a (111) film grown undoped at 850°C,
×270.

discuss all the morphologic features of these films, and we concentrate attention on the layer-growth features.

Films of (111) Orientation. Layer growth was observed under all the conditions used; for instance, Fig. 1 shows a film grown at 850°C without dope, and there is a plateau, which is a part of exact (111) orientation with growth steps emerging from it.

Films of (110) Orientation. Films grown at 680°C without dope show no signs of layer growth (Fig. 2a); the ridged relief indicates rather that the growth was of chain character. Distinct growth steps appeared (Fig. 2b) on raising the temperature to 850°C or adding PCl_3 at a concentration $\geq 5 \cdot 10^{-4}$.

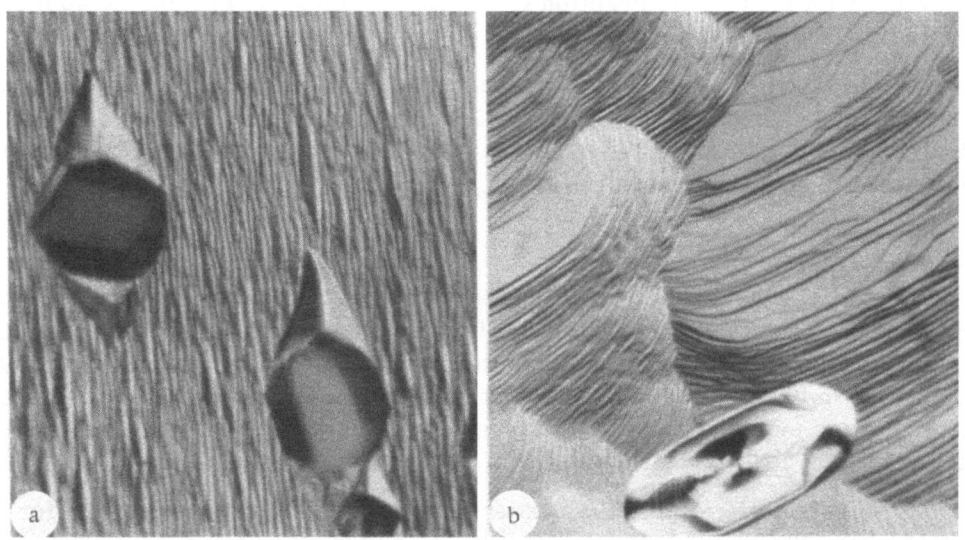

Fig. 2. Surface of a (110) film (a) grown undoped at 680°C, ×550,
and (b) grown doped at 850°C, $5 \cdot 10^{-4}$ M PCl_3, ×270.

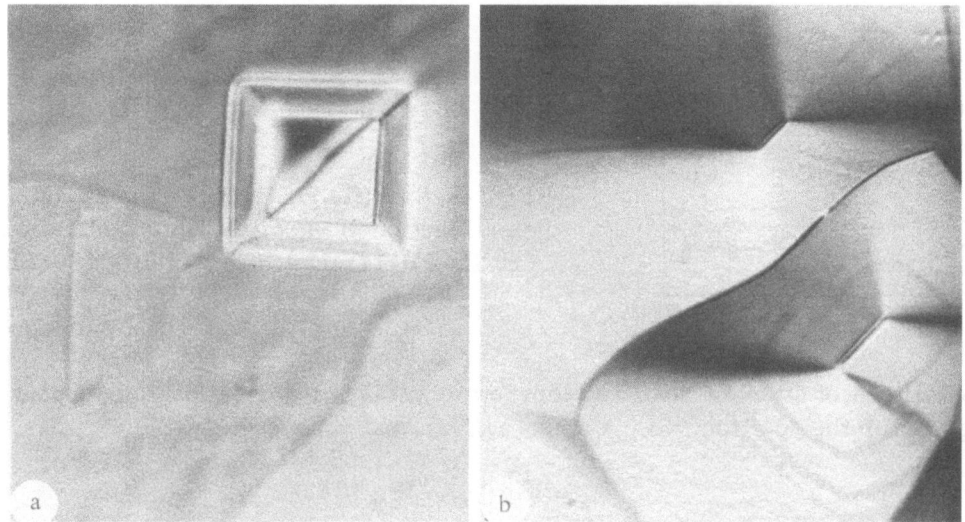

Fig. 3. Surface of a (100) film (a) grown undoped at 850°C, ×1000, and (b) grown doped at 850°C, 0.5 M PCl₃, ×270.

Films of (100) Orientation. Films grown at both temperatures without dope showed no signs of layer growth; for instance, a film grown at 850°C without dope (Fig. 3a) had a uniform rough form, with growth steps not visible, while smooth areas were not seen even on prolonged growth, and around the microtwins there were no humps. Doping facilitated the occurrence of signs of layer growth at both temperatures. Films grown at 850°C without dope (Fig. 3b) had stepped pyramids around the microtwins, which were very active sources of growth steps. Between the pyramids were relatively smooth parts of the (100) face.

The results may be summarized as follows: (a) a trace of PCl₃ facilitates layer growth on (110) and (100) faces, (b) layer growth on (110) faces is favored by raising the temperature, (c) a (111) face grows by the layer mechanism under all conditions that we used.

Discussion

The discussion is based on the molecular-kinetic theory, which involves elementary steps of attachment and detachment of atoms at various positions on the faces. These modes of incorporation of atoms into the lattice will be considered, and we ignore the interaction of these processes with the elementary acts of chemical reaction, on the assumption that these merely supply the atoms of the matrix material and dope to the crystallization medium.

The atomic structures of the (111), (110), and (100) faces in the diamond lattice indicate that (111) faces have fairly strong bond forces between first-order neighbors, which results in a planar net, i.e., in principle, it is possible for this face to grow layerwise. Also, layer growth is possible for (100) faces if we assume that there are also substantial binding forces between second-order neighbors. For (110) faces, one would have to take into account forces between third-order neighbors. The morphology indicates layer growth for (111) faces, and also for (110) and (100) faces under certain conditions, so the subsequent discussion will be based on the forces between neighbors up to the fourth order inclusive.

We introduced the layer growth coefficient C_{la} as the ratio of the probabilities of addition of an atom to a crystal at a kink in the step A_{st} and a smooth surface A_{sur}:

$$C_{la} = A_{st} / A_{sur}.$$

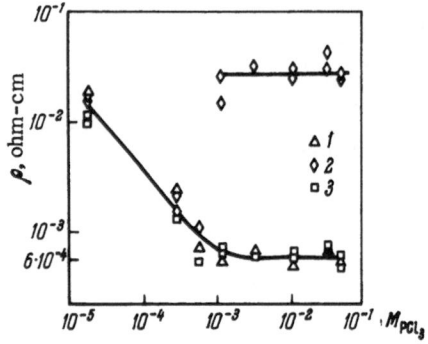

Fig. 4. Specific resistance of germanium films as a function of vapor PCl_3 concentration at 680°C: 1)(111) face; 2) (110) face; 3) (100) face.

The probability A of attachment of an atom to a crystal is exponentially dependent on the binding energy W for the position, i.e., $A \approx \exp (W/kt)$, and consequently

$$C_{la} \approx \exp (W_{st} - W_{sur})/kT.$$

Calculations on C_{la} have been made for (111), (110), and (100) faces of a diamond lattice, which gave the result

$$C_{la}^{(111)} \gg C_{la}^{(100)} > C_{la}^{(110)}.$$

This C_{la} represents the maximal proportion of layer growth in the general growth mechanism, because the expression for C_{la} presupposes rapid exchange of atoms between the steps and the rest of the surface of the crystal. If such exchange is not very rapid, the proportion of layer growth will be less.

Atom exchange between growth steps and the rest of the surface can occur by surface diffusion or via the vapor state involving reversible chemical reactions. This exchange should accelerate as the temperature is raised, and hence also should the layer growth. Under given external conditions (temperature, initial composition of vapor, etc.) the exchange is the more rapid the weaker the bonding of the atoms to the crystal as adsorbed on the surfaces and at kinks in steps. The binding energy of an atom at a kink is large and is approximately the same for all the faces considered, whereas the binding energy for an isolated atom on these faces varies substantially: $W_{sur}^{111} < W_{sur}^{110} \ll W_{sur}^{100}$. Then the coefficient for layer growth of a (100) face is larger than that for (110), but it is more difficult to attain this maximal C_{la} for a (100) face. This appears to be why raising the temperature from 680 to 850°C leads to layer growth only for (110) faces. The trace of PCl_3 also results in more rapid atom exchange, since it facilitates solution of the germanium, at least at low temperatures, as is evident from the reduction in the growth rate in the presence of this substance; this may account for the occurrence of signs of layer growth on (110) and (100) faces above a certain PCl_3 concentration.

Fig. 5. Distributiion of specific resistance over film thickness, 10^{-3} M PCl_3. The ordinate axis corresponds to the free surface of the film.

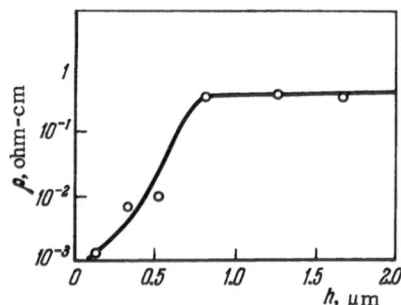

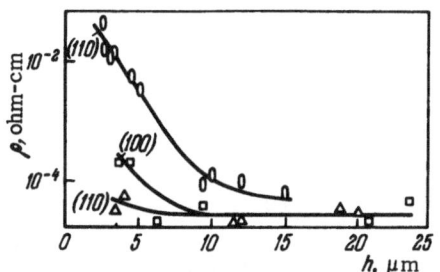

Fig. 6. Specific resistance of a film as a function of thickness, 10^{-3} M PCl_3: $\square$ (100) face; 0 (110) face; $\triangle$ (111) face.

The growth mechanism should influence not only the surface morphology but also the growth rate and details of the structure and composition; layer growth is the most orderly process, and it should lead under these conditions to a crystal with a more perfect structure. We can consider from this viewpoint the dependence of this resistance ρ for films grown at 680°C on the concentration of PCl_3 in the vapor phase (Fig. 4). At PCl_3 concentrations of approximately 10^{-3}, there was a monotonic fall in ρ for films of all three orientations, the minimum value attained being $6 \cdot 10^{-4}$ ohm-cm. At higher PCl_3 concentrations, the ρ for (111) and (100) films remained at this minimal value, which ρ for (110) film increased by nearly a factor of 100, which was not due to less uptake of the phosphorus by the (110) face but to transition of part of the impurity into an electrically active state, as was clear from the distribution of the specific resistance with depth in the (110) film (Fig. 5). The ρ at the surface of the film was practically as for films of other orientations, but at distances of less than 1 μm it increased by more than two orders of magnitude. Figure 6 shows the dependence of the mean specific resistance on the thickness for films of all three orientations grown under identical conditions; there is an increase in ρ only for thin layers, where there is a large contribution from the topmost film and the film near the boundary with the substrate, though the effect varies in magnitude with the orientation of the film; it is largest for (110) faces, weaker for (100), and weakest of all for (111). This electrical neutrality is usually ascribed to uptake preferentially at structure defects; then the degrees of perfection of the film run in a sequence (111), (100), (110), which corresponds to the sequence of layer growth coefficients for these faces. Then the larger the proportion of the layer growth mechanism, the more perfect the film growth under the conditions we have used; this appears to be correct for films of any thickness, but it is most prominent near the boundary with the substrate, where the film structure is more defective. The exact nature of the defects has not been established, but it is clear that they are not the ones usually observed in the optical microscope: packing defects or microtwins, since there was no considerable difference in the concentration of these between the faces. It may be that they are point defects or ones of small volume. The production of them should be dependent on the mode of ordering of the structure during growth, and they may serve as sites for mass uptake of phosphorus atoms.

Conclusions

1. A trace of PCl_3 in the vapor facilitates layer growth on (110) and (100) faces of germanium in the chloride process.

2. Layer growth on (110) is facilitated by elevated temperatures.

3. A (111) face grows by the layer mechanism under all conditions so far examined.

4. The changes in growth mechanism are ascribed to effects on the elementary act of attachment of atoms to different positions on the crystal faces.

5. The (111), (110), and (100) films differ in structural perfection on account of differing proportions of layer growth in the overall mechanism.

We are indebted to L. N. Obolenskaya for assistance in the experiments.

Literature Cited

1.   P. Hartman and W. G. Perdok, Acta Cryst., 8:49 (1955).
2.   P. Hartman and W. G. Perdok, Acta Cryst., 8:521 (1955).
3.   W. K. Burton, N. Cabrera, and F. C. Frank, Phil. Trans., A243:299 (1950).
4.   N. Cabrera, Disc. Faraday Soc., 28:16 (1959).
5.   K. A. Jackson, Growth and Perfection of Crystals (1958), p. 319.
6.   J. W. Cahn, Acta Met., 8:554 (1960).
7.   J. W. Cahn, W. B. Hillig, and G. W. Sears, Acta Met., 12:1421 (1964).
8.   H. C. Theuerer, J. Electrochem. Soc., 108:649 (1961).
9.   E. I. Givargizov, Fiz. Tverd. Tela, 7:1804 (1969).

MORPHOLOGY OF AUTOEPITAXIAL GALLIUM ARSENIDE IN THE CRYSTALLOGRAPHIC RANGE (111)A–(100)–(111)B

L. G. Lavrent'eva, M. P. Yakubenya, O. M. Ivleva, and V. A. Moskovkin

Introduction

It is usual [1] to examine the morphology of autoepitaxial films of gallium arsenide by use of substrates whose surfaces are singular planes.* On the other hand, it is well-known that vicinal surfaces often give layers with smoother vicinal planes often give layers with smoother surfaces, as for silicon [3], germanium [4], and gallium arenide [5, 6]. The only published description of the morphology of gallium arsenide layers on substrates of orientation (112) and (113) is [7], where layers 0.5-0.8 mm thick were used. On the other hand, planes of orientation {hkk} have recently aroused interest in connection with the scope for making films with low carrier concentrations but high mobilities [8].

Here we describe the morphology and general degree of perfection in autoepitaxial films of gallium arsenide made on substrates with orientation deviation from {111} to {100} in the crystallographic range (111)A − (100) − (111)B. The specimens were grown in an open gas-transport system containing iodine as transport agent and hydrogen as carrier gas. The source temperature was 830°C, and the substrate was at 770°C, while the hydrogen had a linear flow speed of 40 cm/min. The supersaturation was varied by adjusting the iodine temperature within the limits 60-75°C, a typical iodine source temperature being 70°C. The substrates were oriented by x-ray methods to better than 20'. We examined the following angles of deviation from (111)A and (111)B: 0; 2; 6; 10; 19; 47; 29.5; 38.9; 43.3; 49; 53; 54.7° in directions to (100), which covered the crystallographic range (111)A − (100) − (111)B. The substrates were polished mechanically and chemically, and were annealed for 20 min at 830°C in hydrogen. No gas etching in iodine vapor was used before the films were deposited.

Morphology of Epitaxial Films

Growth on (111)A and Surfaces Deviating up to 10°. Substrates oriented on (111)A to 20' gave mirror-smooth films with isolated rounded vicinal pyramids (Fig. 1a, b). The side faces of the pyramid were inclined at angles to the base dependent on the supersaturation, the values ranging from a few minutes to 2°. A deviation of 2° from (111)A altered the film surface, which became slightly wavy, and the growth pyramids vanished. The unevenness in the height of the relief did not usually exceed 0.2-0.3 μm, though 0.05 μm was the limit for

---

*We give the name singular to the (111)A, (111)B, and (100) faces, while faces corresponding to small angular deviations from singular are called vicinals and ones with large deviations are called nonsingular [2].

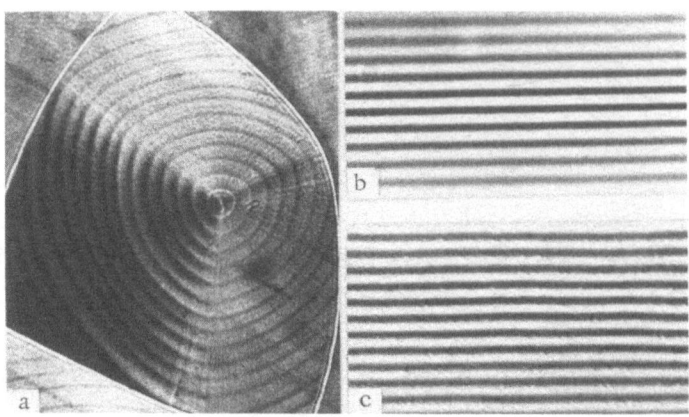

Fig. 1. Growth figures in (111)A. a) High supersaturation, photomicrograph, ×280; b) part between growth figures, interferogram, ×580; c)
interferogram of surface lying at 2° to (111)A,
×580.

certain specimens (Fig. 1c). The films retained this morphology as the angle of deviation
from (111)A increased; for all specimens in this range we found parts oriented on (111)A
(small facets).

Growth on (111)B and Surfaces Deviating up to 10°. Substrates oriented
on (111)B gave films with numerous prominent three-faced pyramids (10^2-10^3 cm^{-2}), which
were also layered in structure; the density of the growth figures increased with the supersaturation (Fig. 2).

If the substrate was a material with uniformly distributed dislocations at entity in the
range 10^4-5 · 10^5 cm^{-2}, a deviation of 2-4° from (111)B caused the pyramidal growth figures to
vanish and new ones of boat type to appear, which were extended along ⟨110⟩ (Fig. 3). These
were usually observed on specimens disoriented from (111)B by 4-6°. The mirror-smooth
part A in such a growth figure is the point of emergence of a plane close to (111)B. The
height unevenness in areas between growth figures did not usually exceed 1 μm. If the sub-

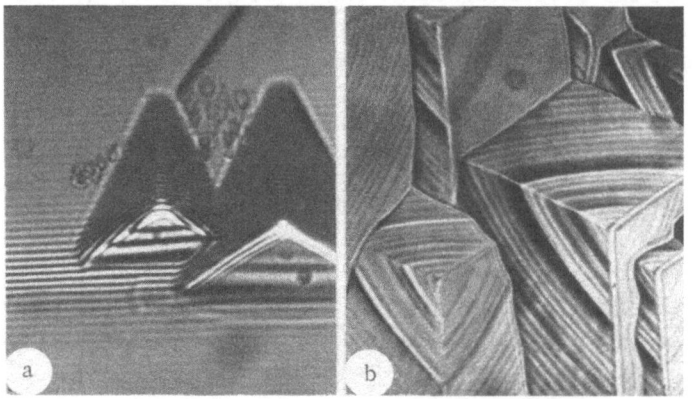

Fig. 2. Interferograms from growth figures on
(111)B. a) Low supersaturation, ×580; b) high
supersaturation, ×140.

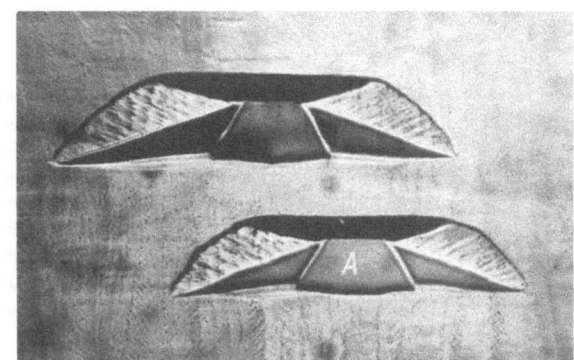

Fig. 3. Growth figures on a surface at 6° to (111)B, ×280.

strate had a high dislocation density (> 10^6 cm^{-2}), and if the dislocations were unevenly distributed, the ranges in the angles for occurrence of pyramidal and boat figures increased to 3-5 and 10°, respectively. The pyramidal growth figures were accompanied by areas of (111)B, plane resembling mica flakes; the number of growth figures decreased as the deviation from (111)B increased, and only a few small figures persisted on 10° specimens.

Growth on (h11). Substrates of orientation (211), (311), (511), and (711) gave smooth films with slightly wavy relief, the height of the unevenness never exceeding 0.2-0.5 μm, and as a rule being less than 0.1 μm. The crests of the waves were usually about 5 μm apart. There were no growth figures in the form of isolated pyramids (Fig. 4). The smoothest films were obtained on substrates of orientation (511) and (711) (Fig. 5a). Specimens of orientation (211) and (311) gave facets of {111} faces together with growth figures. Specimens of orientations (511) and (711) had not only (111) facets but also (100) facet faces on the opposite end of the specimens, which also contained vicinal growth figures (Fig. 5b).

Growth on (100) and Surfaces Deviating up to 8°. Figure 6 is a typical pattern for a film grown on a substrate of (100) orientation. The growth figures here take the form of rounded complete or truncated pyramids; these figures usually clearly revealed their layered structure and the defects at the vertices, which were analogous to those in [1] (the specimens were not etched). This type of growth pyramid persisted on specimens whose surfaces deviated from (100) by 2-6° toward (111)A or (111)B, but the number of growth pyramids fell rapidly as the angle of deviation increased. The height of the unevenness in the relief between them did not exceed 0.3 μm.

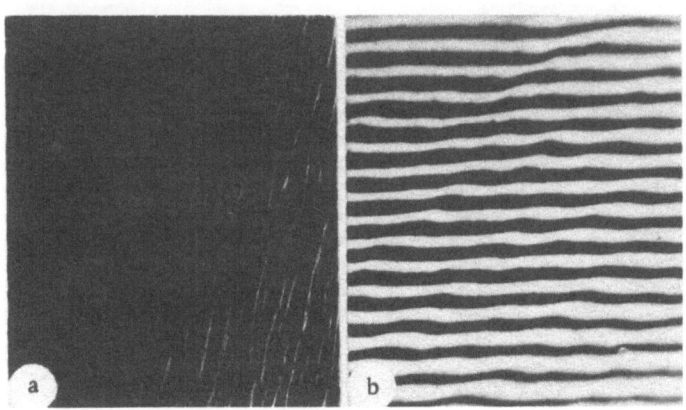

Fig. 4. Surface of a film of (311)B orientation.
a) Photomicrograph, ×50; b) interferogram,
×580.

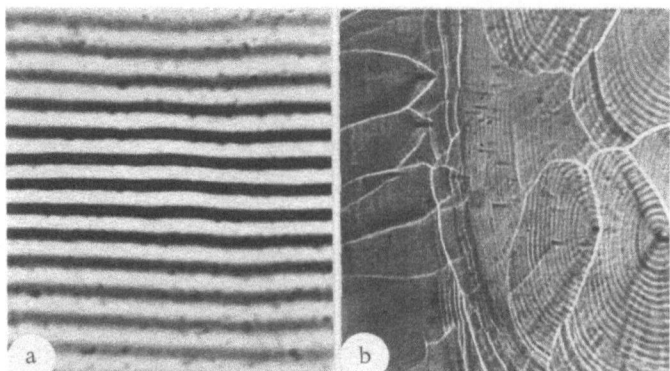

Fig. 5. Surface of a layer of (711) orientation.
a) Interferogram, ×580; b) facetted (100) face
on (711) orientation, ×280.

The number of growth pyramids was maximal on parts of the specimens closest to the
GaAs source, and the number fell away from it; this is true for specimens of all orientations
that gave growth figures.

Film Structure Perfection

The general perfection was examined via the integral intensity of the x-ray at the K
absorption edge of gallium in the Bragg diffraction position [9]. Table 1 gives these discontinuities in the integral intensity for (111)A, (111)B [9], and also (100) (calculated by the method
of [9]). Let F be the ratio of the discontinuities in the integral intensity for the epitaxial film
and perfect single crystals; then we have a quantity characterizing the degree of perfection of
the layer, which may be called the structural perfection factor SPF. Unity on this scale corresponds to an ideal single crystal, while 3.05 corresponds to an ideal mosaic structure (Fig. 7).
Figure 7 shows that this factor for films of (111)A orientation is the same as for a single crystal; small-angle disorientation (up to 6°) makes for a somewhat worse factor, but the increase
is slight. At the same time there is an increase in the dislocation density by a factor of 2-3.
Films of (111)B orientation contain numerous growth figures related to twins, so the factors
for them are substantially larger than unity and vary little in the angular range 0-6°.

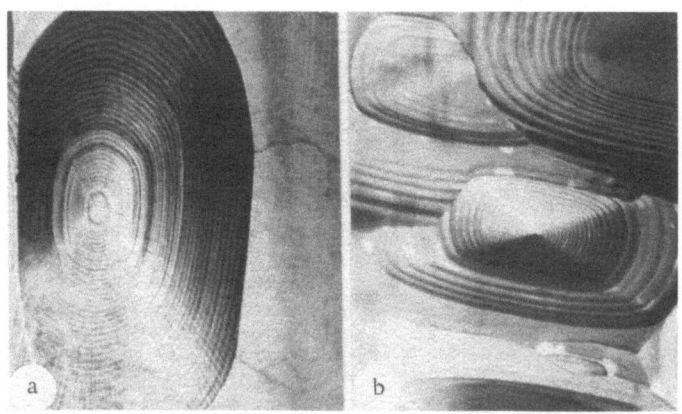

Fig. 6. Growth figures in (100). a) Low supersaturation, ×280; b) high supersaturation, ×70.

TABLE 1. Calculated Values of the Integral-Intensity Step

| Orientation | Step in integral x-ray intensity | | Growth parameters deduced from | | | |
| | | | anisotropy in growth rate | | loss of growth figures | |
	ideal single crystal	ideal mosaic crystal	angle (deg) suppressing spontaneous nucleation	λ_s, Å	angle (deg) for loss of growth figures	λ_s, Å
(111) A	1.44	4.42	3—5	20—30	2	~50
(111) B	1.08	3.29	15—20	3—5	8—12	6—10
(100)	1.23	3.75	10—15	4—7	6—9	8—12

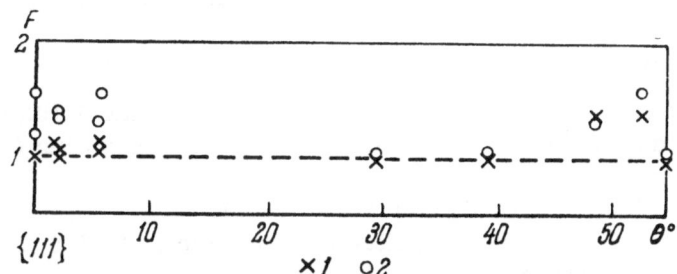

Fig. 7. Variation in layer perfection factor in relation to orientation. a) A-type planes; 2) B-type planes.

Films of {100} orientation also had factors equal to unity; disorientation in this case at first caused the factor to rise and then gradually to fall. The factors for films of {311} and {511} orientations usually corresponded to single crystals of this orientation.

Discussion

We have [10, 11] considered the growth anisotropy of epitaxial GeAs films for the crystallographic range (111)A — (100) — (111)B; we assumed that the (111A), (111)B, (100) singular bases grow by nucleation and subsequent motion of closed steps, while nonsingular surfaces of {hkk} type grow by displacement of rectilinear parallel steps arising from deviation from a singular face, while vicinal faces grow via a mixed mechanism.

Under these conditions we determined the deviations from the singular faces that completely suppressed nucleation in the parts between the rectilinear steps (Table 1). We assumed that we have $y_0 = 2\lambda_s$ for these surfaces, where y_0 is the distance between steps and λ_s is the diffusion length for adsorbed atoms, from which we estimated λ_s. The calculations on y_0 and λ_s were performed without allowance for step coalescence.

The singular faces characteristically give smooth surfaces with isolated growth figures; nonsingular faces with rectilinear parallel growth steps produced surfaces with slight waviness but without growth figures, while vicinal faces have surfaces bearing growth figures. We consider that growth figures occur only on surfaces for which nucleation is important; these nuclei usually arise around imperfections (microtwins, dislocation clumps, etc.). We can therefore assume that structural imperfections arise at the nucleation stage, which means that loss of the growth figures related to imperfections should occur at deviations from singular faces such that $y = 2\lambda_s$. This effect in the nucleation stage has been observed [12] for NaCl, for example. Then this assumption indicates (Table 1) that the λ_s given by morphologic data are somewhat larger than those estimated from growth rate [10, 11].

Nucleation does not produce structural defects responsible for growth figures. Simultaneously, there should be mechanical stresses and other causes of defects, such as imperfections and nonuniformities in the surface of the substrate, as well as elevated supersaturation in the crystallization medium. These factors jointly give rise to the nuclei and, in our opinion, to the growth figures.

Apart from major imperfections responsible for growth figures, one gets conditions for additional structure defects during nucleation and especially when there coexist rectilinear and closed steps, as on vicinal faces; this results in an increase in the structural perfection factor. The wavy relief characteristic of vicinal and nonsingular faces is of the same nature as the layer lines on growth figures and is due to fusion of adjacent steps.

Literature Cited

1.   N. N. Sheftal and H. A. Magomedov, Kristall und Technik, 1:193 (1966).
2.   V. Cabrera and R. V. Coleman, in: Art and Science of Growing Crystals, ed. J. J. Gilman, Wiley, New York (1963), p. 3.
3.   S. K. Tung, J. Electrochem. Soc., 112:436 (1965).
4.   A. Reisman and M. Rerkenblit, J. Electrochem. Soc., 112:315 (1965).
5.   R. C. Taylor, J. Electrochem. Soc., 114:410 (1967).
6.   A. E. Blakeslee, Trans. Metallurg. Soc. AIME, 245:577 (1969).
7.   F. A. Pizzarello, J. Electrochem. Soc., 110:1059 (1963).
8.   L. G. Lavrent'eva and Yu. G. Kataev, Izv. VUZ, Fizika, No. 11, 159 (1969).
9.   H. Cole and N. R. Stemple, J. Appl. Phys., 33:2227 (1962).
10.   L. G. Lavrent'eva and M. P. Yakubenya, in: Gallium Arsenide [in Russian], No. 2, Izd. Tomsk. Univ., Tomsk (1969), p. 40.
11.   L. G. Lavrent'eva, V. A. Moskovkin, and M. P. Yakubenya, Izv. VUZ, Fizika, No. 3, 63 (1970).
12.   H. Bethge, K. W. Keller, and E. Ziegler, J. Crystal Growth, 3-4:184 (1968).

FILMS OF GALLIUM ARSENIDE ON FLUORITE

N. N. Magomedov and R. N. Sheftal'

Considerable advances have been made in producing films of gallium arsenide by gas reaction, especially autoepitaxial films and films on germanium [1-3]. However, the most promising films are ones on insulators such as fluorite CaF_2, as the lattice parameters differ by less than 5% and the lattice symmetries are similar. Our tests on crystallization of GaAs were done by the open hydrogen-iodide method with a vertical two-zone oven [1], with the source and substrate temperatures monitored by chromel—alumel couples. Particular attention was given to stabilizing the substrate temperature, because this largely controls the film quality. The range was 350 to 650°C, while the iodine evaporator was operated between 35 and 55°C. The hydrogen flow rate was 10 liter/hr. The source zone contained large pieces of GaAs cut from an n-type single crystal with a specific resistance of $2.6 \cdot 10^6$ ohm-cm. We examined the film perfection in relation to substrate temperature, source temperature, and partial pressure of iodine vapor. The substrates were of {111} orientation and were made by cleaving immediately before use.

Each substrate was heated to 450°C in hydrogen before the GaAs was deposited. The film thickness and growth rate were recorded with an interference microscope.

Films deposited at 350-420°C were polycrystalline, with random grain orientations. At 450-550°C (Fig. 1) we obtained single-crystal films with the orientation $\{111\}_{GaF_2} \parallel \{111\}_{GaAs}$. Higher temperatures gave isolated crystals with various heights and perimeters. The film perfection was very much dependent on the state of the substrate surface; scratches and other damage resulted in dislocations and other defects.

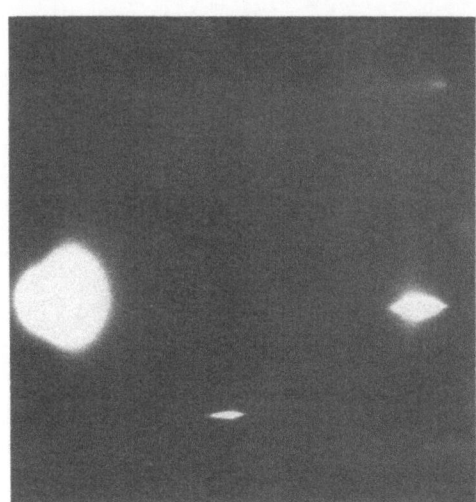

Fig. 1. Electron-diffraction pattern in reflection from a single-crystal gallium arsenide film on fluorite.

We have thus determined the temperatures needed to produce single-crystal gallium arsenide films on fluorite.

Literature Cited

1. Kh. A. Magomedov, "Crystallization of epitaxial gallium arsenide films," Thesis, Institute of Crystallography, Academy of Sciences of the USSR, Moscow (1960).
2. J. A. Amick, RCA Rev., 24:555 (1963).
3. T. Gavor, J. Electrochem. Soc., 111:821 (1964).

VAPOR GROWTH OF LANTHANUM AND NEODYMIUM SULFIDE CRYSTALS

S. I. Uspenskaya, A. A. Eliseev,
and A. A. Fedorov

The high melting points and the various polymorphic forms make it difficult to grow sulfides of the rare earths from melts; x-ray evidence on polycrystalline materials is often erroneous, and the atomic structures of these compounds have not been examined.

The present study was designed to provide single crystals of various phases and modifications, and also to elucidate the conditions for gas-transport reactions, which can affect the habit of the growing crystal. In the second case, we elucidate the scope for growing sulfide crystals of given crystallographic orientation and also the observed anisotropy in the physical properties of rare-earth chalcogenides.

We made single crystals of NdS_{2+x}, the disulfides of lanthanum and neodymium MeS_2, and the α and β modifications of the sesquisulfides Me_2S_3. The initial materials were mixtures of pure lanthanum or neodymium with sulfur, while the carrier gas was iodine of V-5 grade. Table 1 gives the best conditions for the mixture composition and temperature gradient as found by experiment. If the size of the tube is suitable (diameter 15-18 mm, length 150-170 mm) and if the transporter gas has an appropriate concentration (5-6 mg/cm^3), the material is transferred in the diffusion way, and then one gets euhedral crystals of the individual sulfide phases (Fig. 1).

The symmetry classes of the single crystals were established with a goniometer by analysis of stereographic projections and checking of the angles between the faces and directions [1]. The x-ray studies were made with RKOP-A and KFOR-4 cameras with Mo and Cu Kα radiations. The structures of some of the compounds were elucidated by Patterson projections and differential electron-density syntheses. The goniometric result agreed well with the space group derived from the structures (see Table 4 of [2]).

The temperature gradients and the composition of the mixture had marked effects on the crystal habit; one reason for uneven face development was the directional concentration gradient in the gas, which arises during the growth; the gas speed varies from one crystal face to another on account of these flows, and the supersaturations vary markedly near the faces, which leads to dominant growth of some particular faces.

We made single crystals of α-Nd_2S_3 with various shapes: 1) rhombic prisms elongated on [010] and [001] with (100) and (010) as basal planes, 2) needle crystals with the c axis as the principal direction of growth. The increase in transport rate caused by altering the temperature difference by 20°C altered the growth direction of the prismatic crystals from [001] to [110], but temperature differences Δt greater than 30°C led to uneven growth, which resulted in parallel

257

TABLE 1. Results

Composition	Mixture composition, at.% Me	Mixture composition, at.% S	Temperature, °C hot zone	Temperature, °C cold zone	Habit	Preferred growth direction	Dominant faces	Color	Symmetry class	System	Space group	Structure type
NdS$_{2+x}$	~40.0	~60.0	750	650	Plates	[100]	{001}, {110}, {010}	Blue	D_{2h}—mmm	Ortho-rhombic	P_{nmm}	—
NdS$_2$	33.3	66.7	800	720	truncated tetragonal pyramids	[100]	{001}, {101}	Golden-brown	D_{4h}—4/mmm	Tetragonal	$P4/nmm$	Fe$_2$As
LaS$_2$	40.0	60.0	850	780								
	40.0	60.0	750	650								
α-Nd$_2$S$_3$	50.0	50.0	900	800	Prisms	[001]	{110}, {010}, {111}					
					»	[010]	{100}, {010}, {111}					
α-Nd$_2$S$_3$	50.0	50.0	920	800	Needle crystals	[001]		Dark blue to black	D_{2h}—mmm	Ortho-rhombic	P_{nma}	Nd$_2$S$_3$
α-Nd$_2$S$_3$	40.0	60.0	910	800	Prisms	[110]	{100}, {010}, {111}					
α-Nd$_2$S$_3$	47.0	53.0	920	820	»	[110]	{100}, {010}, {111}					
β-Nd$_2$S$_3$	50.0	50.0	1050	980	Prisms Needle crystals	[100] [100]	{311}, {331}	Ruby				
β-Nd$_2$S$_3$	50.0	50.0	980	900	Crystals combining many forms	[110]	{110}, {100}, {311}		O_h—$m3m$	Cubic	Fd 3m	—
β-Nd$_2$S$_3$	50.0	50.0	970	880		[100]	{100}, {110}, {111}					
β-La$_2$S$_3$	50.0	50.0	880	800	The same	[100]	{100}, {211}	Olive				

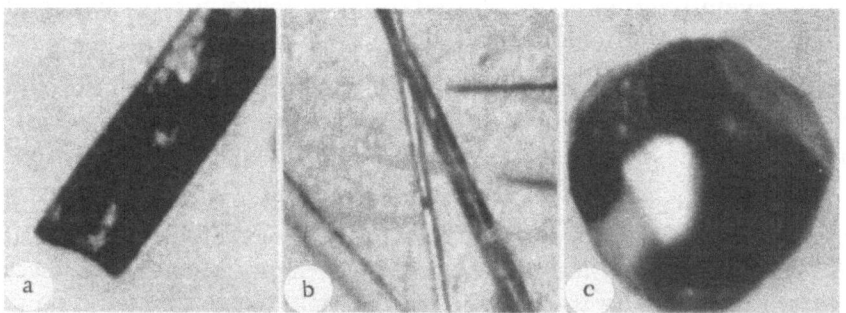

Fig. 1. Single crystals of neodymium sulfides. a) α-Nd$_2$S$_3$ (prisms); b) α-Nd$_2$S$_3$ (needles); c) β-Nd$_2$S$_3$.

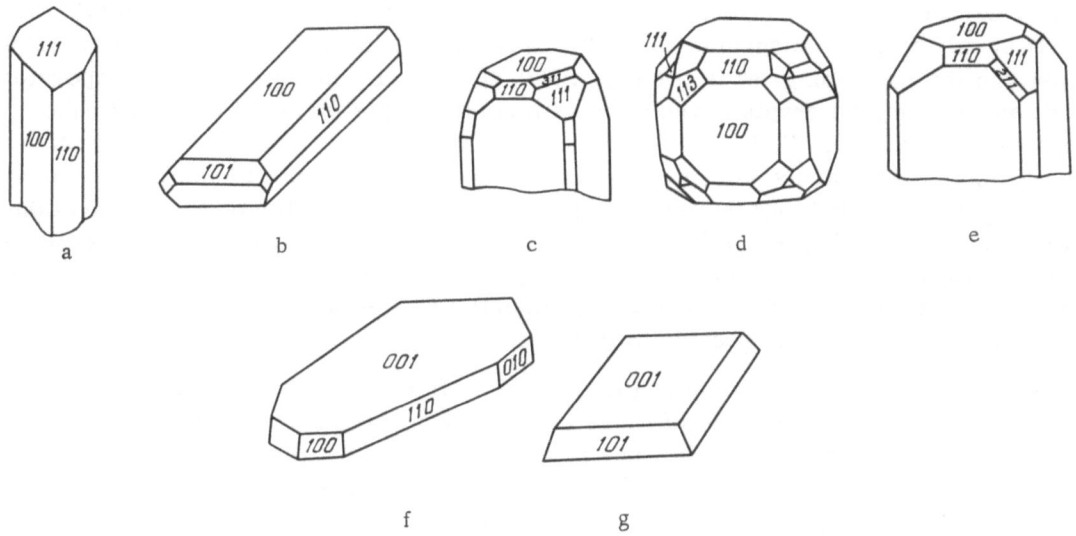

Fig. 2. Growth forms of lanthanum and neodymium sulfide crystals. a) Prisms of α form; b) crystals of β phases of sesquisulfides; c-g) crystals of disulfides and polysulfides.

Fig. 3. Morphology of single crystals of β-Nd$_2$S$_3$ in relation to crystallization temperature.

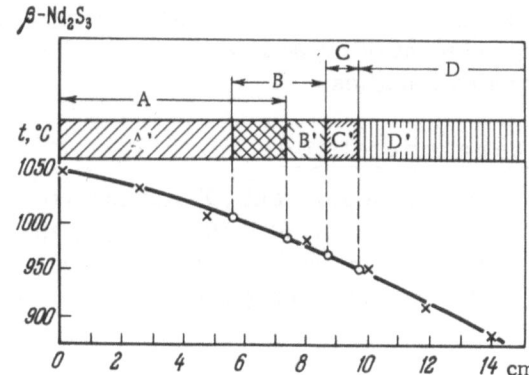

intergrowths. The crystallization range shifted towards lower temperatures as the gradient was increased and with it the diffusion rate.

The (100) face grew rapidly (Fig. 2a, b) when the proportion of sulfur in the mixture was increased to 53.0 at.% as against 50.0 at.%, while the growth rates of the (101) and the (110) faces were reduced. All the single crystals of α-Me$_2$S$_3$ lacked the (001) basopinacoid face.

Crystals of the β forms of Nd$_2$S$_3$ and La$_2$S$_3$ are euhedral polymorphic combinations (Fig. 2c-e). Fig. 3 shows schematically the relation of morphology to crystallization temperature for β-Nd$_2$S$_3$. The regions with different preferred morphologies (for instance A and B) partially overlap, while generally the boundaries are only approximate. The principal growth regions are A, B, C, and D.

Region A gives prisms with (311) and (331) basal planes, needle crystals with [100] directions, and many twins with (110) twin planes. Region B is characterized mainly by (110), (100), (111) faces, the growth directions being [110] and [100], while (311) faces are only slightly developed. Region C gives (100), (111), (110) faces, the growth direction being [100]. Region D gives (110), (111), (110) faces, with [100] as growth direction. Crystals with (311) and (331) basal faces are elongated in [100] directions, and needle crystals with [100] as principal growth direction grow in the range 980-1050°C. If the source zone temperature is reduced, one gets single crystals whose faces include simple forms of the (100) cube, (110) rhombododecahedron, and (111) octahedron. Faces of the (331) trigonotrioctahedron and (311) tetragonotrioctahedron are small or absent. As regards the simple forms of the β modification of La$_2$S$_3$, it was found that the indices of the simple form for the tetragonotrioctahedron alter in the sequence (311) → (211). Their relationship between the principal directions and the reticular density or the dominant face corresponds to the packing-density sequence of the planar grid, and this can be determined only when the atomic structure has been elucidated for the β modifications of the sesquisulfides.

Neodymium and lanthanum polysulfide and disulfide crystallize as tin plate, often with complete sets of (001), (110), (010) of faces for rhombic crystals, or (001), (101) for tetragonal (Fig. 2f, g). If the temperature difference is raised to 25-30°C, the crystals produced in the crystallization zone are thin dendrites, but the direction of preferential growth is still [100]. These crystals readily split on (001) cleavage planes and also cleave on (110). These sulfides have less tendency to produce crystals of various habits, as may be judged from the absence of forms other than plates and heavily truncated tetragonal pyramids.

Conclusions

It has been found that the habit is dependent on the mixture composition and crystallization temperature for the α and β modifications of La$_2$S$_3$ and Nd$_2$S$_3$, the disulfide MeS$_2$, and the polysulfide (LaS$_{2+x}$, NdS$_{2+x}$), i.e., the morphology is affected by the concentration of the transporter gas, and increase in the latter results in oriented flows of crystals and uncontrolled displacement of crystallization zone. The symmetry classes agree well with the results of structural studies.

Literature Cited

1.   A. I. Kitaigorodskii, X-Ray Structure Analysis [in Russian], GITTL, Moscow (1950).
2.   A. A. Eliseev, S. I. Uspenskaya, and A. A. Fedorov, Abstracts for the Seventh All-Union Symposium on the Physical Properties and Electronic Structure of Transition-Metal Alloys and Compounds [in Russian] (1969), p. 80.

POLYMORPHISM AND GROWTH OF NIOBIUM OXYBORIDE CRYSTALS

I. É. Gerasimova and G. M. Safronov

This compound was first made in the Laboratory of Inorganic Polymers, Institute of General and Inorganic Chemistry, Academy of Sciences of the USSR [1]. It is considered [2] that this type of substance can occur in several modifications in the solid state and have very interesting physical properties.

We examined the production of single crystals, which were made by distillation in a sealed tube. The amorphous powder was placed in a quartz boat at the end and the tube was evacuated to 10^{-5} mm Hg before sealing and placing in a two-zone oven, which was gradually brought up to the working temperatures. In all cases the tube volume was 32 cm^3. The tests were done at 800-1150°C in the crystallization zone with a temperature difference of 100°C. The runs lasted from 72 to 200 hr. The amount transferred was deduced from the weight change in the boat.

We obtained crystals of two morphologic types:

1. Dark-gray ones with a metallic luster bearing faces of the {101} tetragonal bipyramid and sometimes those of the {100} prism. The pyramid faces were smooth, while the prism ones had coarse striations perpendicular to the length. Conchoidal fracture. The crystals formed intergrowths and clumps. The average size was 0.5-1 mm, but some attained 2.5 mm (Fig. 1).

2. Dark gray ones with a silky luster, platy, with coarse striations parallel to the length, edges uneven, often dentate, cleavage highly perfect. The plates were not more than 0.1 mm thick. Some intergrowths consisted of twins at 90°, bunches of plates, radiated aggregates, etc. (Fig. 2).

Chemical analyses showed that both types had the formula NbOB. The two types occurred together in some runs. Preliminary x-ray data showed that these are distinct modifications:

Fig. 1. Tetragonal bipyramidal crystals of the rutile-type modification, × 4.

261

Fig. 2. Crystals of the platy modification, × 4.

Fig. 3. Intergrowth of the two crystal modifications, × 4.

1) with a rutile-type structure and $a = 13.68$ Å, $c = 5.95$ Å, $z = 28$; 2) with a layered tetragonal structure consisting of chains intersecting at right angles: ...−O−Nb−O−Nb−... and ...−B−Nb−B−Nb−..., which lie in the same plane. Here $a = 3.98$ Å, $c = 3.68$ Å (V. I. Pakhomov). The temperature limits for the existence of the two phases were not established; both occur in the entire range from 800 to 1100°C.

We used Knudsen's formula $p - p_e = \sigma \sqrt{2\pi RT/M}$, where σ is evaporation rate (g/cm²-sec) to calculate the pressure difference $\Delta p = p - p_e$, which varied from 2.6 to 14.3 · 10⁻⁵ mm Hg. The rutile-type phase was produced at Δp of 2.6-5.2 · 10⁻⁵ mm Hg, while the layered one needed higher Δp. Intermediate Δp produced the two together, with the rutile one deposited later.

Times of over 200 hr at 800-1100° caused the layered form to pass gradually into the rutile one, with a definite orientation for the new phase determined by its internal structure. The faces of the {101} pyramid were parallel to the surfaces of the plates of the layered phase, with the plate axis perpendicular to the [010] edge of the dipyramid (Fig. 3).

The crystallization of NbOB is a typical case of crystallization of a substance with a high enthalpy for transition from the vapor state to the solid one [3]. Different transport mechanisms are probably involved in the production of the two phases. The rutile-type phase grows by addition of single NbOB molecules, whereas the platy one may require addition of chain fragments.

Prolonged heating of the platy crystals may cause moelcules to evaporate followed by nucleation of the rutile phase.

Literature Cited

1. E. M. Fedneva, Yu. A. Buslaev, and V. I. Alpatova, Izv. Akad. Nauk SSSR, ser. neorg. mat., 3:1942 (1967).
2. E. M. Shustorovich, Electronic Structures of Polymer Molecules with Multiple Bonds in the Main Chain [in Russian], Nauka, Moscow (1967).
3. K. A. Jackson, J. Phys. Chem. Solids, Suppl. 1, 17 (1967).

GROWTH OF SOME SEMICONDUCTOR CRYSTALS VIA TRANSPORT REACTIONS

L. I. Bezrodnaya, N. I. Makarova, E. P. Strukova, Yu. S. Kharionovskii, and S. G. Yudin

We have examined the use of chemical transport reactions in the growth of single crystals of germanium heavily doped with elements giving deep traps, and also in producing PbTe crystals and PbTe–SnTe solid solutions.

Growth of Doped Germanium Crystals

Most of the dopes in germanium have retrograde solubility, and there is no particular difficulty in carrying out the crystallization at a temperature corresponding to the limiting solubility of the dope, so chemical transport is most convenient for producing heavily doped crystals. As doping elements we tested Au, Zn, Cd, Hg, and Te.

The single crystals were produced by growing in a closed volume in the systems Ge–Br and Ge–I. A quartz tube 1-2 cm in diameter and 13-15 cm long had a constriction to divide it into two zones, one of which contained germanium, the doping material, and the solvent component; this was evacuated to 10^{-5} mm Hg and placed in a resistance furnace having two temperature zones. The temperature in the evaporation zone was 600-960°C in accordance with the nature of the dope and the degree of doping required, while that in the crystallization zone was 600-880°C, both values being stabilized to ±5°C. The runs lasted from 1 to 5 days.

We used the theory of diffusion via a vapor [1] to calculate the transport rate for germanium as a function of the solvent component pressure for the Ge–Br system, and we confirmed this for bromine pressures from 1 to 6.5 atm. Figure 1 shows the Ge transport rate as a function of bromine pressure; the points are average experimental results from three or

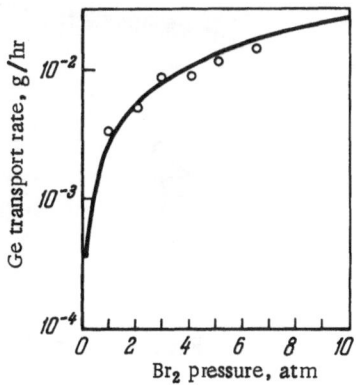

Fig. 1. Ge transport rate as a function of bromine pressure.

263

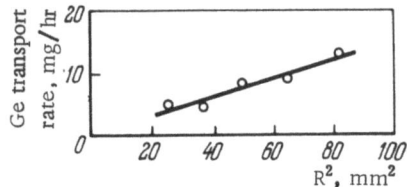

Fig. 2. Ge transport rate as a function of tube cross section.

four runs. It is clear that the diffusion state prevailed in our experiments because there was a linear dependence of the transport rate on the cross section of the reaction tube (Fig. 2). If the tube diameter was more than 2.5 cm, convection set in, which greatly increased the transport rate. The tube as a rule contained numerous crystals, whose sizes in some cases were several mm; these doped crystals resembled those described in [2, 3], and they were verified as being single crystals and the indices of the faces were determined by x-ray methods. The gold-doped crystals grew along $\langle 111 \rangle$ and took the form of hexagonal prisms. The faces of heavily doped crystals ($N_{Au} = 5 \cdot 10^{15} - 3 \cdot 10^{16}$ cm^3) were rough and had distorted growth steps.

Doping with Zn, Cd, or Hg always resulted in crystals in the form of octahedra or cube-octahedra, as well as needle and platy crystals. When the Ge−I system was used, the {111} faces of some cube octahedra showed defects in the form of growth or etch figures. Tellurium doping usually produced cubes. Parts a to c of Fig. 3 show typical doped crystals. The dislocation structure was examined only for {111} faces; no dislocations were found in whisker crystals, and they were observed only in crystals with imperfect surfaces, kinks, humps, and other defects. The dislocation densities in such specimens were as high as 10^4 cm^{-2}. Table 1 gives the electrical parameters of some of the crystals; the dope concentration determined by radioactivation analysis is higher than that calculated from the Hall effect, because the former gives the total concentration in the germanium, which includes some electrically neutral atoms.

The dope concentrations attained with gold, mercury, and tellurium approximate to the limiting solubility in germanium recorded by other methods [4-6]; crystallization at 600°C not only gave a perfect structure but also fairly high dope levels.

Most of the doped crystals had impurity photoconduction in the spectral region corresponding to the ionization energy of the dope; this was particularly so in the case of germanium crystals doped with gold to $N_{Au} = 2 \cdot 10^{16}$ cm^{-3} [7].

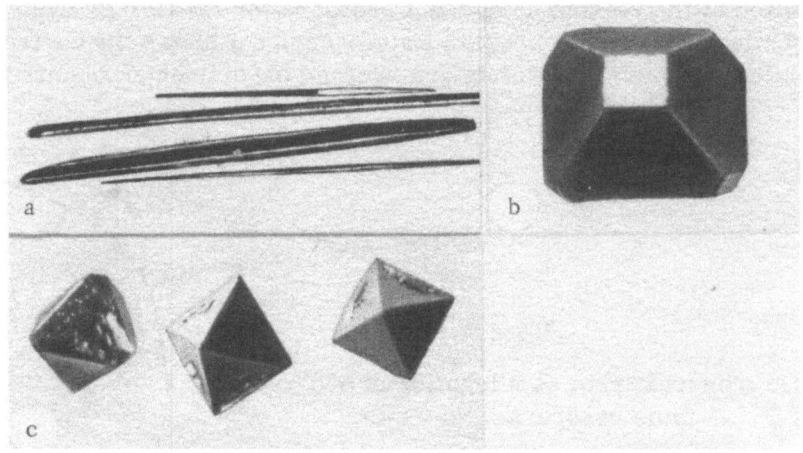

Fig. 3. Doped Ge crystals, ×10: a) Ge : Hg; b) Ge : Au;
c) Ge : Zn.

TABLE 1. Growth Conditions and Electrical Properties
for Doped Ge Crystals

Initial Ge	Pressure (atm) of solvent		Dope	Temperature, °C		Final properties			
	I_2	Br_2		evaporation zone	crystalliza-tion zone	conduction type	specific resist-ance, Ω-cm	dope core, cm^{-3}	
								from Hall effect	radioactiva-tion
p-type $\rho = 45$ ohm-cm, poly-crystal		1,5	Au	850	800	p	0.4—0.5	$1.7 \cdot 10^{16}$ $3.2 \cdot 10^{16}$	$1,5 \cdot 10^{17}$
n-type $\rho = 40$ ohm-cm, single crystal	2		Au	650	600	p	3.0	$2.0 \cdot 10^{15}$	$8 \cdot 10^{15}$
n-type $\rho = 40$ ohm-cm, single crystal	3		Zn	950	880	p	1.0—2.0		
n-type $\rho = 40$ ohm-cm, single crystal	4		Cd	950	880	p	1.5—2.5		
n-type $\rho = 40$ ohm-cm, single crystal		3	Hg	960	880	p	2.5	$3.0 \cdot 10^{15}$	$3 \cdot 10^{16}$
n-type $\rho = 40$ ohm-cm, single crystal		3	Hg	960	880	p	2.8	$1.5 \cdot 10^{15}$	$1.7 \cdot 10^{16}$
p-type $\rho = 45$ ohm-cm, poly-crystal		3	Te	650	600	n	3—3.5	$2.0 \cdot 10^{15}$	

Growth of Crystals of PbTe and PbTe — SnTe

Crystals of these compounds and solid solutions have been produced; a characteristic feature of PbTe—SnTe solid solutions is that the width of the forbidden band is dependent on the balance between the lead and tin. Crystals of these compounds are made mainly by Bridgman's method and from the vapor, but we used chemical transport in a closed volume with the system PbTe—I and PbTe—SnTe—I, with evaporation-zone temperatures of 780-890°C and crystallization temperatures of 450-610°C. The vapor pressure of the transport component (iodine) did not exceed 1.0 atm, because higher pressures resulted in extensive deposition of the intermediate compounds PbI_2, SnI_2, PbI_4, SnI_4 in the crystallization zone. The PbTe and PbTe—SnTe crystals grew as octahedra and cube-octahedra (Fig. 4) with smooth perfect faces; there were also whisker crystals with rectangular cross sections. We obtained crystals with n and p conduction types and specific resistances of 0.001-0.01 ohm-cm. By varying the balance be-

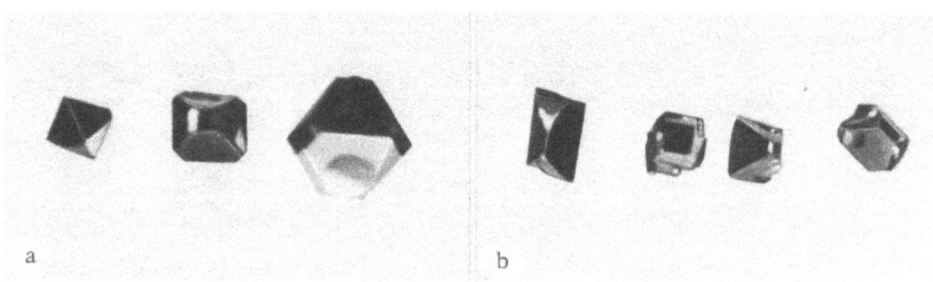

Fig. 4. Crystals of PbTe and PbTe—SnTe.

tween lead and tin in the evaporation zone we were able to vary the composition of the crystals within wide limits. In certain cases we used diffusion to prepare p—n junctions, whose voltage—current characteristics showed that the crystals were of high quality.

Conclusions

Chemical transport has been used to grow single crystals of germanium doped with materials giving deep traps, and also crystals of the semiconductor compounds PbTe and PbTe—SnTe.

The dope concentrations in the germanium crystals approximate to the limiting solubility as found by other methods.

Literature Cited

1.   R. F. Lever, J. Chem. Phys., 37:1174 (1962).
2.   A. V. Sandulova, A. I. Andrievskii, and M. I. Dronyuk, in: Growth of Crystals, Vol. 4, Consultants Bureau, New York (1966), p. 98.
3.   A. V. Sandulova, A. I. Andrievskii, and M. I. Dronyuk, in: Growth of Crystals, Vol. 4, Consultants Bureau, New York (1966), p. 101.
4.   V. M. Glazov and V. S. Zemskov, Physicochemical Principles of Semiconductor Doping [in Russian], Nauka, Moscow (1967).
5.   S. B. Borello and H. Levinstein, J. Appl. Phys., 33:2947 (1962).
6.   W. W. Tyler, J. Phys. Chem. Solids, 8:59 (1959).
7.   B. I. Beglov, Yu. S. Kharionovskii, and S. G. Yudin, Fiz. Tekh. Poluprov., 3:288 (1969).

MORPHOLOGY AND STRUCTURE OF
EPITAXIAL GaAs$_{1-x}$P$_X$ FILMS

M. A. Konstantinova, E. S. Kopeliovich,
V. N. Maslov, and R. L. Petrusevich

We have used microscopic methods to examine the structure of epitaxial films in this system grown on (111)B substrates of gallium arsenide in moist hydrogen, with the source placed close to the substrate (sandwich method) [1].

We examined the morphology of the freshly grown layers and also the etch patterns on (111)A and (111)B surfaces in various sections parallel to the plane of the substrate and at 4° to it [2]. The film morphology was very variable (Fig. 1), which was due to deviation of the substrate from (111)B and change in the conditions of crystallization; most of the films had structural features that were independent of the crystallization process, composition of the solid solution, and concentration and type of dope, as well as other such factors. The edges of the layers showed different states, and they were smoother than the central part, while the middle of the surface was the most uniform, and it occupied 70-95% of the total area. The morphologic variations indicate structural nonuniformity. The distribution of the dislocation pits is related to the growth figures (Fig. 2). The largest number of dislocation etch pits occurs in the points of junction between layers growing on different centers, e.g., region A.

The design of the reaction vessel and graphite block are amongst the possible causes of the macroscopic variation over the area of the film; there was a temperature difference between the demountable parts of the blocks, and in the gap between the source (at the bottom) and the substrate (at the top) there could be gradients perpendicular to the substrate and longitudinal and transverse in relation to the flow, i.e., the overall gradient was at a certain angle to the flow. Substrate etching and film growth on it in various places occurred at various rates.

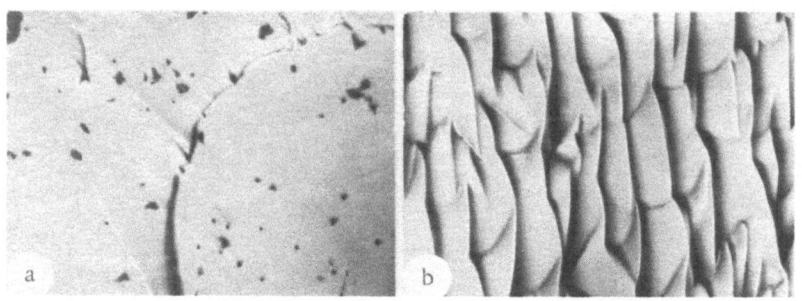

Fig. 1. Morphology (×70) of gallium phosphide films grown on the following substrates: a) (111)B; b) 3° from (111)B.

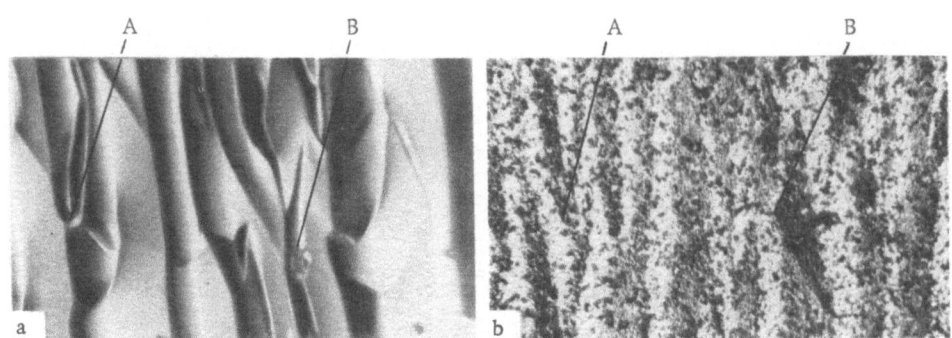

Fig. 2. Relation of morphology to dislocation etch pits, ×70. a) Freshly grown surface; b) the same after selective etching. The letters indicate identical parts.

The motion of the gas in the gap also affected the layer morphology; the screening insert was not very firmly attached to the graphite block, so the diffusion in the gap was accompanied by a directional flow of hydrogen. The suspension holes in the screen were perpendicular to the plane of the substrate and hindered the escape of gas and other material from the gap, which led to accumulation at the exit; to obtain a more uniform layer, we used an insert with a flow around an internal rim (Fig. 3). The distribution of the material in the gap was more uniform, because gas escape from the gap was facilitated. The layer produced in this case had no rim at the hydrogen inlet and outlet and was substantially more uniform in thickness.

Structure of Epitaxial Layers in Thickness

It has been observed [3, 4] that epitaxial films of gallium phosphide and solid solutions as used here are variable throughout their thickness; when cross sections are etched, one sees characteristic banding, which reflects the relief of the growth surface, and the structure becomes more perfect as the layers thicken. Our $GaAs_{1-x}P_x$ epitaxial films had structural and morphologic defects, as well as dislocations revealed by etching. The structural defects were observed in the 30–50 μm film adjoining the substrate (Fig. 4). They extended from the latter in the [111] direction. If they grew right through the layer, the defects or groups of defects terminated in growth pits. Most frequently, they did not pass right through the layer, but were overlapped by single-crystal films from adjacent regions. The side surfaces of an etch pit were bounded by {110} planes.

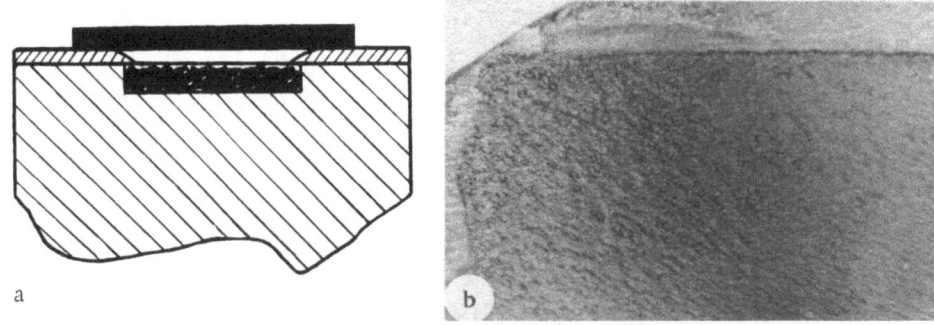

Fig. 3. a) Pyrocarbon screening layer; b) morphology of GaP film without droplets, ×6.

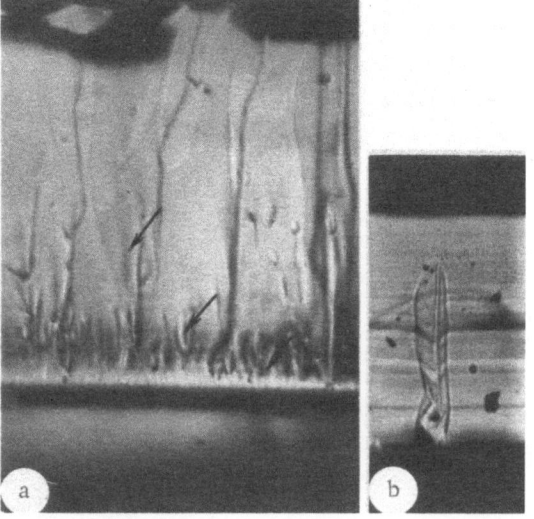

Fig. 4. Typical defects revealed by etching on cleaved layers, ×120. a) Defects in GaAs$_{0.7}$P$_{0.3}$ films; b) in GaP films. Defects indicated by arrows.

We examined the given area by successive grinding parallel to the substrate and treatment in a polishing etching agent [2], which showed that a defect has a triangular cross section, in accordance with the face symmetry. Within the triangle one could see clearly three symmetrical sections, whose boundaries were crystallographically oriented at 120° one to another (Fig. 5). The sectors could be traced down to the substrate.

The preliminary standard treatment of the substrate had a particular effect on the formation and density of the defects; plates after orientated cutting were ground with Al$_2$O$_3$ powder (M5-M7) and were polished in a mixture of 3H$_2$SO$_4$: H$_2$O$_2$: H$_2$O. Before the growth commenced, the substrates were gas etched by moist hydrogen directly in the reaction vessel. Very high growth pit densities (up to 10^4 cm^{-2}) were observed on substrates that were not etched after grinding, i.e., that had damaged surface layers. If the substrates were chemically polished but not etched in the reaction vessel, the etch pit densities were considerably lower than those of mechanically polished substrates. Finally, films on substrates prepared in the standard fashion usually had no growth pits.

The dislocation densities were highest in the part of the film adjoining the substrate; then there was a marked fall, with an approximately constant value for thickness of about 150 μm. Figure 6a shows the distribution in thickness for the dislocations in gallium phosphide films grown on substrates treated in various ways. Substrates with deformed layers (curve 1) had maximal dislocation densities; in the other cases, differences in dislocation density did not exceed a factor 3 or so.

It is clear that the general tendency for the dislocation density to fall is accompanied by peaks in certain parts of the thickness; Fig. 6b shows the relation between the dislocation density in the layer on (110) substrate (upper curve) and the growth bands, which were revealed by etching cleaved section. The dispositions of the parts were determined with an MII-4 interferometer; the growth bands are readily revealed because of change in the arsenic concentration entering the epitaxial film during the etching of the substrate at the time of growth [4, 5] and also on account of pressure change in the system [3]. The arsenic atoms replace phosphorus in the gallium phosphide and distort the lattice, which results in dislocations.

It has been found [6] that there is a relation between the growth bands and the dislocation distribution, the composition of fluctuations, and the pressure in the system. The peaks on the

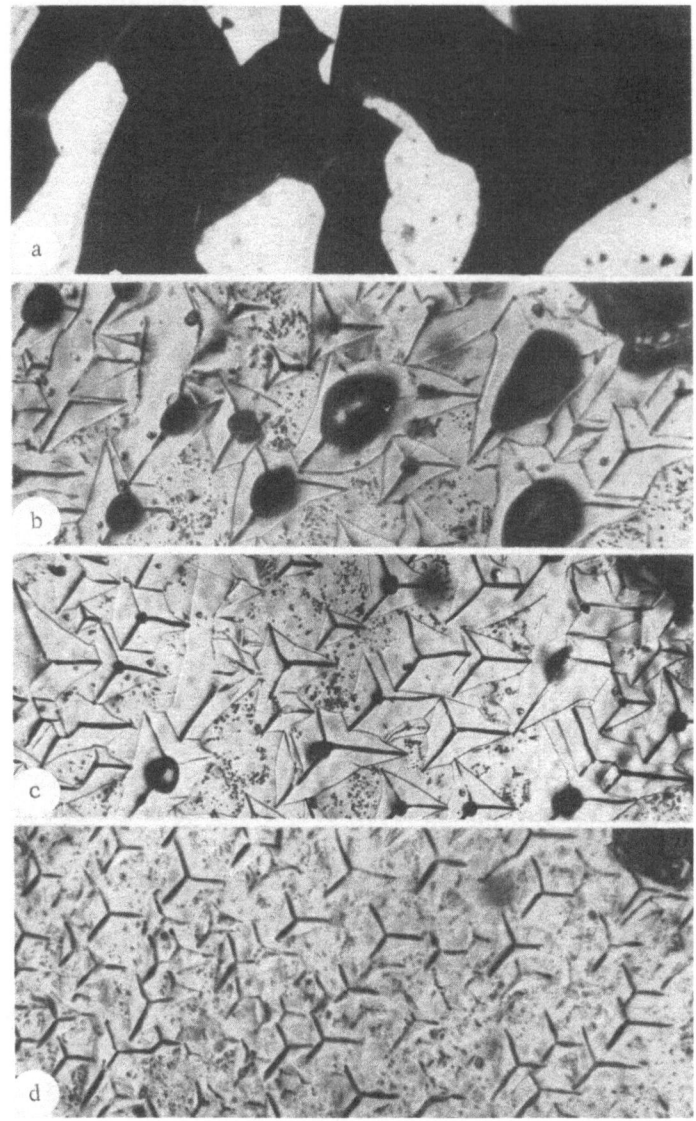

Fig. 5. Formation of growth pits, ×70. a) Fresh
760 μm film; b) the same area after cleaving 520
μm by grinding and etching; c) 380 μm; d) 300 μm.

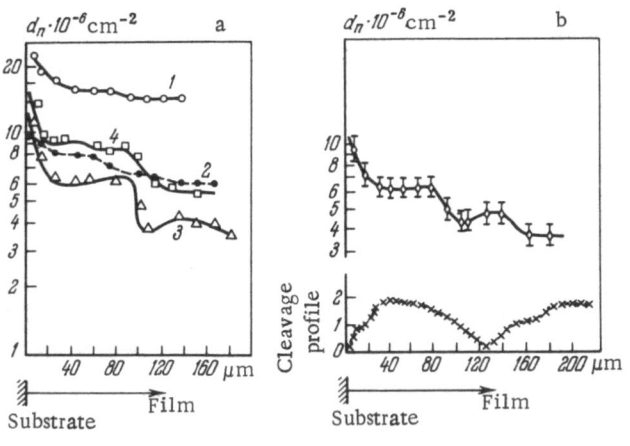

Fig. 6. a) dislocation distribution in thickness of GaP films grown on substrates prepared in various ways; b) relation in profile
and etched cleavage surface: 1) substrate
ground, without subsequent gas etching, 2)
mirror-smooth substrate after etching in
$3H_2SO_4:H_2O_2:H_2O$; 3) the same followed by
gas etching, 4) mirror facetted substrate
after chemical polishing 6 $HF:HNO_3:H_3PO_4$:
CH_3COOH without subsequent gas etching.

dislocation distribution curve correspond to rises in arsenic content to 7.8%, as against a general content of 5.5 wt.%.

These sandwich $GaAs_{1-x}P_x$ films thus show macroscopic nonuniformity, which is related to the growth conditions and to the thickness distribution of the structural defects.

Literature Cited

1. L. I. D'yakonov, V. N. Maslov, and B. A. Sakharov, Dokl. Akad. Nauk SSSR, 183:76 (1965).
2. M. A. Bendik and R. L. Petrusevich, Izv. Akad. Nauk SSSR, ser. neorg. mat., 4:1488 (1968).
3. W. G. Oldham, J. Appl. Phys., 36:2887 (1965).
4. A. V. Lishina, V. N. Maslov, R. L. Petrusevich, and T. V. Troneva, in: Growth of Crystals, Vol. 8, Consultants Bureau, New York (1969), p. 224.
5. L. I. D'yakonov and V. N. Maslov, Izv. Akad. Nauk SSSR, ser. neorg. mat., 7:1147 (1967).
6. F. A. Gimel'garb, É. S. Kopeliovich, V. N. Maslov, R. L. Petrusevich, and V. I. Fistul', Kristallografiya, 14:1104 (1969).

VAPOR GROWTH AND MORPHOLOGY OF CdO AND ZnO SINGLE CRYSTALS

G. A. Ivanov and Ya. S. Savitskaya

The literature [1-10] gives only the optimal conditions for producing mixtures of CdO and ZnO crystals of various shapes, while studies have been made [11, 12] of effect of some experimental parameters on the shape of CdO crystals. Here we consider the relation between growth conditions and morphology for CdO and ZnO single crystals in mass spontaneous crystallization and in growth on seeds. We have examined also the effects of indium In as a dope as regards the shape of ZnO crystals.

The CdO and ZnO crystals were obtained by vapor-phase reaction in a flow; the essence of the method was to oxidize the vapor of the metals as supplied to the crystallization zone by transporting gases, which gave CdO and ZnO (Fig. 1).

Figure 2 shows CdO single crystals, and a schematic representation of these has been published previously [12]. The goniometer was used to index the faces of the CdO crystals, and it was found that the side faces of needle crystals (part 1 of Fig. 2) are faces of the $\{1\bar{1}0\}$ rhombododecahedron, and they are often covered by striations, which indicate that the direction of preferential growth coincides with $\langle 111 \rangle$ direction. The vertex of a needle shows $\{100\}$ cube faces, which may develop (part 2 of Fig. 2) and form the habit of the finished crystal (part 4 of Fig. 2). If the crystallization temperature is raised, the cubic crystals acquire a skeletal form (part 5 Fig. 2). Some CdO crystals take the form of octahedra (part 6 of Fig. 2).

We related the morphology of the CdO and ZnO crystals to the growth parameter via calculation of the quantity α, which characterizes the bulk degree of supersaturation and which itself is defined by the equation [12, 13]

$$\alpha = K_p P_{Me} PO_2^{1/2}, \tag{1}$$

where K_p is the equilibrium constant of the reaction

$$Me_{gas} + {}^1/_2 O_2 \rightleftarrows MeO_{solid},$$

which can be calculated via published data [14, 15], where P_{Me} and P_{O_2} are the partial pressures of Me and O_2 under the working conditions. The α for the various conditions of spontaneous crystallization or growth on seeds may be compared with the morphology of the CdO crystals to show that in both cases the shape changes from needles bearing rhombododecahedron faces to octahedron and cubed forms as α decreases, i.e., the habit changes in the direction $\{110\} \rightarrow \{111\} \rightarrow \{100\}$. There are substantial differences in the α for identical crystal shapes under conditions of mass spontaneous crystallization and growth on seeds (the α for cubes are 10^4-10^5 and 4-5, while for prisms they are 10^5-10^7 and about 10^2); this appears

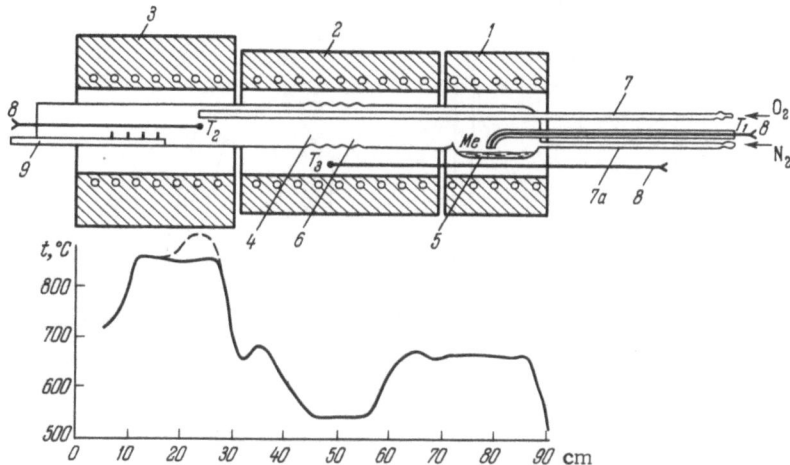

Fig. 1. Apparatus for growing single crystals of CdO and
ZnO, and temperature distribution in the ovens. 1) Element
for heating metal; 2) for crystallization zone; 3) for zone of
partial condensation; 4) quartz reaction vessel; 5) recess for
metal; 6) zone of partial metal condensation; 7, 7a) quartz
gas tubes; 8) PP and KhA thermocouples; 9) alundum tube
with CdO seeds.

to arise because in the first case the number of growing crystals is very large, as is the total
growth surface, while the actual supersaturation at the surface of each such crystal is probably rather different from the α as calculated above. There are major difficulties in determining the actual surface supersaturation, especially because the CdO and ZnO crystals are
not stoichiometric.

For ZnO we examined in the main the growth under conditions of bulk crystallization;
Fig. 3 shows schematically the crystals that were examined by x-ray, goniometry, and polarization microscopy. The needles were 10-15 mm long with thicknesses of 1 and 2 mm respectively, while the plates had areas of 0.5 cm^2. From (1) we calculated α for the various crys-

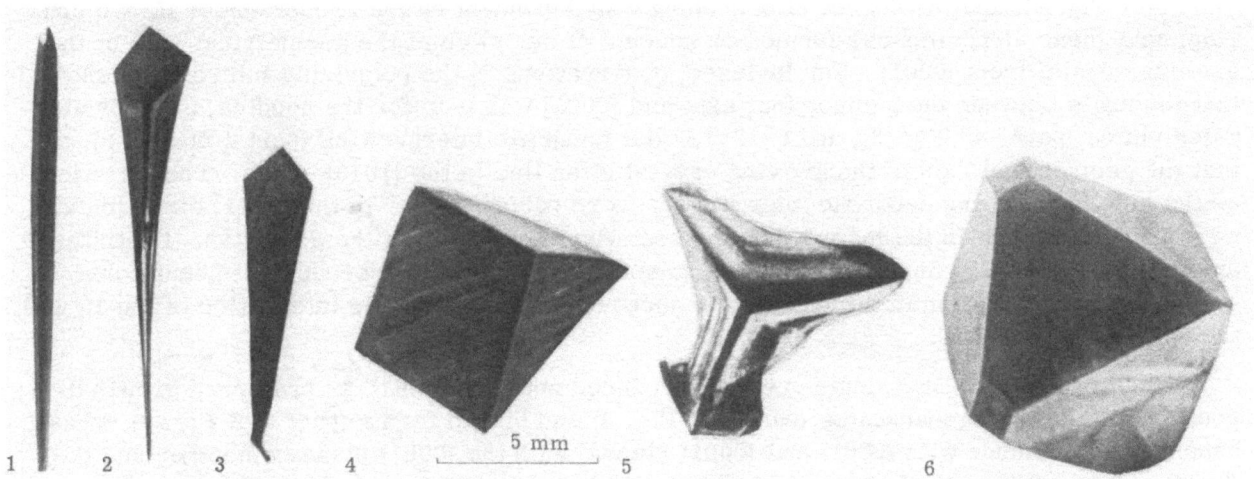

Fig. 2. CdO crystals. 1) Prismatic six-faced needle; 2) pyramidal needle; 3) needle
bearing elongated plate; 4) cube; 5) skeletal crystal; 6) octahedron.

Undoped ZnO

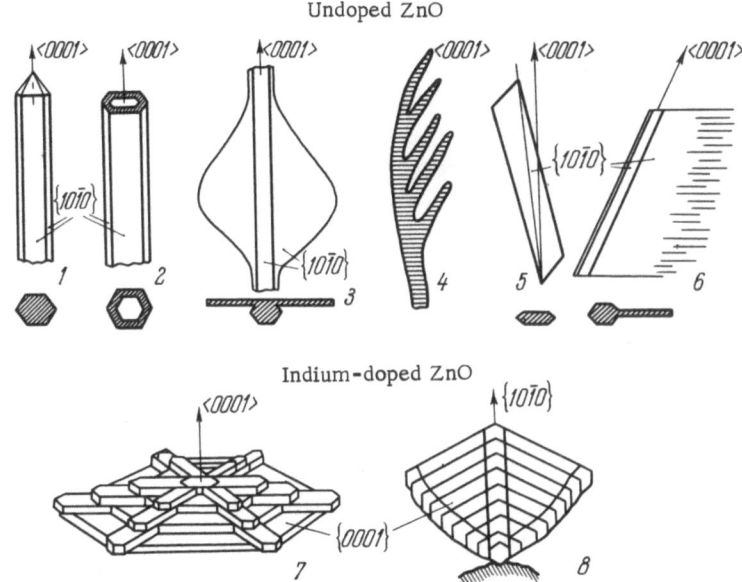

Indium-doped ZnO

Fig. 3. Schematic representation of ZnO crystals. 1) Prismatic six-faced needle; 2) hollow needle; 3) plate grown from needle; 4) dendritic intergrowth; 5) platy crystal; 6) comb; 7) hexagonal plate; 8) fan.

tallization conditions for ZnO and found that it was related to the morphology, namely reduction in α from 10^3 to 10 represented a fall in the relative growth rate along [0001]. The $\langle 111 \rangle$ direction for CdO and the $\langle 0001 \rangle$ for ZnO had alternation of layers of metal and oxygen; the preferred shape of the ZnO crystals remained that of a prismatic six-faced needle (part 1 of Fig. 3), whose side faces were indexed as $\{10\bar{1}0\}$, while the vertex took the form of a pyramid whose faces could not be indexed. Often the needles had lateral platy overgrowths (part 3 of Fig. 3). If the crystallization was conducted at 1400-1450°C, we obtained hollow ZnO prisms (part 2 of Figure 3), which indicate skeletal growth, which agrees well with results on CdO.

We examined the effects of transport gas composition (N_2 or a mixture of N_2 with H_2) on the shape of ZnO crystals and found that many dendrites were produced when H_2 was employed (part 4 of Fig. 3), and the degree of branching was dependent on the proportion of H_2; we consider that these dendrites are formed on account of deviation of the geometrical axis of the growing crystal from [0001]. For instance, observations in the polarizing microscope showed that the angle between the geometrical axis and [0001] was 0-2° for the needles, 3-6° for elongated plates (part 5 of Fig. 3), and 6-12° for flat dendritic intergrowths (part 4 of Fig. 3), and that the geometrical axis of the growing crystal often lies in the $\{10\bar{1}0\}$ plane. The striations on the side faces of the dendritic intergrowths were perpendicular to the [0001] direction and were parallel to this in the main trunk and secondary outgrowth of the dendrites. Dentrites are formed with ZnO when H_2 is used as transporting gas probably because of nonuniform temperature distribution in the medium on account of the exothermic interaction of the H_2 with the O_2 and also the ZnO.

Indium doping of ZnO single crystals produced not only $\langle 0001 \rangle$ as preferred growth direction but also hexagonal plates (part 7 of Fig. 3) and bladed forms (part 8 of Fig. 3), whose basal planes coincide with $(000\bar{1})$ and (0001) planes, with the $(000\bar{1})$ planes smoother and the (0001) planes consisting of steps. The difference between $(000\bar{1})$ and (0001) has been observed for other $A^{II}B^{VI}$ crystals with a wurtzite lattice, and is due to polarity of these compounds [16, 17].

Literature Cited

1. S. van Hauten, Nature, 195:484 (1962).
2. R. H. Fahig, J. Appl. Phys., 34:234 (1963).
3. J. G. Marinace, IBN J. Res. Devel., 4:248 (1960).
4. S. Mayashi, Oyo Butsuri, 37:825 (1965).
5. E. Scharowsky, Z. Phys., 135:318 (1953).
6. G. Bogner and E. Mollwo, J. Phys. Chem. Solids, 6:136 (1958).
7. I. T. Drapak, Nauch. Zap. Chernovits. Univ., No. 53, 88 (1961).
8. I. W. Nielson and E. E. Dearborn, J. Phys. Chem., 64:1762 (1960).
9. R. A. Laudise, E. D. Kold, and A. J. Caparaso, J. Amer. Ceram. Soc., 47:9 (1964).
10. I. N. Kuz'mina and A. F. Antonova, in: Growth of Crystals, Vol. 4, Consultants Bureau, New York (1966), p. 125.
11. G. A. Ivanov and Ya. S. Savitskaya, in: Semiconductor Physics [in Russian] Izd. AN SSSR, Novosibirsk (1968), p. 49.
12. G. A. Ivanov, Ya. S. Savitskaya, and L. A. Velikzhanina, Neorg. Mat., 5:1915 (1964).
13. F. A. Kuznetsov, Yu. G. Siderov, and I. K. Maranchuk, Fiz. Tverd. Tela, 6(10):2981 (1964).
14. Ya. S. Gerasimov, A. N. Krestovkikov, and A. S. Shakhov, Chemical Thermodynamics in Nonferrous Metallurgy [in Russian], Vol. 1, Metallurgizdat (1960); Vol. 4 (1966).
15. D. R. Stull and G. Sinke, Thermodynamic Properties of the Elements, Washington, 1956.
16. G. A. Wolff, Z. Phys. Chem., 31:1 (1962).
17. A. A. Simanovskii, Abstracts for the Second All-Union Symposium on the Growth of Crystals and Films [in Russian], Novosibirsk (1969).

CONTROLLED GROWTH OF ORIENTED SYSTEMS
OF WHISKER CRYSTALS

E. I. Givargizov and Yu. G. Kostyuk

Until recently, there has been much uncertainty about the growth mechanisms of whisker crystals; the growth has been largely uncontrollable. However, the VLS mechanism was recently discovered [1-4], and it has been described fully in a series of papers [5-7], which provide a detailed description of many aspects of the growth of whiskers and can serve as the basis for controlled growth.

Here we report some characteristics of whisker crystals grown on single-crystal substrates by the VLS mechanism; we have examined the characteristics of the oriented crystal systems and the detailed structure of the individual crystals. The main results relate to silicon whiskers on a silicon substrate, but some results are given also for germanium on germanium.

Experimental Method

The crystallization was performed in a vertical quartz vessel [8], as used in epitaxial technology. The heating was by induction. We used the chloride−hydrogen process in a flow system. For silicon on silicon, the temperature was 900-1050°C, most often 1000°C. The H_2 flow rate was 60 liter/hr, and the $SiCl_4$ concentration in the H_2 was 0.01-0.02. The germanium was crystallized at 700-800°C (750°C) and a $GeCl_4$ concentration in the H_2 of about 0.01. The systems of silicon whiskers were grown on (111) silicon plates, which were first ground, mechanically polished, and chemically polished. The germanium crystals were grown on (111) germanium.

The solvent material was usually gold, but we also used other metals such as silver, indium, gallium, and aluminum; the gold was deposited on the substrate in a continuous film 1000 Å thick. We used two methods of depositing the gold: electrolytic from acetamide solution and a vacuum (evaporation with preliminary ionic etching of the substrate or without this). When the temperature was raised above the eutectic point, the gold formed a liquid alloy with the substrate, and this split up into islands in response to surface tension, which then served as basis for whiskers. These islands varied in size with the melting conditions and the local properties of the substrate: from ones just visible in the microscope to 5-10 μm in size. When the substrate had been kept at the crystallization temperature for 3-5 min, the growth proper began, and this usually lasted 10-20 min.

The systems of points grown on the substrates were examined to determine the uniformity and filling density in individual areas; we also determine the main crystallographic growth directions (for the principal systems) and the relative importance of each, together with the uniformity of the whiskers as regards height and relative growth rate in relation to cross

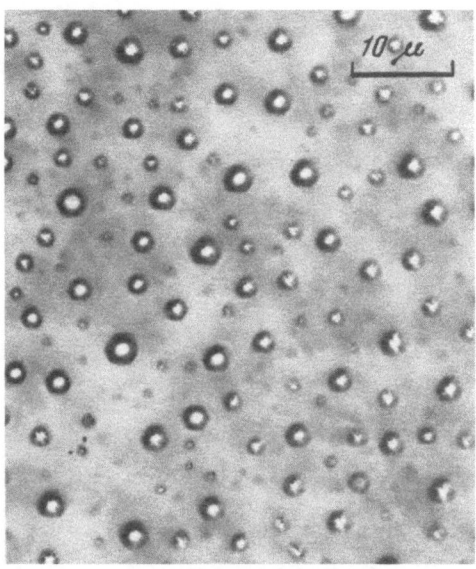

Fig. 1. Si whisker crystals perpendicular to Si substrate (photomicrograph).

section. In addition, we examined the form and facets of the individual crystals. These measurements were made by optical microscopy, shadow electron microscopy, scanning electron microscopy, and light figures; for this last purpose we made a special goniometer.

Results and Discussion

The growth rate of a silicon whisker is dependent on the crystallization temperature, the concentration of the $SiCl_4$ in the H_2, and the crystal diameter; the growth rate was 1–3 μm/min under typical conditions, so times of 10–20 min gave crystals 10–60 μm high. Figures 1 to 9 give photomicrographs of such systems.

The features of these crystals are such that it was essential to combine the various microscopic methods in order to draw definite conclusions.

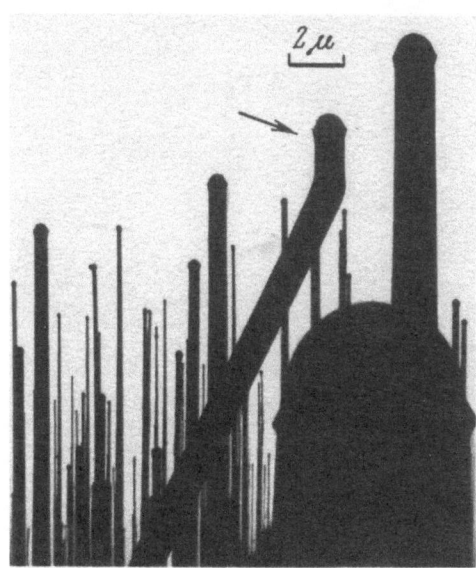

Fig. 2. Shadow electron micrograph of Si whiskers grown on a (111) substrate.

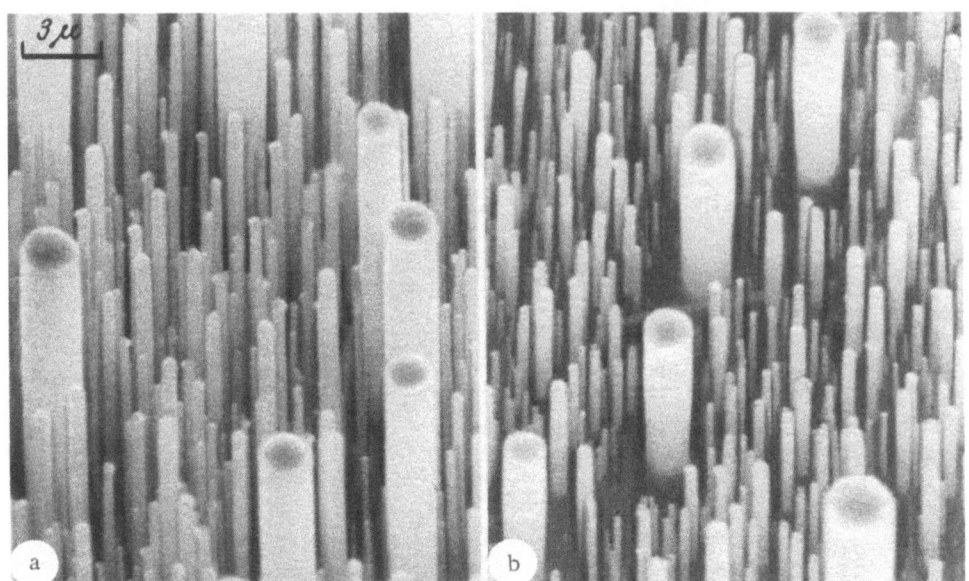

Fig. 3. Si whisker crystals seen in the scanning electron microscope with angles of incidence of (a) 45° and (b) 22°30'.

The uniformity of substrate covering was determined by optical and scanning electron microscopy at low magnifications (50–100). The value was dependent on the local properties of the substrate and the thickness of the initial gold film; the thinner the film, the more uniform the whisker system. Figure 1 gives some idea of the uniformity, where the microscope was focused on the vertices of the crystals, and one can see light patches of the crystallized silicon—gold alloy. The whisker density here was about 10^6 cm^{-2}. Away from the focal plane there are the vertices of thinner crystals, which grow more slowly.

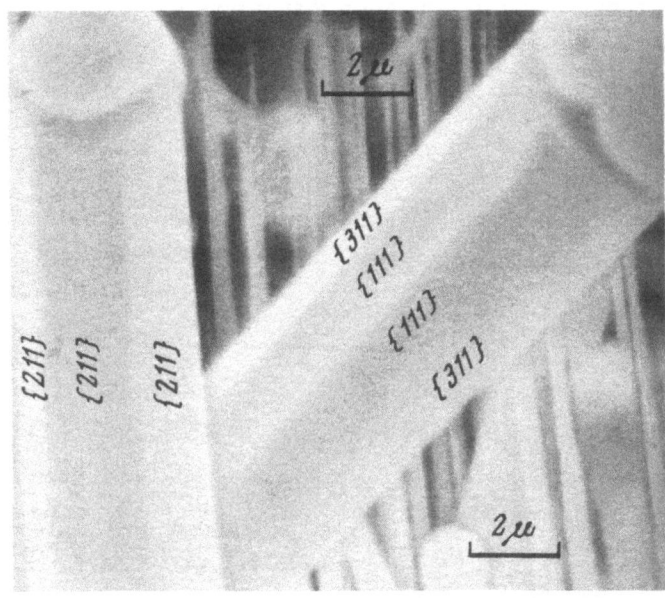

Fig. 4. Faces of Si whiskers growing along [111] and ⟨110⟩ as seen in the scanning electron microscope.

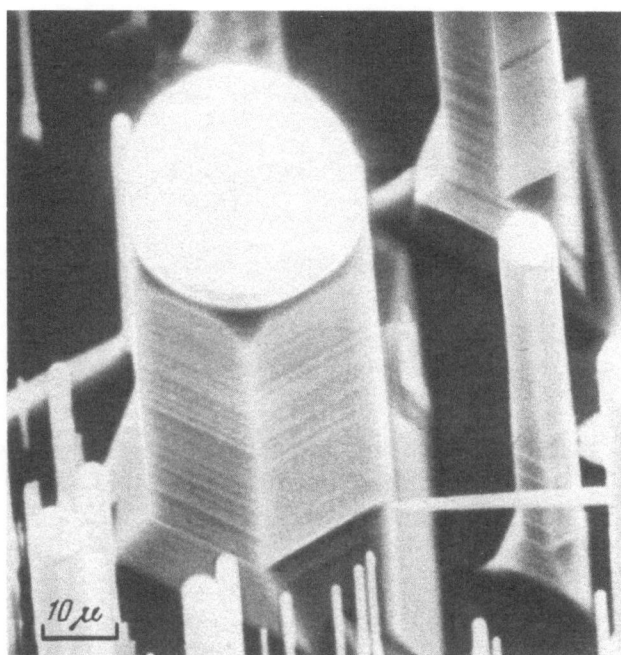

Fig. 5. Facets and traces of growth layers on sides of Si crystals as seen in the scanning electron microscope.

The main growth directions of the crystals of silicon grown by gold are $\langle 111 \rangle$, $\langle 110 \rangle$, and $\langle 211 \rangle$ [2-4], which continue the corresponding directions in the substrate. The goniometer showed that the main systems in our case grew on $\langle 111 \rangle$ and $\langle 110 \rangle$. Crystals perpendicular to the substrate ([111]) were found in all the specimens, as were the three $\langle 110 \rangle$ systems, which form an angle of 70°32' with the first. These systems gave in the goniometer reflections that were similar in intensity, so the thicknesses were roughly equal. The crystals grew in $\langle 110 \rangle$ directions as a rule from ones initially growing on [111]; if circumstances were favorable, they straightened out, as shown in Fig. 2 by the arrow, and continued to grow on [111].

The results on the numerous specimens show that most of the whiskers (at least 95%) grew perpendicular to the (111) silicon substrate in our cases; the others grew mainly in the

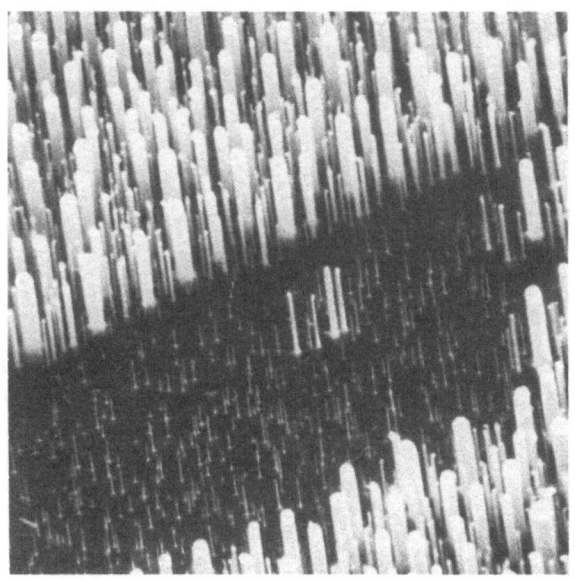

Fig. 6. Si whiskers seen from side, with gap.

Fig. 7. Whiskers of various diameters grown
under identical conditions of rapid supply.

$\langle 110 \rangle$ directions, and there were only a few in the $\langle \bar{1}11 \rangle$ directions, which form an angle of
19°28' with the substrate; we found no [211] strip crystals in our specimens. The growth was
preferentially perpendicular to the substrate because inclined crystals are in unfavorable
supply conditions from the start and are readily suppressed.

There is a close connection between the u n i f o r m i t y i n d i a m e t e r a n d i n h e i g h t;
the thicker crystals grew more rapidly than the thin ones, and some conclusions can be drawn
on the crystallization mechanism from this. The growth rate of a whisker in general is determined by several successive states:

1) supply of material via the gas;
2) deposition on the surface of the drop;

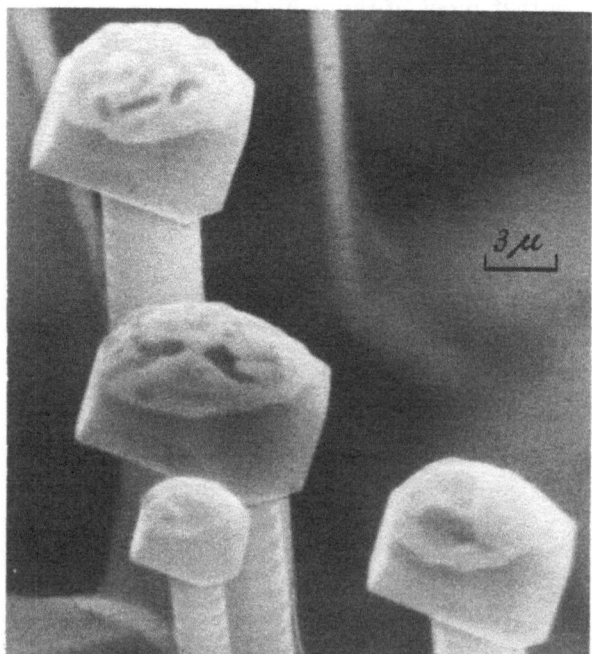

Fig. 8. Ge whiskers grown in (111) substrate
as seen in the scanning electron microscope.

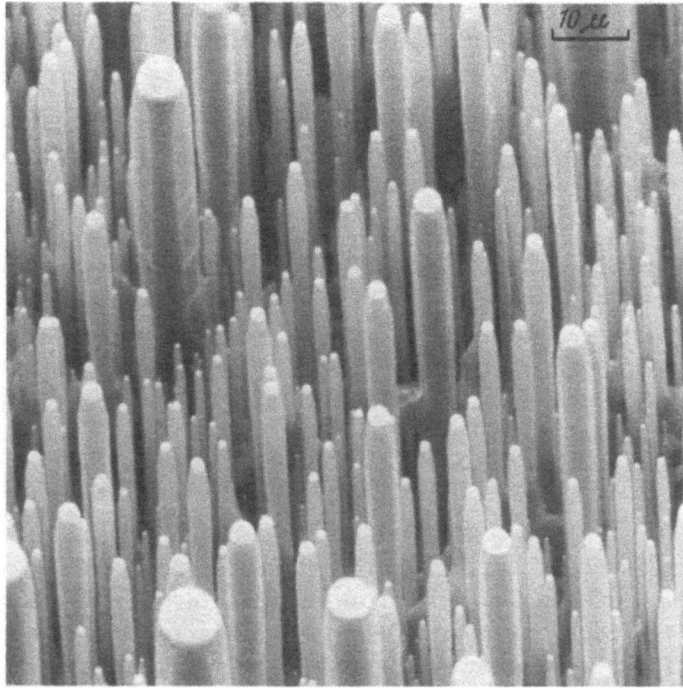

Fig. 9. Si whiskers deliberately given a conical
shape.

3) diffusion of the solute via the liquid;
4) deposition at the crystallization front.

If stage 3 is the slowest one, the thicker crystals, which have the higher hats, would lag appreciably behind the thin ones; however, Figure 1 and Fig. 3b show that thick crystals grow at least no more slowly than thin ones. Moreover, a detailed study in the optical microscope showed that the growth rate actually increases with the crystal diameter, so stage 3 was not the rate-limiting one in our experiment.

Consider stage 2. The ratio of the liquid−vapor area to the liquid−crystal area is much the same for crystals of all diameters (the contact angles of the caps on the crystals of various diameters are roughly equal, Fig. 2). Diffusion through the liquid does not restrict the crystal growth; if stage 2 were the rate-limiting one, all the crystals, whether thin or thick, should have the same height, which was not the case, so stage 2 is not the rate-limiting step.

Consider now stages 1 and 3, which can be invoked to explain the observed differences in the growth rates, but the rate-limiting mechanism differs as between the two. First of all we consider the possible role of supply through the vapor. Consider two parts of the substrate, on which the original liquid film has split up randomly into different numbers of islands. We assume that the gold film was uniform in thickness, and then the volume of a drop is proportional to the cube of the diameter, while the area of the crystallization front is proportional to the square of the diameter, though an area with smaller islands has a relatively large crystallization front area, and so a given area receives a relatively little material when there is a restricted supply from the vapor and the growth rate should be less. Figure 3b clearly shows crystallization gaps around thick crystals, and if there are any thin crystals near a thick one, they scarcely grow at all.

Also, these conditions may also involve an additional mechanism of positive feedback type, which tends to accentuate the inequality of the thick and thin crystals: if the thick crystals begin to run ahead during growth, they subsequently take up more of the supplied material.

Then competition between crystals differing in diameter can cause differences in height when there is a restricted supply from the vapor; this effect is characteristic of whisker growth under diffusion-controlled conditions.

Consider now the crystallization proper in stage 4. To determine the effect here, we provided relatively rapid supply so that the vapor stage was not the rate-limiting step. To this end, the gold was removed locally from the substrate, but leaving a little on the bared part. Then during growth one got a gap (Fig. 6) in which the supply was generous. Under these conditions of generous supply, the thin crystals grew much more slowly then the thick ones (Fig. 7). This means that it is not the gas supply stage that restricts the growth but the small size of the growing crystals.

There are at least two reasons that can be invoked to explain the reduced growth rate of thin crystals:

1. There are reasons for supposing that whiskers grow by a layer mechanism, i.e., via two-dimensional nucleation; this is clear from many features, especially the prominent tendency to grow on the close-packed (111) face, which is characteristic of crystals with a diamond lattice growing by a layer mechanism. Even when the whiskers grew in $\langle 110 \rangle$ directions, the crystallization front was bounded by two $\{111\}$ planes, one perpendicular to the substrate and the other inclined at 19°20'. The bundles of growth layers are clearly visible on the sides of the whiskers, as in Fig. 5.

We may suppose that the small whisker diameters characteristic of our experiments cause there to be an effect from the nucleation frequency as a function of the liquid—crystal interface area; the less the crystal diameter, the lower the probability of two-dimensional nucleation there, this being proportional to the square of the diameter. This tendency should be particularly prominent in the thinnest crystals. We observed crystals of diameter ≤ 300 Å, and Fig. 2 shows that these are largely transparent to the electron beam with an accelerating voltage of 50 kV.

The ratio of the cross-section area to the height for a whisker is a characteristic of this process; we measured these parameters and related the one to the other to determine the critical diameter below which whiskers grow by one-nuclear mechanism, and also the nucleation frequency for this growth region.

2. For whiskers of submicron size one should apply the Thomson—Gibbs formula, which relates the minimum crystal size to the supersaturation; this limiting factor has been pointed out previously [2]. We plotted the growth rate as a function of radius and thus determined the minimum radius, from which one can calculate the saturation.

It is at present uncertain which of these factors is the decisive one; we are at present processing quantitative results from which to draw a final conclusion and also to provide quantities that cannot be measured directly.

All the illustrations indicate the details of the individual crystals, and it is clear that the top of a whisker has a peculiar cap, which consists of two parts: a layer of silicon deposited from the melt in accordance with the phase diagram after the VLS process has ended [7], together with a spherical cap consisting of a mixture of crystals of silicon and gold.

The facets on silicon whiskers have been described in detail already [2-4]; here we note that one usually sees six side faces, which differ somewhat in nature, because they form three pairs of identical faces, clear in the optical microscope (Fig. 1), and even more prominent in the electron micrographs (Fig. 5), which corresponds to the observation [9] on the difference between ⟨1$\bar{1}$0⟩ directions at (111) surfaces of crystals with diamond lattices. The faces were identified by light figures in the goniometer, and it was found that the side surfaces of crystals growing along [111] make up six {211} faces, in accordance with [2-4], whereas inclined crystals growing in ⟨110⟩ directions have four {111} faces and four {311} ones. Figure 4 shows the indices of the faces.

The germanium whiskers (Figure 8) have an unusual form: the trunk is cylindrical, and on it one can see regular traces of growth layers, while the cap is much larger than the cross-section of the trunk, and it is facetted and, as for silicon, consists of a mixture of crystallites.

One can adjust the crystallization temperature to control the crystal diameter [3]; Fig. 9 shows a silicon whisker system where the whiskers have deliberately been given a conical form by programmed temperature variation.

Conclusions

Experimental studies have been made on the VLS growth of oriented whiskers of silicon on silicon and of germanium on germanium. Particular attention has been given to the characteristics of the system as a whole; measurements have been made on the main growth directions, the variations in diameter and height, and the relative growth rates in relation to cross section. Conclusions have been drawn on the most likely rate-limiting steps. Details have been given of the structure of the individual crystals such as the shape and facets.

We are indebted to N. N. Sheftal and A. N. Stepanova for discussion of the results and to V. I. Muratova and L. N. Obolenskaya for assistance in the experiments.

Literature Cited

1.   R. S. Wagner and W. C. Ellis, Appl. Phys. Lett., 4:89 (1964).
2.   R. S. Wagner and W. C. Ellis, Trans. AIME, 233:1053 (1965).
3.   R. S. Wagner and C. J. Doherty, J. Electrochem. Soc., 113:1300 (1966).
4.   R. S. Wagner and C. J. Doherty, J. Electrochem. Soc., 115:93 (1968).
5.   D. W. F. James and C. Lewis, Brit. J. Appl. Phys., 16:1089 (1965).
6.   P. R. Thornton, D. W. F. James, C. Lewis, and A. Bradford, Philos. Mag., 14:165 (1966).
7.   J. D. Filby, S. Nielsen, G. J. Rich, and G. R. Booker, Philos. Mag., 16:565 (1967).
8.   E. I. Givargizov, Fiz. Tverd. Tela, 6:1804 (1964).
9.   F. C. Frank, K. E. Pattick, and E. M. Wilks, Philos. Mag., 3:1262 (1958).

EFFECTS OF SUBSTRATE MATERIAL ON THE MORPHOLOGY
OF SINGLE-CRYSTAL FILMS OF ZINC TELLURIDE

N. N. Magomedov and N. N. Sheftal'

In research on epitaxial growth of gallium arsenide films [1] we made several attempts to grow these films on sapphire substrates with various orientations by the iodide and chloride methods. However, the compound either did not deposit on the sapphire or was deposited as patches, or else as a firmly adhering polycrystalline film. Subsequently, GaAs films on sapphire were obtained by the use of arsine and organic gallium compounds [2]. We have made single crystal zinc telluride films on GaAs substrates with good reproducibility in the structure and orientation. Such films were also made on ZnSe, ZnS, mica (fluorophlogopite), and sapphire. Here we consider only the effects of the substrate on the morphology of ZnTe films.

We used a closed two-zone system made of fused quartz with molybdenum heaters for source and substrate. The temperature in the crystallization zone was kept constant to ± 0.2°C. The reaction was between hydrogen and a ZnTe source via $ZnTe_{(s)} + H_{2(g)} \rightleftharpoons H_2Te_{(g)} + Zn_{(g)}$, which occurs to an appreciable extent even at 600°C and shifts substantially to the right as the temperature is raised. We adjusted the crystallization conditions to give single-crystal p-type films on the above substrates. One of the methods of [3] was used in preliminary treatment of n and p types of GaAs with (111), (110), and (100) orientations, and also 3° from (111). The sapphire substrates of basal orientation were polished mechanically (13-14∇) and fired under vacuum at 1850°C for 3 hr [4]. The GaAs and sapphire substrates were brought to a mirror finish with few etch pits (at $\times 320$). The substrates of mica, ZnS, and ZnSe were made by cleaving single crystals on (001) and (110) cleavage planes, respectively. The ZnTe films were red and had a cubic structure.

We examined the film morphology with Normanski attachments, types MIN-8 and MeF (by Reichert), with a JOL-2 scanning electron microscope, and with an MII-4.

Parts a-c of Fig. 1 show the micromorphology of ZnTe films on A(111) and B(111) (to ± 5) and also deviating 3° from B(111) toward [110].

The basic surface of ZnTe on A(111) (Fig. 1a) is smooth with parallel curvilinear elongated ridges. There were also larger curvilinear positive growth figures and small triangular vicinal pyramids. The height of the large figures was reduced at the higher crystallization temperatures. Films on B(111) (Fig. 1b) were covered with regular vicinal pyramids, which appear to have arisen from increase in supersaturation at the end of crystallization. The difference between the films of Figs. 1a and 1b arises in particular because ZnTe on A(111) grows via its zinc side, while on B(111) it grows via its tellurium one. The orientation of the ZnTe deviated correspondingly when the substrate deviated from (111). Dentate regeneration steps were formed in the film with their vertices towards [111] (Fig. 1c).

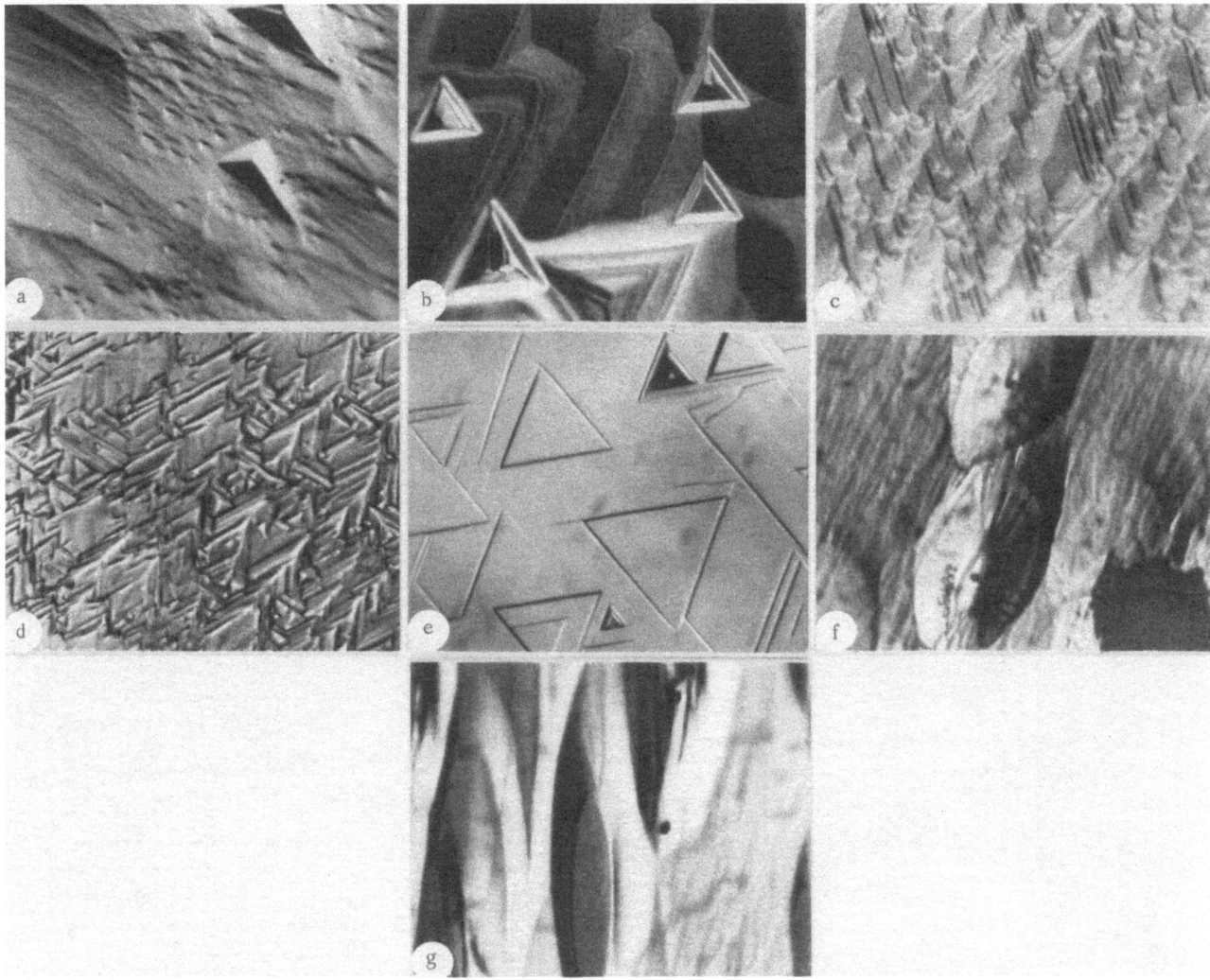

Fig. 1. Films (×320) of zinc telluride. a) On A(111) of GaAs; b) on B(111) of GaAs; c) on GaAs deviating from B(111) by 3°; d) on (0001) of sapphire; e) on mica (fluorophlogopite); f) on zinc sulfide; g) on zinc selenide.

ZnTe films on sapphire inexactly oriented on (0001) (Fig. 1d) were morphologically very similar to the film of Fig. 1c. The substrate had very precisely a (111) orientation when the pseudohexagonal cleavage plane of mica was used, and the films had mirror-smooth surfaces with regular flat triangular growth figures, which were due to two-dimensional structures formed around the most active growth centers (Fig. 1e). Some of these structures had twinned positions, which indicates differences in growth conditions (supersaturation).

ZnTe films on {110} cleavages of ZnS and ZnSe had a morphology whose symmetry corresponded to the symmetry of these planes, with distinctive conical figures of elliptical cross section. On ZnSe these figures terminated at the top in tablets, while on ZnS they had vertices showing clearly the effects of the active growth centers (Figs. 1f and 1g).

The morphology thus indicates a single-crystal state and correlation with the orientation and structure of the substrate, though the latter has comparatively little effect, since the morphology is much the same for ZnSe, ZnS, sapphire, and slight deviation from B(111) in GaAs.

ZnTe grows quite readily on sapphire, whereas GaAs substrates require the use of organic compounds and arsine [2].

The film state is dependent on the composition and structure of the particles in the gas and on the correlation in geometry between the film and the substrate, as well as on the forces in the two. The film continues the atomic structure of the substrate, and this occurs the more readily the greater the similarity. Gallium iodide and chloride are less readily brought into combination with arsenic and attached to the substrate than are organic Ga compounds interacting with AsH_3. The structure discrepancy between ZnTe and α-Al_2O_3 (Δa = 0.44 Å) is less than that for GaAs and α-Al_2O_3 (Δa = 0.76 Å), while the bonds in ZnTe are more ionic than those in GaAs. This probably accounts for the easier growth of ZnTe.

Literature Cited

1.   Kh. A. Magomedov and N. N. Sheftal', Kristallografiya, 9:902 (1964).
2.   Electronics, 14:34 (1968).
3.   Kh. A. Magomedov and N. N. Matomedov, Kristallografiya, 12:2 (1967).
4.   T. A. Zeveke, L. N. Kornev, and V. A. Tomasov, Kristallografiya, 14:579 (1969).

PART VI

THEORY OF CRYSTAL GROWTH

MORPHOLOGY AND FACE GROWTH-RATE RELATIONS FOR CRYSTALS OF DIAMOND STRUCTURE

L. A. Borovinskii and R. P. Vorontsova

Octahedron and cube faces are the main ones for artificial diamonds [1-4] and crystals of silicon and germanium [5, 6]. These faces begin to predominate at relatively low crystallization temperatures; rhombododecahedron faces are less developed or entirely absent [7]. Natural diamonds have roughly the same relative balance between faces [1, 4, 8, 9].

The growth-rate ratios also indicate the relative stability of crystal faces; data have been published [10-13] on the growth rates as a function of substrate orientation for silicon and germanium films. If the growth is restricted by the attachment of particles to the crystal, one usually finds the least film growth rate for substrate orientations of (111) or (100). The growth-rate ratios for these substrates are dependent on the growth conditions; if the temperature is relatively low, films on (100) substrates grow most slowly, while ones on (111) do so at higher temperatures.

To explain the high stability of octahedron faces, it is sufficient to consider the crystal structure and the valencies of the elements for the diamond-lattice crystals, while to explain the relative stabilities of cube and rhombododecahedron faces, one needs to know the nature of the bonds between the surface atoms. A simplified analysis can be based on the assumption that the energy of the bonds between atoms is made up of the energies of the pair interactions, while the interaction energy for each pair of particles is dependent only on the relative disposition; this simplified approach, however, can lead to results in clear conflict with experiment. For instance, Ansheles [14, 15] concluded that cube faces should be displaced by octahedron and rhombododecahedron ones on the basis of the addition of atoms to {100}, {110}, and {111} faces of a diamond crystal, and also that {110} faces, which grow by formation of one-dimensional nuclei, could occur only as narrow bands. This conclusion does not agree with experiment, because Ansheles neglected the differences in bond character at the surface, especially for atoms lying on {100} faces, from the type of bonds found within the crystal.

Also, representation of the energy as the sum of pair interactions leads to the result that the basic faces on the equilibrium form for a diamond lattice should be octahedron ones, while the occurrence of cube and rhombododecahedron faces is ascribed to interaction with second and third neighbors [16-18].

A qualitative explanation has been given for the relative stabilities of (111), (100), and (110) faces on diamond-lattice crystals via the directions of the free valencies at these faces (Fig. 1a-b). Dispositions have been given [19-21] for the atoms in the (111), (100), and (110) surface net; the free valencies at the surfaces of (111) faces are parallel to one another and perpendicular to the face, while the free valencies on a (110) face can be divided into two groups: those parallel to the face and also the crossed ones. When a crystal is divided on a

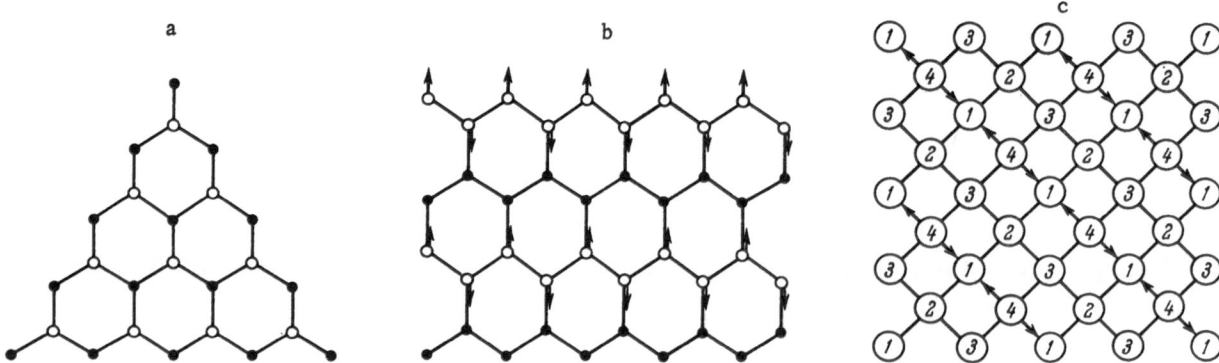

Fig. 1. Atomic arrays in surface grids of diamond. a) (111) face; b) (110) face; c) (100) face.

(100) face, two valency bonds are broken for each surface atom; the nearest atoms in the surface grid for a (100) face are second-order neighbors. The extensions of the lines for the broken bonds for the two nearest surface atoms intersect at 109°27', at points which are the centers of atoms before division of the crystal (Fig. 1c). Up to this instant, the exchange interaction of the electron pairs with compensated spins in the two adjacent bonds is small by comparison with the interaction of the electrons forming a single bond; on separating the crystal, each broken bond is left with one electron with an uncompensated spin. The intersections of the lines for the broken bonds result in mutual compensation of the free valencies in the surface systems, and one gets surface curved bonds, whose energy is, of course, less than the energy of a valency bond deep within the crystal, since the surface atoms are second-order neighbors and the valencies are mutually at angles. However, the energy of the bond between surface atoms may be larger than or comparable with the energy of the thermal motion, so it influences the addition of new atoms and thus affects the equilibrium and growth shape. Strictly speaking, the surface bonds are not isolated one from another, so their interaction via the bulk of the crystal leads to splitting of the electron energy levels for the electrons that form the bonds between the surface atoms, and one gets an energy band for the surface state. The electronic surface states for (100) face of diamond have been considered [22-24]. So far as we are aware, nothing has been published on the application of quantum theory for surface states to problems of crystal growth and form. A quantum-mechanical calculation of the binding energy between surface atoms cannot be undertaken here, and we will assume the abiding energy between surface atoms in (100) face as defined by an empirical parameter, denoting the work required to break such a bond by φ_1. Then one calculates the surface energy via the bond breakage energy on dividing a crystal on (100) planes, subtracting the energy released on formation of the bonds between the surface atoms. The number of newly formed surface bonds on the two parts of the separated crystal is twice the number of surface atoms, while the number of surface bonds formed on each surface is equal to the number of surface atoms, so we get the following expressions for the specific surface energy of a (100) face:

$$\sigma_{100} = \frac{4}{a^2} \left(\varphi_I + 4\varphi_{II} + 5\varphi_{III} - \varphi_1 \right),\qquad(1)$$

where φ_I, φ_{II} and φ_{III} are the interaction energies of neighbors of orders I, II, and III.

The free valencies of the surface atoms in a (111) face cannot mutually compensate, because the directions are parallel; the directions of the free valencies for surface atoms on (110) cross (Fig. 1b), so there can only be weak interaction between the unpaired electrons in the surface. It seems likely that the significance of this interaction is much less than that of surface-bond formation for (100) faces. It has been pointed out [23] that a (111) surface in diamond behaves as a polyradical with unsatisfied bonds, whereas a (100) surface is not a polyradical. This essential difference in the electronic structure of (111) and (100) faces is decisive in explaining the relative stability. It is an open question whether a (110) face can be considered as a polyradical, because it appears that the corresponding quantum-mechanical calculations have not been performed. If we assume that the exchange interaction between free (unpaired) electrons in (111) and (110) faces is slight, then we may suppose that the expressions for σ_{111} and σ_{110} in [16-18] are approximately correct.

We use these expressions and (1) for σ_{100}, which shows that the equilibrium form should contain (111) and (100) faces, whose relative balance is dependent on the relationship between φ_{I}, φ_{II}, φ_{III} and φ_1.

Consider now the growth rate of a (100) face during growth of the crystal or film from the vapor; we assume that the growth rate is limited only by the mode of addition of atoms to this face. The surface bonds between (100) face atoms are responsible for the potential barriers to attachment of atoms; we denote the height of these barriers by φ_2, which is close to φ_1. The exact heights of the potential barriers and their relationships between them may be derived from quantum-mechanical calculations.

We determined the rate as the difference between the number of atoms incident on 1 cm^2 and the number leaving this face or reflected from this area in 1 sec; we assume that the face takes up all atoms that collide with it whose kinetic energy is φ_2. Consider the growth occurring by distillation of the crystalline material into vacuum. Let the source and substrate be parallel plates at a distance small by comparison with the plate size and the mean free path. Then the following is the number of atoms evaporating from 1 cm^2 in the source 1 sec:

$$\nu(T_{\mathrm{svp}}\varphi_2 P_{\mathrm{svp}}) = \frac{\alpha P_{\mathrm{svp}}}{kT_{\mathrm{svp}}\sqrt{2\pi mkT_{\mathrm{svp}}}}\exp\left\{-\frac{\varphi_2}{kT_{\mathrm{svp}}}\right\}, \tag{2}$$

where P_{svp} is the saturation vapor pressure of the material at the source temperature, while $\alpha < 1$ is a factor that takes account of the nonequilibrium evaporation from the source. Then the growth rate is put in the form

$$v_{100} = \frac{\alpha P_{\mathrm{svp}}}{kT_{\mathrm{svp}}\sqrt{2\pi mkT_{\mathrm{svp}}}}\exp\left\{-\frac{\varphi_2}{kT_{\mathrm{svp}}}\right\} - \frac{P_\infty}{kT_{\mathrm{cub}}\sqrt{2\pi mkT_{\mathrm{cub}}}}\exp\left\{-\frac{\varphi_2}{kT_{\mathrm{cub}}}\right\}. \tag{3}$$

Of course, $\alpha \to 1$ for $T_{\mathrm{cub}} \to T_{\mathrm{svp}}$. The P_∞ of (3) is the equilibrium vapor pressure at the substrate temperature.

We have neglected the interaction between the atoms in the surface net of the (100) face; the probabilities of attachment and detachment of atoms at each point in the grid are assumed to be independent of whether there is or is not an atom at any adjacent adsorption site. In this approximation, the growth rate is limited by the energy barriers to addition of atoms, i.e., to formation of nuclei of zero dimensions.

This mechanism is correct for high supersaturations; at low supersaturations, the probabilities of atom detachment are comparable with the probabilities of attachment, and in this case there may be an important part for the interaction between the atoms in the growing layer and hence for the related increase in attachment probability at adsorption sites adjacent to ones already filled, with reduced probability of detachment from such sites. For this reason, one can get a transition to a growth mechanism for (100) via two-dimensional nucleation.

Literature Cited

1. G. N. Bezrukov, V. P. Butuzov, and D. Korolev, in: Growth of Crystals, Vol. 7, Consultants Bureau, New York (1969), p. 91.
2. V. S. Petrov, in: Growth of Crystals, Vol. 7, Consultants Bureau, New York (1969), p. 105.
3. G. G. Lemmlein, M. O. Kliya, and A. A. Chernov, Kristallografiya, 9:231 (1964).
4. S. Tolansky, Proc. Roy. Soc., A270:443 (1962).
5. G. A. Wolff, Amer. Mineralogist, 41:60 (1956).
6. A. V. Sandulova, A. I. Andrievskii, and M. N. Dronyuk, in: Growth of Crystals, Vol. 4, Consultants Bureau, New York (1966), p. 98.
7. H. P. Bovenkerk, Amer. Mineralogist, 46:952 (1961).
8. A. E. Fersman, Crystallography of Diamond [in Russian], Izd. AN SSSR, Moscow (1955).
9. Yu. L. Orlov, Morphology of Diamond [in Russian], Izd. AN SSSR, Moscow (1963).
10. N. N. Sheftal' and E. I. Gwargizov, Kristallografiya, 9:686 (1964).
11. M. Takabayashi, Japan. J. Appl. Phys., 1:22 (1962).
12. S. K. Tung, J. Electrochem. Soc., 112:436 (1965).
13. B. R. Glang and E. S. Wajda, in: Art and Science of Crystal Growing, Wiley, New York (1963), p. 80.
14. O. M. Ansheles, Dokl. Akad. Nauk SSSR, 101:1109 (1955).
15. O. M. Ansheles, Uch. Zap. LGU, No. 215, ser. geol., issue 8, 84 (1957).
16. R. Lacmann, Ber. Bunsenges. phys. Chem., 67:632 (1963).
17. I. N. Stranski and R. Kaischew, Z. Kristallogr., 78:373 (1931).
18. I. N. Stranski, Disc. Faraday Soc. 5:13 (1949).
19. T. A. Smorodina and N. N. Sheftal', in: Growth of Crystals, Vol. 8, Consultants Bureau, New York (1969), p. 301.
20. R. Sangster, $A^{III}B^{V}$ Semiconductor Compounds [Russian translation], Metallurgiya (1967), p. 344.
21. G. Bliznakov and S. Delineschew, Phys. Status Solidi, 13:101 (1966).
22. J. Kontecky, Czech. Fiz. Zh., 12(series B):184 (1960).
23. J. Kontecky, Kin. Kat., 2:319 (1961).
24. J. Kontecky, Phys. Status Solidi, 1:554 (1961).

EQUILIBRIUM SHAPE AND NUCLEATION ENERGY
IN KOSSEL'S MODEL

V. V. Voronkov

In homogeneous nucleation, a nucleus passes through a critical state in which it is in unstable equilibrium with the medium; the work A for formation of a critical nucleus is the most important parameter that governs the nucleation rate. To calculate A we need to know the shape of the critical nucleus, i.e., the equilibrium shape of the crystal, which is defined by Wulff's rule in conjunction with the Gibbs—Thomson equation [1, 2]. Each surface with its normal $\mathbf{n}$ and specific surface free energy $\sigma(\mathbf{n})$ must be placed at a distance $h(\mathbf{n}) \approx \sigma(\mathbf{n})$ from some fixed construction center, with

$$h_{(\mathbf{n})} = \frac{2\sigma(\mathbf{n})}{\rho f}, \tag{1}$$

where ρ is the density of the atoms in the crystal and f is the increment in the bulk free energy when one atom dissolves.*

A similar problem on the form of nuclei arises in production of two-dimensional nuclei on crystal faces; a two-dimensional nucleus is bounded by steps, and to construct a critical nucleus requires each step with its normal $\mathbf{m}$ and specific edge free energy $\alpha(\mathbf{m})$ to lie at a distance $h'(\mathbf{m}) \approx \alpha(\mathbf{m})$ from some construction center

$$h_{(\mathbf{m})} = \frac{\alpha(\mathbf{m})}{\rho_s f}, \tag{2}$$

where ρ_s is the surface concentration of atoms in a two-dimensional nucleus.

Formulas (1) and (2) completely define the shapes of three-dimensional and two-dimensional critical nuclei if $\sigma(\mathbf{n})$ and $\alpha(\mathbf{m})$ are known; of course, for (1) and (2) to be applicable, the critical nucleus must be of macroscopic size, i.e., f must be sufficiently small. When the equilibrium form is constructed via Wulff's rule, some surfaces may lie outside it; in particular, the equilibrium form may contain only a finite number of surfaces (polyhedral form). We envisage the special case where the critical nucleus is a polyhedron with slightly rounded edges and vertices, i.e., when we have not only polyhedron faces but all other crystal surfaces belonging to the equilibrium form. If we neglect the rounding, one would get a finite number of equations such as (1) for the faces of the polyhedra; but if there are several types of faces, calculation of the shapes and areas of each of these involves tedious and geometrical computa-

---

*If the medium is a vapor or solution, we have $f = kT \ln(1 + s)$, where s is the relative supersaturation; if the medium is a melt, $f = H_m \Delta T / T_m$, where T_m is the melting point, H_m is the heat of melting per atom, and ΔT is the supercooling.

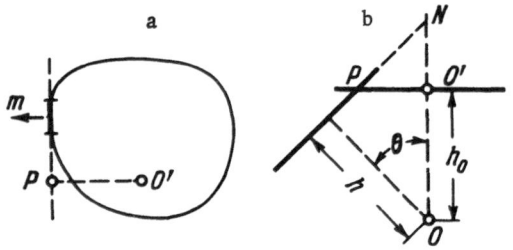

Fig. 1. a) Face contour (plan view); b) inter-section with adjacent surface along contour element.

Fig. 2. Stepped surface.

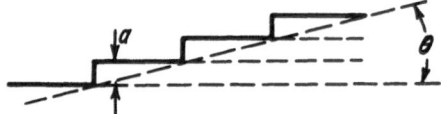

tion, which can be simplified by bearing in mind that all the surfaces belong to the equilibrium form. Each face on a crystal is separated from the others by a certain planar curve (Fig. 1a). Consider an element on this curve with the normal vector **m**, which lies in the plane of the face; an element of the face contour may be considered as the intersection of the face by the adjacent surface, which deviates from the face by a small angle θ in the direction of vector **m**. Figure 1b shows the two intersecting surfaces (the plane of Fig. 1b is normal to the contour element). The distances h_0 and h from the construction center O to the face and to the adjacent surface are given by (1); we project point O on the face and calculate the distance from the projection O' to the straight line passing through the element of the face contour, i.e., distance O'P. Figure 1b shows that O'P = O'N $\tan \theta$, while O'N = ON $- h_0 = h/\cos \theta - h_0$; we substitute h and h_0 from (1) to get

$$O'P = \frac{2}{\rho f} \frac{\sigma/\cos \theta - \sigma_0}{\tan \theta},$$ (3)

where σ_0 and σ refer respectively to the face and to the adjacent surface, which is inclined to the former at an angle θ. Such a surface is stepped (Fig. 2), and the normals to the steps coincide with the normal **m** to the face contour element: the σ for a stepped surface is* given by

$$\sigma = \sigma_0 \cos \theta + \alpha(\mathbf{m}) \sin \theta / a,$$ (4)

where a is the height of a step, i.e., the distance between adjacent atomic layers (Fig. 2). We substitute for $\sigma(\theta)$ from (4) into (3) to get

$$O'P = \frac{2\alpha(\mathbf{m})}{\rho_s f},$$ (5)

where $\rho_s = \rho a$ is the surface concentration of the atoms in an atomic layer parallel to the face, i.e., in a two-dimensional nucleus on this face.

Comparison of (5) and (2) shows that the edge of a face is distant from the fixed point O' by twice the amount of an edge of a two-dimensional critical nucleus, so the face on a three-dimensional critical nucleus is geometrically similar to two-dimensional critical nucleus on this face and exceeds the size of the nucleus by a factor 2 in linear dimension. This result reduces the construction of the three-dimensional equilibrium form to the construction of

---

* It is stated [3] that the choice of $\sigma(\mathbf{n})$ is not unambiguous, since there is the physically un-important term (**cn**), where **c** is an arbitrary constant vector, so that (4) is one of the many physically equivalent representations for $\sigma(\mathbf{n})$ in the neighborhood of this face.

two-dimensional equilibrium forms for all faces of the crystal. Also, the work of formation A for a three-dimensional critical nucleus is 1/3 of the free surface energy [2]. The area of a face is four times greater than the area S of a two-dimensional nucleus, so that

$$A = {}^{4}/_{3} \sum_i \sigma_i S_i, \tag{6}$$

where the summation is taken over all faces. The work of formation for a two-dimensional critical nucleus A' is expressed via the area S [2]:

$$A' = \rho_s f S. \tag{7}$$

We substitute for S from (7) into (6) to get the final formula for A:

$$A = {}^{4}/_{3} f \sum_i \tilde{\sigma}_i A_i, \tag{8}$$

where $\tilde{\sigma} = \sigma/\rho_s$ is the free energy of the face per surface atom.

If the shape of the two-dimensional critical nucleus is nearly a polygon with slightly rounded corners, the length of the sides and A' may be found by analogy with the above. Intersections of a side of the polygon with a step of similar orientation give a result as in Fig. 1b, while a step deviating from the side of the polygon constitutes an echelon of atoms and an analogous stepped surface (Fig. 2). A step can deviate in one of two directions, so there are two types of kink, positive and negative. We denote the energies of formation of these by β_+ and β_-; also, $\alpha(\theta)$ is given by (4), in which one replaces $\alpha(\mathbf{m})$ by β_+ or β_-, while a is replaced by the kink depth b, where b is the distance between the close-packed atomic rows parallel to the side of the polygon. Then the distance from O' to the corner of polygon P is

$$O'P = \beta_{+, -}/\rho_s b f. \tag{9}$$

The total length L of a side is made up of two of these O'P, so that

$$L = 2\beta/\rho_s b f. \tag{10}$$

The energy $\beta_+ + \beta_-$ of a pair of kinks is denoted for convenience as 2β; from the side lengths, we readily get A', which is half the edge energy of a critical nucleus [2]:

$$A' = {}^{1}/_{2} \sum_k \alpha_k L_k = {}^{1}/_{f} \sum_k \tilde{\alpha}_k \beta_k. \tag{11}$$

Here the summation is taken over all sides of a nucleus; by $\tilde{\alpha}$ we denote $\alpha/\rho_s b$, i.e., the free energy of a step per atom in a close-packed atomic series parallel to the step, because $\rho_s b$ is the linear density of the atoms in this row.

Formulas (8) and (11) define the work of formation for three-dimensional and two-dimensional nuclei; the results are applicable, however, only to a special class of crystals, where the polyhedral equilibrium form contains all other surfaces. This condition is never met in models for ionic crystals of NaCl type [2], whereas Kossel's model for crystals with short-range forces belongs to this class. To prove this, we first consider the stability as an important characteristic of the surface. A surface with a given orientation is stable if any deviations from planar form result in an increase in the surface free energy F_s. If shape variations reduce F_s, such a surface is unstable, and it cannot form part of the equilibrium form of the crystal, because this form gives minimum F_s for a given volume. On the other hand, if the given surface lies outside the equilibrium form as constructed via Wulff's rules, such a surface is also unstable [1], so there is unique correspondence between the stability of a surface and the presence of that surface on the equilibrium shape of the crystal.

Consider now the surface stability in Kossel's model for a simple face lattice; we denote the vectors joining an internal atom in the crystal with its partners by $\mathbf{v}_i$, while the corresponding interaction energies are denoted by φ_i; if two vectors $\mathbf{v}_i$ and $\mathbf{v}_k$ differ only in direction, but not in length, then $\varphi_i = \varphi_k$. The interaction radius and the number of vectors $\mathbf{v}_i$ are both finite. To each $\mathbf{v}_i$ from a given atom there is an energy $\varphi_i / 2$, because the energy φ_i is common to the pair of atoms. Some of the bonds are broken for surface atoms, and we denote the number of broken bonds of $\mathbf{v}_i$ by N_i, and so the energy per surface atom is as if such atoms were in the volume of the crystal less the quantity $\sum_i N_i \varphi_i/2$, but the surface energy E_s of the crystal is

$$E_s = -{}^1\!/_2 \sum_i N_i \varphi_i. \tag{12}$$

The energy φ_i is negative, on account of the attraction between atoms, though $E_s > 0$. In (12) we neglect the contribution to E_s from the external medium, which is justified for a crystal−vapor system, and also probably for some crystal−solution systems.

If $T = 0$, F_s is the same as E_s, and in this case the stability of the surface is determined by how N_i varies on altering the surface shape. We take a very simple configuration for the surface, namely one where it coincides with the corresponding crystallographic plane $P(\mathbf{n})$, and consider the broken bonds of type $\mathbf{v}_i$ starting from this plane. Each broken bond $\mathbf{v}_i$ corresponds to the end of an atomic row parallel to $\mathbf{v}_i$ (Fig. 3). The number of such rows, of course, is not dependent on the shape of the surface, so the number of broken bonds $\mathbf{v}_i$ does not alter on varying the shape, or else it increases if the atomic rows have gaps, as in row $a-a$ in Fig. 3. Further, if the plane $P(\mathbf{n})$ has a bond vector $\mathbf{v}_k$, one gets lateral broken bonds on varying the surface (Fig. 3). Then the variation in surface shape either does not alter the surface energy or increases it, so Kossel's model is the limiting case for the class of crystals on which all surfaces are stable and belong to the equilibrium form. In fact, the only stable surfaces are those in which the $P(\mathbf{n})$ plane has at least two bond vectors $\mathbf{v}_k$ that are not mutually parallel, and in this case any variation increases E_s. All other surfaces are indifferent as regards arbitrary or special variations; for instance, if there is only one $\mathbf{v}_k$ in the $P(\mathbf{n})$ plane, then E_s does not alter on adding to the surface unbounded rows of atoms parallel to $\mathbf{v}_k$. Indifferent surfaces can be present on the equilibrium form only as edges and vertices, because indefinitely small perturbations in $\sigma(\mathbf{n})$ would make these surfaces unstable and take them outside the equilibrium form. Then the equilibrium form at $T = 0$ is a polyhedron. Similarly, the two-dimensional equilibrium form is a polygon, and the sides of this are atomic rows parallel to the bond vectors $\mathbf{v}_k$. The corners of the polyhedron, and also the edges, become rounded for $T > 0$; however, usually kT is small compared with energy of the bonds between nearest neighbors, so this rounding is slight, and in this case σ and α have nearly the values for $T = 0$.

Then the equilibrium form in Kossel's model is a polyhedron, but in principle it contains all other surfaces, so the above results apply to it. In particular, each face of a three-dimen-

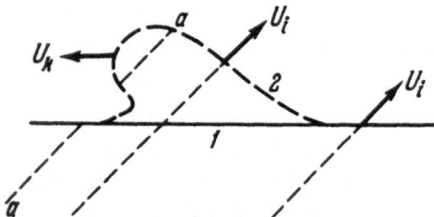

Fig. 3. Surface shape variation. 1) Initial shape; 2) altered shape.

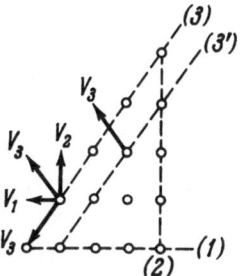

Fig. 4. Atomic array on {011} in a free lattice.

sional critical nucleus is geometrically similar to a two-dimensional critical nucleus on this face, and is twice as great as it in the dimensions. Stranskii and Kaishev [4] derived this conclusion for the particular case of a {001} face in a simple cubic lattice. We now apply (8) and (11) to calculate A and A_i' for some types of simple lattices with allowance for interactions between neighbors of the first, second, and third orders. The mean-energy method [4] has been used [5, 6], but this is rather tedious for use in calculating A_i', and it involves considering the angles of a two-dimensional nucleus, whereas A_i' and A are expressed in (8) and (11) via the elementary quantities $\tilde{\sigma}_i$, $\tilde{\alpha}_k$, β_k, and these formulas contain no geometrical dimensions, which greatly simplifies the calculation. The quantity $2\tilde{\sigma}$ equals the energy needed to separate the crystal into two parts along a given plane, as reckoned per surface atom; similarly, $2\tilde{\alpha}$ is the energy required to separate a planar atomic layer into two parts along a given line, again reckoned per edge atom. Finally, 2β is the energy for separating an atomic series. An equivalent but more convenient method for calculating the boundary energies $\tilde{\sigma}$, $\tilde{\alpha}$, β consists in calculating N_j, the number of broken bonds starting from a surface, step, or kink; the boundary energy is given by (12). To illustrate this calculation, we consider a {011} face in face-centered cubic lattice. We denote by V_k the set of bond vectors of order k, while ψ_k is the bond energy between neighbors of order k; ψ_k is the interaction energy with the sign reversed, i.e., $\psi_k > 0$. There are three types of vectors (Fig. 4) in a {011} plane and consequently there are three types of step. We denote these types by the subscripts 1, 2, and 3 in the order of decreasing β (Fig. 4). The number of steps of type i on the equilibrium form is n_1, and Fig. 4 shows that $n_1 = 2$, $n_2 = 2$, and $n_3 = 4$, while the kink energies are $\beta_1 = \psi_1/2$, $\beta_2 = \psi_2/2$, $\beta_3 = \psi_3/2$. Now we consider the $\tilde{\alpha}_i$; in the case of a step of type 1, there is one broken bond V_2 from an edge atom in a series and two broken bonds V_3, so that

$$\tilde{\alpha}_1 = \frac{1}{2}(\psi_2 + 2\psi_3).$$ (13)

Similarly,

$$\tilde{\alpha}_2 = \frac{1}{2}(\psi_1 + 2\psi_3).$$ (14)

In the case of a step of type 3, each atom in an extreme row (row 3 in Fig. 4) gives rise to one each of the broken bonds V_1, V_2, and V_3; also, from each atom in the next row in depth (row 3' in Fig. 4) there is one V_3 broken bond, so that

$$\tilde{\alpha}_3 = \frac{1}{2}(\psi_1 + \psi_2 + 2\psi_3).$$ (15)

We substitute for n_i, β_i, $\tilde{\alpha}_i$ in (11) to get the work of formation of a two-dimensional critical nucleus on a {011} face:

$$A_{011}' = \frac{1}{J}\sum_{i=1}^{3} n_i \beta_i \tilde{\alpha}_i = \frac{1}{J}(\psi_1\psi_2 + 2\psi_1\psi_3 + 2\psi_2\psi_3 + 2\psi_3^2).$$ (16)

TABLE 1. Face Energy Parameters and Kossel's Model

Lattice type	Face type	Number of faces of equal form	$\tilde{\sigma}_{hkl}$	B_{hkl}
fcc	$\{111\}$	8	$^3/_2\,\psi_1 + {}^3/_2\,\psi_2 + 6\,\psi_3$	$3\psi_1^2 + 12\psi_1\psi_3 + 9\psi_3^2$
	$\{001\}$	6	$2\,\psi_1 + \psi_2 + 8\psi_3$	$\psi_1^2 + 4\psi_1\psi_2 + 2\psi_2^2$
	$\{011\}$	12	$3\psi_1 + 2\psi_2 + 10\psi_3$	$\psi_1\psi_2 + 2\psi_1\psi_3 + 2\psi_2\psi_3 + 2\psi_3^2$
	$\{113\}$	24	$^7/_2\psi_1 + {}^5/_2\psi_2 + 12\psi_3$	$2\psi_1\psi_3 + \psi_3^2$
	$\{012\}$	24	$5\psi_1 + 3\psi_2 + 16\psi_3$	$2\psi_2\psi_3 + \psi_3^2$
	$\{135\}$	48	$^{13}/_2\psi_1 + {}^9/_2\psi_2 + 21\psi_3$	ψ_3^2
bcc	$\{011\}$	12	$\psi_1 + \psi_2 + 3\psi_3$	$\psi_1^2 + 2\psi_1\psi_2 + 2\psi_1\psi_3 + 2\psi_2\psi_3$
	$\{001\}$	6	$2\psi_1 + \psi_2 + 4\psi_3$	$\psi_2^2 + 4\psi_2\psi_3 + 2\psi_3^2$
	$\{112\}$	24	$2\psi_1 + 2\psi_2 + 5\psi_3$	$\psi_1\psi_3$
	$\{111\}$	8	$3\psi_1 + 3\psi_2 + 6\psi_3$	$3\psi_3^2$
Simple cubic	$\{001\}$	6	$^1/_2\psi_1 + 2\psi_2 + 2\psi_3$	$\psi_1^2 + 4\psi_1\psi_2 + 2\psi_2^2$
	$\{011\}$	12	$\psi_1 + 3\psi_2 + 2\psi_3$	$\psi_1\psi_2 + 2\psi_1\psi_3 + 2\psi_2\psi_3 + 2\psi_3^2$
	$\{111\}$	8	$^3/_2\psi_1 + 3\psi_2 + 3\psi_3$	$3\psi_2^2$
	$\{112\}$	24	$2\psi_1 + 5\psi_2 + 4\psi_3$	$\psi_2\psi_3$

The calculation is analogous for other faces and other lattices; Table 1 gives the results for fcc and bcc lattices, and also a simple cubic lattice. The $\tilde{\sigma}$ were taken from [2]. Instead of A'_{hkl} the table contains $B_{hkl} = f\,A'_{hkl}$, which were introduced in [5, 6]. The B_{hkl} differ somewhat from the results of [6] because in the latter incorrect estimates were used for the contribution from the third-order bonds as regards the energies of some steps. The table now a allows one to calculate from (8) the work of formation for each of the three-dimensional critical nuclei:

$$
\begin{aligned}
A_{\text{fcc}} &= 64\psi_1^2/f^2\,(\psi_1 + {}^{21}/_8\psi_2 + 12\psi_3),\\
A_{\text{bcc}} &= 16\psi_1^2/f^2\,(\psi_1 + 3\psi_2 + 9\psi_3),\\
A_{\text{sc}} &= 4\psi_1^2/f^2\,(\psi_1 + 12\psi_2 + 12\psi_3).
\end{aligned}
\tag{17}
$$

In (17), we have retained in the terms for the second and third order bonds only the largest ones of the form $\psi_1^2\psi_2$ and $\psi_1^2\psi_3$ (we have discarded $\psi_1\psi_2^2$ etc.). To obtain numerical estimates, we make the usual assumption about the r^{-6} decrease in ψ_k, where r is the distance between neighbors. We also use the standard relation between the sublimation energy H per atom and the energies ψ_k [2]:

$$
H = {}^1/_2\sum_k m_k\psi_k,
\tag{18}
$$

where m_k is the number of bonds of order k from an internal atom in the crystal. If initially we neglect the second- and third-order forces, and if we express ψ_1 in terms of H, we get

$$
A_{\text{fcc}} = {}^{8H^3}/_{27f^2}, \qquad A_{\text{bcc}} = {}^{H^3}/_{4f^2}, \qquad A_{\text{sc}} = {}^{4H^3}/_{27f^2}.
\tag{19}
$$

If now we incorporate the second- and third-order forces via (17) and (18), we have to multiply (19) by a correction factor, which in the three cases takes the value of 1.21, 1.01, and 1.34, respectively. Then the correction introduced in A by the second- and third-order forces is small if we calculate this for constant H; on the other hand, the correction introduced directly into (17) with constant ψ_1 is fairly considerable and results in an increase by a factor of 2-3.

Literature Cited

1. C. Herring, Phys. Rev., 82:87 (1951).
2. B. Honigmann, Crystal Growth and Form [Russian translation], IL (1961).
3. V. V. Voronkov, Kristallografiya, 12:831 (1967).
4. I. N. Stranski and R. Kaishev, Usp. Fiz. Nauk, 21:408 (1939).
5. R. Kaishew and G. Bliznakow, Compt. rend. Acad. Bulg. Sci., 1(2/3):23 (1948).
6. N. A. Pangarov, Electrochim. Acta, 7:139 (1962).

FILM GROWTH MECHANISM IN LIQUID EPITAXY

Yu. M. Kozlov, L. N. Aleksandrov, and A. G. Cherevko

Here we consider the kinetics of nucleation and growth for semiconductor epitaxial films produced from solution in molten metals, with allowance for convection and diffusion. This topic is of interest because it is unlikely that material is supplied to the growing crystal by diffusion alone. A temperature gradient in the liquid, and consequently a concentration gradient, results in convection, and cases where convection is absent arise only when the solution has a vertical density gradient on account of the temperature gradient: however, there are also density gradients produced by the composition change resulting from displacement of impurities, so it is clear that convection is very likely to accompany diffusion.

General concepts of phase transitions in crystallization from solution in metals allow one to write equations, in which the basic parameters are the nucleation rate $J = dN/dt$ and the growth rate of the nuclei $R = dr/dt$. The nuclei arise only at a certain supersaturation, i.e., when the atoms or molecules undergo random density fluctuations and give rise to independent thermodynamically stable groups; the critical supersaturation here acts as thermodynamic pressure, which causes the soluent to begin to be deposited from the solution.

The rate of nucleation by fluctuation may [1] be put in the following form on the basis of the kinematic characteristics of the solution and solute:

$$J_i = \frac{2D}{\lambda^h} \exp\left(-\frac{\Delta G_i^*}{kT}\right),\tag{1}$$

where $i = 2$ and $h = 4$ for two-dimensional nuclei and $i = 3$ and $h = 5$ for

$$\Delta G_{i=2}^* = \frac{\beta' a' (\sigma')^2}{kT \ln S},\tag{2}$$

where $\beta' a'$ is area, β' is the geometrical form factor, λ is the distance through which molecules diffuse, $\sigma' = \sigma_3 d$ is the surface tension of a two-dimensional nucleus, σ_3 is the same for a three-dimensional one, D is the diffusion constant of the coefficient of the soluent, and $S = C_r/C_\infty$ is the relative supersaturation.

This last determines the scope for stable two-dimensional nuclei and may be calculated from

$$\ln S = \frac{\beta' a'}{k \ln\left(\frac{2D}{\lambda^4 J_2}\right)} \left(\frac{\sigma'}{T}\right)^2.\tag{3}$$

The critical supersaturation can be found if we assume in (3) that the lowest nucleation rate that can be observed by experiment is one per square centimeter per second.

The following equation characterizes the formation of three-dimensional nuclei:

$$\Delta G_{i=3}^{*} = \frac{\beta v_0^2 \sigma_3^3}{(kT)^2 \ln^2 S} \, , \tag{4}$$

where v_0 is the volume of a molecule in the crystalline phase and β is the geometrical form factor.

The relative supersaturation for this case is

$$\ln S = \left[\frac{\beta v_0^2}{k^3 \ln \left(\frac{2D}{\lambda^5 J_3} \right)} \right]^{\frac{1}{2}} \cdot \left(\frac{\sigma_3}{T} \right)^{\frac{3}{2}} . \tag{5}$$

This nucleation mechanism relates to homogeneous nucleation; in the heterogeneous case, we have to introduce into the expression for ΔG_i^{*} terms that reflect the interaction of the deposited atoms with the substrate. In the Gibbs−Volmer theory, this is taken into account via the specific surface energies σ_{fj} in the function for the contact angle θ (f, j = 1, 2, 3). The σ_{fj} are related by Yang's equation; in general,

$$\Delta G_{\text{het}}^{*} = \Delta G_{i \, \text{hom}} \Phi_i (\theta),$$

and for i = 3

$$\Delta G_{i=3}^{*} = \frac{4}{3} \pi \, (\sigma_{f,j})^3 \, v_0^2 \, \frac{(2 + \cos \theta)(1 - \cos \theta)^2}{(\Delta G_v)^2} \, ,$$

which for $\theta = 180°$ (complete failure to wet) becomes $\Delta G_{i=3\text{hom}}^{*}$, while for $\theta = 0°$ (complete wetting) it is $\Delta G_{i \, \text{het}}$, and nucleation should be very rapid.

However, if the production of critical nuclei is due to concentration fluctuations in the solution, the subsequent growth will be limited by transport processes.

It is usually assumed that the supply of soluent to the phase boundary is governed not only by diffusion but also by motion of the solute; the differential equation has been considered [2] for this case where there will be a diffusion layer δ near the growing surface. Then

$$\delta = D^{\frac{1}{3}} \xi^{\frac{1}{6}} \left(\frac{l}{U_0} \right)^{\frac{1}{2}} , \tag{6}$$

where ξ is the kinematic viscosity of the solution, l is the characteristic dimension, U_0 is the speed of the incident flow.

Then the nucleation rate in the symbols of [2] is

$$\frac{1}{F_r} \cdot \frac{dm}{dt} = \Omega \, (C_r - C_\infty), \qquad \Omega = \frac{D}{\delta} = D^{\frac{2}{3}} \xi^{-\frac{1}{6}} \left(\frac{U_0}{l} \right)^{\frac{1}{2}} , \tag{7}$$

where dm/dt is the change in weight of a nucleus and F_r is the surface area of the nucleus. We express m in terms of the density γ and volume L, and use the condition that γ = const, dm/dt = $F_r \gamma$ dr/dt; substitution in (7) gives

$$R = \gamma^{-1} D^{\frac{2}{3}} \xi^{-\frac{1}{6}} \left(\frac{U_0}{l} \right) \Delta C. \tag{8}$$

We assume that $U_0 = dr/dt$ to get

$$R = \gamma^{-2} D^{\frac{4}{3}} \xi^{-\frac{1}{3}} l^{-1} (\Delta C)^2. \tag{9}$$

In relation to the use of the dynamic method, it is of interest to consider the supply of solute to rotating substrate; in this case [2]

$$\delta = 1.61 \left(\frac{D}{\xi}\right)^{\frac{1}{3}} \left(\frac{\xi}{\omega}\right)^{\frac{1}{2}}, \tag{10}$$

where $\omega = 2\pi n$ is the speed of rotation of the substrate in the solution. Use of (10) corresponds to changing (9) for the velocity.

We assume that an epitaxial film begins to grow with the production of two-dimensional nuclei, which expand into planar clumps and come together, which enables us to determine the area of the growing surface of the new phase, S_t. The substrate exerts its orienting influence during the growth of the epitaxial layer in this case via the structure of the two-dimensional nuclei, and these can only take the form of the planar net for the equilibrium faces of the crystal and the set of these is restricted, so there can only be a finite number of possible forms for the two-dimensional nuclei. The probability of formation for the various types on an orienting substrate is determined for a given supersaturation by the energies of formation, so the actual form of the nuclei will be determined by the detailed growth conditions. The actual overgrowth on the substrate can occur in various ways, and much will be determined by the external conditions and by the quality of the substrate. At those points on the substrates where there are contamination or defects, the nuclei will be deformed, which may lead to a change in the growth mechanism and to structure defects in the film. In the general case [3], the surface S_t of the film is as follows for growth by the two-dimensional mechanism:

$$S_t = S_0 \eta_{i=2} = S_0 \left\{ 1 - \exp\left(-\int_0^t J_2(\tau) S(t-\tau)\, d\tau\right) \right\}. \tag{11}$$

We have as follows for isothermal crystallization if we neglect the time dependence of $J_2(\tau)$ and constant R:

$$\eta_{i=2} = 1 - \exp\left(-\frac{\pi}{3} J_2 R^2 t^3\right). \tag{12}$$

Then the following is the increase in the film area in unit time:

$$\frac{dS}{dt} = \pi S_0 J_2 R^2 t^2 \exp\left(-\frac{\pi}{3} J_2 R^2 t^3\right), \tag{13}$$

while the kinetic curve for the increase in coverage of the substrate surface is described by

$$t = \left\{ -\frac{3 \ln(1 - \eta_{i=2})}{\pi J_2 R^2} \right\}^{\frac{1}{3}}. \tag{14}$$

At some stage the surface of the epitaxial monolayer produces new crystallization centers and the process is repeated; the production rate for the new nuclei may be different, and the calculations in this case become much more complicated. The matter is considerably simplified if we assume that the film consists of very thin similar layers, each of which grows in the same way; in this case, the change in the monolayer volume $V(t)$ in a time t less than the coverage time t_1 is put as

$$V_{i=2}(t) = \int_0^t \pi dS_0 J_{i=2} R^2 (t-\tau)^2 d\tau = \frac{\pi}{3} dS_0 J_2 R^2 t^3 \qquad (15)$$

(here R = constant). If the layers are comparatively thick $t > t_1$, the total volume of a layer is defined by

$$V_2(t, t_1) = \frac{t}{t_1} V_2(t_1) = \frac{\pi}{3} dS_0 J_2 R^2 t^2 t_1. \qquad (16)$$

The effective normal growth rate of such a film will be

$$\tilde{w}_{i=2} = \frac{V_2'}{S_0 t} = \frac{\pi}{3} dJ_2 R^2 \left[-\frac{3 \ln (1-\eta_2)}{\pi J_2 R^2} \right]^{-\frac{2}{3}}, \qquad (17)$$

for $\eta_2 \to 0.99$

$$\tilde{w}_{i=2} = A_1 (J_2 R^2)^{\frac{1}{3}}, \qquad A_1 = \text{const.} \qquad (17a)$$

When the three-dimensional mechanism is involved in coverage of the substrate (nucleation via domes), the volume of a finished monolayer is given by the following (r = constant and $R_r = R_n = R$):

$$V_{i=3}(t) = \int_0^t \pi S_0 J_{i=3} R^3 (t-\tau)^3 d\tau = \frac{\pi}{6} S_0 J_3 R^3 t^4. \qquad (18)$$

In a time $t \gg t_1$, we have for specially thick layers

$$V_{i=3}(t) = \frac{t}{t_1} V_{i=3}(t_1).$$

We finally take account of the fact that t_1 is determined by the kinetics of two-dimensional nucleation, which gives us the normal growth rate of the epitaxial film as

$$\tilde{w}_{i=3} = \frac{\pi}{6} R^3 J_{i=3} t_1^3 = \frac{\pi}{6} R^3 J_3 \left[-\frac{3 \ln (1-\eta_2)}{\pi R^2 J_3} \right]; \qquad (19)$$

and for $\eta_3 \to 0.99$

$$\tilde{w}_{i=3} = A_2 R \approx (\Delta C)^2, \qquad A_2 = \text{const.} \qquad (19a)$$

We see from (17a) and (19a) that the effective normal growth rate of the epitaxial film from solution is a power function of the supersaturation whose power is dependent on the growth mechanism; the above relationships have been used in calculations for nucleation and growth of epitaxial semiconductor films of gallium stibnide and arsenide from solution in gallium.

The measurements on the nucleation and growth were made under static and dynamic conditions in an apparatus as in [4]; the substrates were plates of gallium stibnide or arsenide oriented on (111). These crystallographic directions were chosen for their crystallochemical features: slowly growing faces on which it is convenient to examine nucleation kinetics. The substrates have been treated with a polishing etching agent, then they were plunged into the melt and kept for a certain time at the set temperature of about 650°C; the contact with the solution caused a surface layer of the substrate to dissolve to a depth of 15-20 μm. This removed most of the structure defects, inclusions, and oxides from the surface. Then, when the steady state had been reached, and the substrate was no longer dissolving, the layer was allowed to grow; when the overgrowth was complete, the substrate was extracted from the solution and rapidly rotated to remove the residual drops of solution, since any residual solution caused deposition of more solute, and the morphology was substantially affected.

The films were examined optically and by electron microscopy employing replicas.

We used the values of [5] for the solubilities of the compounds in gallium; the following quantities were the basic parameters in the calculations:

$$d^3 = 10^{-22} \text{ cm}^3; \qquad a \approx d^2; \qquad \lambda \approx d;$$
$$\delta \approx 10^{-4} \text{ cm}; \qquad l = 10^{-4} - 10^{-5} \text{cm}; \quad D = 10^{-4} - 10^{-5} \text{ cm}^2/\text{sec};$$
$$k = 1.38 \cdot 10^{-16} \text{ erg/deg}; \quad \gamma_{Ga} \approx 6.09 \text{ g/cm}^3; \qquad \xi_{Ga} = 0.165 \text{ cm}^2/\text{sec}.$$

We found that the calculated nucleation rate for gallium stibnide tended to a limit of 10^{25}-10^{30} cm^{-2} sec^{-1} as the supersaturation increased; the actual limiting value was substantially less than the calculated one, which corresponds to a lower value for ΔG_i^*; estimates from (1) gave ΔG_i^* of 40-50 kT. The discrepancy arises because in the deduction we neglected the loss of solute to the growing nuclei, which is important at high supersaturations. Theoretical estimates were made for the surface tension of gallium stibnide via [6, 7], which gave for σ_{fj} values of 200-300 ergs/cm^2; for these values of σ, the experimental results indicate that the rate of two-dimensional nucleation is higher than the three-dimensional rate at relatively low saturations, but that the latter rate increases with the supersaturation, and it may exceed the two-dimensional rate at a suitable point. Then the nucleation kinetics is determined by the supersaturation. In practice, the supersaturation varies within considerable limits in the production of epitaxial films, which leads to change in the growth dimensions during the growth, which ultimately affects the properties of the films.

We are indebted to Professor N. N. Sheftal' for direction in the experiments and for comments on the theoretical study.

Literature Cited

1. A. E. Nielsen, Kinetics of Precipitation, Pergamon Press (1964).
2. V. G. Levich, Physicochemical Hydrodynamics [in Russian], Moscow, Fizmatgiz (1959).
3. L. N. Aleksandrov and Yu. G. Siderov, in: Growth of Crystals, Vol. 8, Consultants Bureau, New York (1969), p. 197.
4. Yu. M. Kozlov and N. N. Sheftal', Proceedings of the Second All-Union Symposium on Crystallization of Single Crystals and Films of Semiconductor Compounds [in Russian], Novosibirsk, Izd. SO AN SSSR (1970).
5. R. N. Hall, J. Electrochem. Soc., 110:385 (1963).
6. M. Ya. Dashevskii, G. V. Kukuladze, V. G. Lazarev, and M. S. Mirgalovskaya, in: Growth of Crystals, Vol. 8, Consultants Bureau, New York (1969), p. 85.
7. S. N. Zadumkin and A. A. Karashev, in: Surface Phenomena in Melts and the Resulting Solid Phases [in Russian], Kabarda-Bulkar Book Publishers, Nal'chik (1965), p. 85.

THE STATISTICAL THEORY OF NUCLEATION
AND GROWTH FOR CONTOUR REPRESENTATION
OF AN INTERFACE

V. F. Dorfman

Crystal face growth has repeatedly been discussed from the viewpoint of random formation and intergrowth of individual centers, in particular via kinetic equations; this method has been used [1] to describe the competition between formation of centers in coherent and defective positions, with allowance for the internal degrees of freedom of the nuclei differences in particle composition at the surface (consisting of one or more atoms, etc.), and other factors. However, a difficulty arising in this approach is the increasing variety of nuclei as regards size and also effects from fusion, which also results in a variety of configurations. These difficulties have [2] been overcome via a method that includes description of the grain boundaries, which enables one to give a logical statistical description of the crystallization via the mechanisms of the microscopic processes. A rough face is considered as superposition of partially completed layers, and in each layer there are two processes of nucleation: a form in which the nuclei arise and grow without mutual interference, and the actual one in which the nuclei fuse and fill out the layer (Fig. 1). The description is given in terms of the generalized contour P, the total perimeter of all the nuclei for each layer i, and degree of layer filling S_i. The equations in their final form contain the actual parameters:

$$\frac{dP_i}{dt} = -\frac{P_i}{1-S_i} + (1-S_i)\,\delta\,(r_0 S_{i-1} g_i^* + \sigma_i \int_{t_{i-1}}^{t} S_{i-1} g_i^* dt), \tag{1}$$

$$\frac{dS_i}{dt} = P_i \sigma_i, \tag{2}$$

where g^* is the nucleation frequency per unit free surface, while σ is the linear (tangential) growth rate, r is the radius of a nucleus, which is related to the perimeter p by $p = \delta r$, and $r = r_0$ for a critical nucleus. The dynamics of the two-dimensional gas on the active surface of layer i, which is $S_{0i} = S_{i-1} - S_i - aP_i$ (a is the lattice parameter), includes the following microscopic processes: the flux to the surface $v_1 = S_{0i}\omega$, the formation of subcritical nuclei [1] $v_2 = \sum_{k=2}^{k_{cr}} k\dot{G}_i^k$, where G_i^k is the concentration of k-atom associations in layer i, while the flux to the contour of a given layer and the one lying below is $v_3 = R_i(c_1 P_i + c_2 P_{i-1})$, and the detachment from the contours is $v_4 = c_3 P_i + c_4 P_{i-1}$; with the re-evaporation $v_5 = S_{0i}c_5 R_i$ ($c_1 - c_5$ are the probabilities of the corresponding microscopic processes). Then we have for the density of the monomer per unit active surface that

$$\frac{dR_i}{dt} = S_{0i}^{-1}\sum_{l=1}^{5} v_l - \frac{d}{dt}(\ln S_{0i}). \tag{3}$$

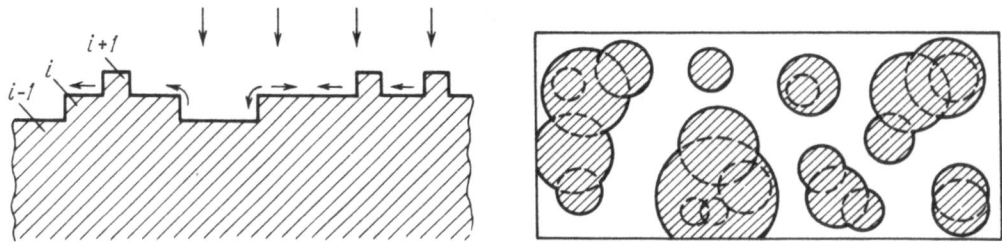

Fig. 1. Production of a general shape (the actual shape is shown by the solid
line, while the arrows indicate microscopic processes).

The system of linked groups of equations of the form (1)–(3) describes the growth of a rough
face, and the number of groups is determined by the maximum height of the roughness expressed in atomic layers. If there is macroscopic roughness, whose height is much greater
than the lattice parameter, one can transfer from a discrete description to a continuous one
by replacing subscript i by the continuous linear coordinate along the growth axis, i.e., one
transfers to one group of equations and partial derivatives (time is the normal coordinate). If
the shape of the surface has been established statistically, we have for any growth parameter
m that

$$\frac{\partial M}{\partial t} = \text{const} \, \frac{\partial M}{\partial x} \, ,$$

which may be used to analyze the kinetically stable form of growth in this model.

A deficiency of this approach is that it gives only a macrokinetic description of the
growth, and so it does not take account of the fluctuations, which are particularly important in
crystallization near critical points. This restriction is completely eliminated by combining
the contour representation of a face with the microkinetic equations [1] for subcritical nuclei,
which has formed the basis of growth description in [2] and in the present paper.

Consider the nucleation kinetics for an initially atomically smooth surface. The activation energies of the microscopic processes will be determined from the nearest-neighbor
model, and the energy per bond will be expressed in units of E/kT. Figure 2 shows the characteristic relationships for the nucleation rate and integral nucleus concentration in dimension-

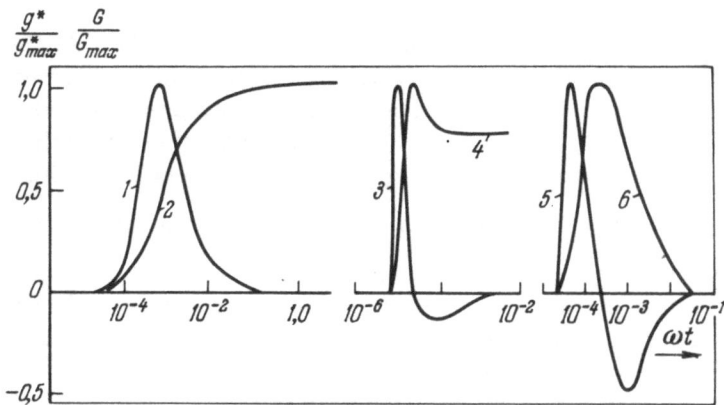

Fig. 2. Three characteristic cases of nucleation. 1, 3,
5) Nucleation rates; 2, 4, 6) integral center densities.

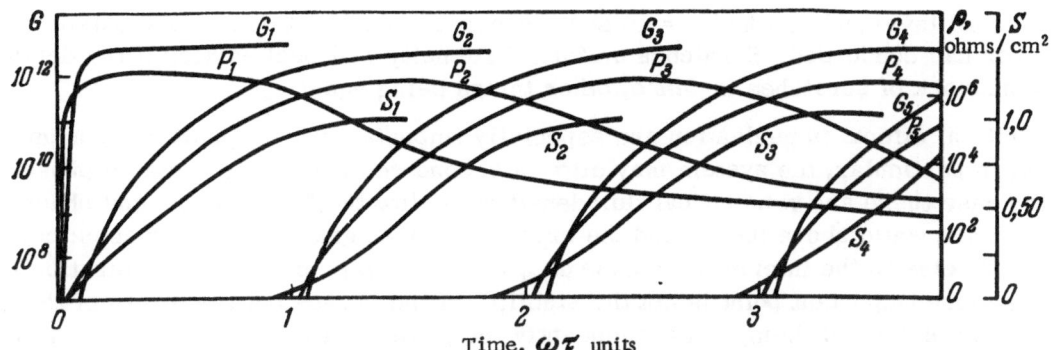

Fig. 3. Formation of the first five layers.

less units as functions of the dimensionless time ωt. Curves 1 and 2 refer to high supersaturations ($E/kT = 10$, $\omega = 10^3$ sec^{-1}), while curves 3 and 4 relate to low ones ($E/kT = 4, 6$, $\omega = 1$ sec^{-1}), while the case of 5 and 6 can be realized when a metastable second phase is formed (see below for details). At low supersaturations, nucleation occurs for a very short time in the initial growth period, and there is a critical monomer density below which stable nuclei are not formed. The expression $R_{cr} \approx 2N \exp(-2E/kT)$ describes approximately the critical density of the monomer as a function of temperature within the nearest neighbor model, where N is the atom packing density per unit area of the layer.

This feature of crystallization at low supersaturations is extremely important; when the nucleation is complete, the monomer density is determined only by the interaction with the generalized contour and with the flux to the surface, with $\omega \gg dR_i/dt$ and $\omega \gg (d/dt)(\ln S_0)$. Then R_i is simply defined by (1); in particular, we have $R_i \approx \omega S_{0i}/(c_1 P_i + c_2 P_{i-1})$ if detachment from the steps can be neglected.

This may be taken in conjunction with the condition for the critical monomer concentration to show that there is a critical radius for a nucleus below which the next layer cannot arise:

$$r^* = \frac{2\nu_D \exp(-3E/kT)}{\omega N^{1/2}},$$

where ν_D is the Debay frequency for the lattice. If the flux density is $\omega = 1$ layer/sec and $E/kT = 10$, the second layer begins to form at nuclei with sizes of the order of 10 atoms, while for $E/kT = 7$ the radius r^* is of the order of 1 μm. If the characteristic temperature is higher, one cannot generally neglect atom detachment from the steps, but this approximation is correct for crystallization of the first layer on a substrate with which the material has a high binding energy. In particular formation of a second layer is unlikely until the entire initial surface has been completely covered by adsorbed material when $E/kT = 4.6$ and the size of the substrate is about 1 cm^2.

Figure 3 [2] gives an example of successive growth of five layers on an initially smooth surface; here the characteristic ratio of the frequency of the diffusion jumps to the flux density is $\gamma = 5 \cdot 10^5$, and the growth becomes steady from the third layer onwards (virtually smooth growth). If $\gamma = 5 \cdot 10^7$, the steady state is attained only for the fourth or fifth layer, i.e., we get transition to rough growth.

The surface initially has mechanical defects, there is a critical flux density for healing of a scratch of given depth in layer i; for instance, if $E/kT = 7$ and $\omega = 10^3$ sec^{-1}, the healing of a scratch with a grid of pitch 1 μm and a cross section of 10^2 atomic positions occurs only in the second layer, while when $\omega = 10 \ sec^{-1}$ it occurs when material to the extent of 0.07 monolayer has condensed. Scratches uniform in density will heal earlier if the depth is small and the network of scratches is dense, other things being equal.

Critical effects in nucleation are especially important when a multicomponent flux crystallizes; in particular, the system may allow a second comparatively unstable phase to form, in which case there are two critical flux densities; below the first, the second phase is not formed at all, while above the second the crystal includes grains of both phases; curves 5 and 6 in Fig. 2 refer to the intermediate case $\omega_{1cr} < \omega < \omega_{2cr}$ [2]. If the supersaturation is very high, the more important point is not the stability of the phases but the length of the generalized contour of each of them, and consequently the grain configuration is the relevant point. Under these conditions, one can get preferential formation of the metastable phase if the nuclei of this have a larger ratio of perimeter to radius; for instance, the stable phase may have tetragonal nuclei, while the metastable one has trigonal ones. Similarly, one of the components may dissolve completely in any of the phases, in which case at low supersaturations it will enter preferentially the phase to which it is more firmly bound, and only that phase below the critical supersaturation. At high supersaturation, it will enter preferentially the phase with the larger generalized contour. If the system has a normal eutectic, low supersaturations cause the overgrowth of the successive layers to produce inheritance of structures from previous ones, while this structure correlation becomes only slight at high supersaturations. The conditions for smooth or rough growth for the different components in a complex system in general do not coincide, which leads to the component with the weaker bond entering the lower layers and being depleted in the projecting parts. If some property of a nucleus is dependent on the size (for instance, the melting point), then at low supersaturations the critical size will be attained mainly at the earlier stages of condensation (as a result of free expansion of nuclei) than is the case at high supersaturation (then fusion of nuclei is the important point). For instance, the critical size might be 1000 atoms, in which case $E/kT = 3$ and $\omega = 1 \ sec^{-1}$ would give this as attained by practically all nuclei even when 0.001 monolayer has condensed, as against only for 0.2 when $E/kT = 10$ and $\omega = 10^3 \ sec^{-1}$.

The method considered here is applicable to crystallization with a mainly random distribution of centers; as such centers we may have: 1) polyatomic particles of the soluent adsorbed directly from the medium or arising by collision in the two-dimensional gas, 2) adsorbed atoms of impurities, 3) active centers related to surface defects and electronic structures in the surface, 4) elements of the surface microrelief. The random distribution of the first two types of center is obvious; experiment shows that point centers also lead to a random distribution of the nuclei [3], and so the center distribution is always random except for certain conditions of crystallization of cleavage surfaces and planes with high indices [3]. Correspondingly, measurements on nucleation and layer growth, where these are possible, agree with the conclusions derived from statistical methods as described here. For instance, there is complete quantitative agreement with the observed distribution function for sizes of nuclei [4], while recent studies [5-7] have confirmed that these functions have second and subsequent peaks at the stages where the nuclei fuse. Our arguments agree with direct observations on nucleation in the electron microscope [3] in showing that one suddenly gets 10^{10}-10^{11} cm^{-2} particles at the very start of condensation, while in the subsequent stages there are practically no new centers. Metastable phases at high growth rates have also been observed [8]; see [9] for a comparison of the theoretical and experimental conclusions on the kinetic effects in growth defect formation. Then the present method is effective for many cases of crystallization of practical importance.

Literature Cited

1.   V. F. Dorfman and M. B. Galina, Dokl. Akad. Nauk SSSR, 182:372 (1968).
2.   V. F. Dorfman, Kristallografiya, 15:435 (1970).

3.   A. J. W. Moore, J. Austral. Inst. Metals, 11:220 (1966).

4.   M. Perrot, J. Phys., 25:104 (1964).

5.   B. T. Boiko, A. T. Puguchev, and V. M. Bratsykhin, Fiz. Tverd. Tela, 10:35 (1968).

6.   F. A. Krivorotov, L. N. Aleksandrov, and Yu. G. Sidorov, Izv. Akad. Nauk SSSR, neorg. mat., 5:287 (1969).

7.   L. S. Palatnik and M. N. Naboka, Surface Diffusion and Growth [in Russian], Nauka, Moscow (1969), p. 208.

8.   D. L. Ageeva, Zh. Neorg. Khim., 3:605 (1958).

9.   V. F. Dorfman, M. B. Galina, and L. I. Trusov, Kristallografiya, 14:71 (1969).

APPENDIX

MEMORIAL TO ALEKSEI VASIL'EVICH SHUBNIKOV

N. N. Sheftal'

Academician Aleksei Vasil'evich Shubnikov died on April 27, 1970; he was one of the senior Soviet crystallographers and the editor of the early volumes of Growth of Crystals.*

A. V. SHUBNIKOV
March 29, 1887 — April 27, 1970

In May 1969 he participated in the symposium from which the present volume derives; his film on crystal growth aroused much interest.† Only 11 months later he died.

Shubnikov made profound surveys in all regions of crystallography; one of his most notable productions was the book "How Crystals Grow," which has now appeared in a new edition.‡ This was used in training several generations of Soviet crystallographers, mineral-

---

*There are several articles [1-3] on Shubnikov's life and work.
† Some good frames from this film appear in the recently published book by Shubnikov and Parvov "Crystal Nucleation and Growth," Moscow, Nauka, 1969.
‡ To this we may add a small but excellent and fundamental work "Crystal Formation," published in 1948.

ogists, physicists, and chemists. This book was also highly regard by English readers, and Buckley, the author of an outstanding English monograph on crystal growth (1951), refers to it as a standard work in his foreword.

In addition, he published books on optical crystallography, and also piezoelectric quartz, and he wrote a series of manuals on the working and mechanical properties of crystals. However, among all these activities there was one that particularly occupied him throughout his life, which was the study of symmetry. He showed that this was a vast untapped field in the world of crystals, and he extended it far beyond the limits of the 230 space groups, which had long been considered as the final development in the edifice of geometrical crystallography. He introduced four-dimensional sign-varying operations into three-dimensional space; the symmetry of black-and-white figures or antisymmetry, which extended the framework of symmetry and encompassed the physical properties of crystals. Here we find his tendency to give symmetry an extended meaning, in the way that he did at an earlier stage of his life in introducing the limiting symmetry groups, which relate the shape and physical properties of crystals. He devised the symmetry of similarity and the symmetry of textures. His book "Symmetry" was a book a joy to read and profound in content. All these works brought forward ideas that the study of symmetry can be applied to real crystals as a major working method in crystallography.

He was also greatly attracted to the philosophical side of symmetry principles; it is no accident that he had an enduring interest in the activities of two other scientists related to him in spirit, Pierre Curie and Louis Pasteur, who provided crystallography with the principle of dissymmetry and the principles of right- and left-handedness in the world of crystals, which are features closely related to the physical properties.

He was interested in crystal growth from 1911 onwards; his first paper in this region was concerned with the symmetry of potassium dichromate crystals. He examined the effect on the crystal shape from supersaturation, temperature fluctuations, electric fields, and other such, and he also examined the crystallization pressure, geometrical selection, orthotropism, and rhythmic growth. He examined the production of vicinals, spherulities, dendrites, and skeletal forms, as well as formation of twins and equilibrium forms, and he examined the relation between the morphologic and structural symmetries of crystals in relation to diamond, rock salt, sylvene, and chromium trioxide. In recent years he collaborated with Gordeeva on enlarging crystals by dynamic recrystallization under variable temperature conditions.

He examined always real crystals and profoundly understood the importance of such study, in that he sought to simulate crystal growth, which is particularly well seen in his work on geometrical selection and equilibrium forms. He defined the theory in words, taken from a paper concerned with crystal growth: we need a theory that will enable us to grow single crystals.

In 1932 he organized the Laboratory of Crystal Growth for Rochelle salt at the Leningrad Institute of Technical Physics, because this material had aroused interest on account of its unique properties. On his initiative, work was started in the USSR on the synthesis of piezoelectric quartz and diamonds, as well as studies on liquid crystals.

The Institute of Crystallography devotes great attention to the growth of single crystals, and this naturally became the leading institute for this problem, and in the 1930s it became one of the centers for the new technology. Here were developed industrial methods of growing single crystals of Rochelle salt (1941), gem corundum in Boules (1937), and rods (1949), as well as scintillation crystals (1959), single-crystal semiconductor films (1954), and many other such examples. Many research centers arose directly or indirectly from the Institute of Crystallography: laboratories, institutes, plants, and factories.

A few words must be said on his style of exposition; he was a master of the written word, and he wrote in an original style with exceptional clarity and brevity, but accurately and clearly. This simplicity concealed much work on his writings. The clarity of expositions was aided by the excellent illustrations, which he did himself.

Although concerned with scientific problems and research on real crystals, he was not detached from real life; he valued everything in his surroundings and aided the success and advances of the Institute and society. Better than any other he knew crystallography and could work in any area of it with anyone.

Shubnikov always expected good work from his colleagues and students, and he knew how to value work, and was firm in his decisions and cooperative when the team expressed an opinion to him, and often took note of it. A characteristic feature of the man was his direct approach to the solution of problems and the directness in formulation of his opinions.

He was a persistent and careful worker, flexible in his approach to work and careful in his relationships with colleagues, whom he regarded as invaluable in the principal work of his life. In the Institute of Crystallography he found all the necessary conditions for successful work, and he went on from success to success. Simultaneously, his popularity and authority increased.

In spite of progressive weakness, he continued to work to the end of his life, and in that respect remains an example to all scientists.

His memory will always be treasured in the history of crystallography and in the hearts of all those who were close to him in his work.

Literature Cited

1. Kristallografiya, 2(1):5 (1957).
2. Kristallografiya, 12(2):179, 180 (1967).
3. Akust. Zh., 13(3):468 (1967).

Chronological List of Shubnikov's Papers on Crystal Growth

1. Über die Symmetrie der Kristalle von Kaliumdichromat, Z. Kristallogr., 50(1):19-23 (1911).
2. Effects of solution supersaturation on the shape of deposited alum crystals, Izv. Ros. Akad. Nauk, series 6, 7(14):817-828 (1913).
3. Über den Einfluss der Temperaturschwankungen auf die Bildung der Kristalle, Z. Kristallogr., 54(3-4):261-266 (1914).
4. Effects of temperature variations on crystals, Zh. Russ. Fiz.-Khim. Obshch., ch. fiz., 50(1-3):39-44 (1918).
5. Growth rates of crystal faces in relation to crystal size, Izv. Ros. Akad. Nauk, series 6, 13(16-18):1135-1142 (1919).
6. Geometrical theory of crystal growth, Izv. Gorn. Inst. Ural. Univ., No. 5, 1-15 (1920).
7. Salol crystals with curved faces, Izv. Ural. Univ., 2:1-10 (1921) (jointly with S. G. Mokrushin).
8. A Statistical study of vicinals of an alum octahedron, Trudy Min. Muz. Akad. Nauk SSSR, 1:1-34 (1926) (jointly with O. M. Shubnikova).
9. Orthotropism in crystal growth, Dokl. Akad. Nauk SSSR, No. 4, 61-64 (1927) (jointly with G. G. Lemmlein).
10. Die Kristallisation auf der Oberfläche der Schmelz, Z. Kristallogr., 67(2):329-338 (1928).
11. Nature of vicinal faces on an alum octahedron (answer to O. M. Ansheles), Zap. Ros. Min. Obshch., 58(1):143-149 (1929) (jointly with O. V. Shubnikova).

12. Patent: Description of a temperature regulator, No. 11,719, Class 42,15, certificate No. 31,583, deposited August 20, 1928, published 1929.

13. Crystallization at the surface of a melt, Nauch. Zap. Tsukr. Prom., 9(3-4):349-354 (1930) (jointly with G. G. Lemmlein).

14. Über die Natur der Vizinalflächen des Oktaeders des Aluminium-alauns, Z. Kristallogr., 77(3-4):336-345 (1931).

15. Artificial production of regular intergrowths of potash alum crystals, Trudy Lomonosov Inst. Geokhim., Krist., i Mineral., Krist. ser., No. 3, 51-66 (1933) (jointly with M. I. Shaskol'skaya).

16. Untersuchung der Vizinalflächen des Alaunoktaeders wahrend der Kristallisation, Z. Kristallogr., A, 88(4):336-342 (1934).

17. Vorlaufige Mitteilung über die Messung der sogenannten Kristallisationskraft, Z. Kristallogr., A, 88(5-6):466-469 (1934).

18. How Crystals Grow [in Russian], Izd. AN SSSR, Moscow and Leningrad (1935), p. 175.

19. Vicinal faces of an alum octahedron during growth, Trudy Lomonosov Inst. Geokhim., Krist., i Mineral., Krist. ser., No. 6, 5-11 (1935).

20. Preliminary measurements on crystallization force, Trudy Lomonosov Inst. Geokhim., Krist., i Mineral., Krist. ser., No. 6, 17-21 (1935).

21. The equation for Popov's two-sheet form, Trudy Lomonosov Inst. Geokhim., Krist., i Mineral., Krist. ser., No. 8, 5-9 (1936).

22. Supersaturation, Physics Dictionary [in Russian], Vol. 4 (1938), pp. 108-109.

23. Recrystallization, ibid., pp. 646-648.

24. Law of geometrical selection in formation of a crystal aggregate, Dokl. Akad. Nauk SSSR, 51(9):679 (1946).

25. Crystal Formation [in Russian], Izd. AN SSSR, Moscow and Leningrad (1947).

26. Crystals in science and technology, Nauka i Zhizn', No. 3, 25-28 (1951).

27. Aspects of the crystallization of diphenylamine, Kristallografiya, Vol. 1, No. 3 (1956).

28. A lecture experiment to demonstrate the rhythmic growth of salol crystals, Kristallografiya, 1(5):606-607 (1956).

29. Spherulite production, Kristallografiya, 2(3):424-427 (1957).

30. Nuclear forms of spherulites, Kristallografiya, 12(5):584-589 (1957).

31. Rhythmic growth of triphenylmethane spherulites, Kristallografiya, 3(4):499-501 (1958).

32. Mutually parallel disposition of ammonium chloride dendrites in a drop of solution, Kristallografiya, 6(2):244 (1961).

33. Which is warmer: the growing crystal or the mother liquor? Kristallografiya, 6(2):315 (1961).

34. Production of nuclei in a drop of ammonium chloride solution in response to an electric field, Kristallografiya, 6(3):443 (1961) (jointly with V. F. Parvov).

35. Address at the Second Conference on Crystal Growth, in: Growth of Crystals [in Russian], Vol. 3, pp. 5-3 (1961). [English translation: Consultants Bureau, New York (1962), p. xi.]

36. Floating magnets and floating crystals, Priroda, No. 3, 17 (1962).

37. Crystallization, Physics Dictionary [in Russian] (1962), Vol. 3, p. 523 (jointly with A. A. Chernov).

38. Faces of zero growth rate on a $K_2Cr_2O_7$, Kristallografiya, 9(3):435 (1969) (jointly with V. F. Parvov).

39. Enlargement of Rochelle salt grains in a solution in response to temperature fluctuations, Kristallografiya, 12(2):186-190 (1967) (jointly with N. V. Gordeeva).

40. Crystal Nucleation and Growth [in Russian], Nauka, Moscow (1969) (jointly with V. F. Parvov).
    Small items in Physics Dictionary 1963-1966 (Vols. 3-5) on orthotropism, growth pyramids, solution of crystals, skeletal forms of crystals, crystal spherulites, and epitaxy.

THE EQUILIBRIUM SHAPE OF POTASH ALUM CRYSTALS *

N. M. Shchagina

The equilibrium shape of a crystal continues to attract theoretical and experimental attention; our object has been to repeat earlier experiments [1, 3] and also to find the equilibrium form for alum crystals with a modified method of production; in addition we have examined the effects upon this form from added dye. A crystal of equilibrium form is one whose shape is close to that produced in response to temperature fluctuations.

Recent theoretical work on equilibrium forms has been based on the principle of minimum free surface energy of the faces for constant volume. The equilibrium forms have been deduced not only theoretically but also by experiment for many ionic and other crystals. Shubnikov proposed a method of finding by experiment the equilibrium form by means of small temperature variations, which cause the crystal to grow and dissolve alternately.

The apparatus for this purpose has been described in detail [2].

This apparatus has two horizontal hollow shafts rotating around their axes in ball bearings. The crystallizers have a device with a cylindrical glass vessel containing solutions and a crystal, the latter being in the form of a sphere, which is firmly attached to a rubber bung. The sphere as it begins to form faces is not allowed to slide on any dominant face in the rotating vessel by being held in a spherical container made of plastic 55 mm in diameter having holes 5 mm in diameter, which is placed in the solution.

The solution is heated periodically by an electric heater, while a fan provides the cooling. The apparatus is kept in a thermostatic room, which was maintained at 24.3°C to ±0.1°. The amplitude of the temperature variations was 0.6-0.7°, while the period was 1.3 hr. The solution for the purpose was prepared with temperature variations from 26 to 26.7°C; then it was divided into two equal parts, to one of them was added a dye called direct blue (color index KI 518). Preliminary tests had shown that the presence of this dye solution altered the shape of the growing crystal by increasing the growth rates of the octahedron faces. The proportions of this dye by weight (0.03 and 0.06%) were such as to suppress completely the {111} faces.

We made 8 spheres from potash alum crystals to have diameters of 15, 18, 21, and 24 mm, two spheres of each diameter; these were cut from a single-crystal block of {111} orientation, which was grown in a centrifugal system in a holder that restricted the lateral growth. The seed was an alum plate parallel to {111} cut from a large crystal. The spheres were placed in two vessels; one batch contained spheres of all diameters and was placed in the pure water solution of the alum, while the other was placed in a solution containing dye 518 at the level of 0.03% by weight. Each vessel contained 0.77 liters of solution. The experiments lasted 5 months.

---

*This paper is published as one of the last studies on crystal growth done under A. V. Shubnikov's direction.

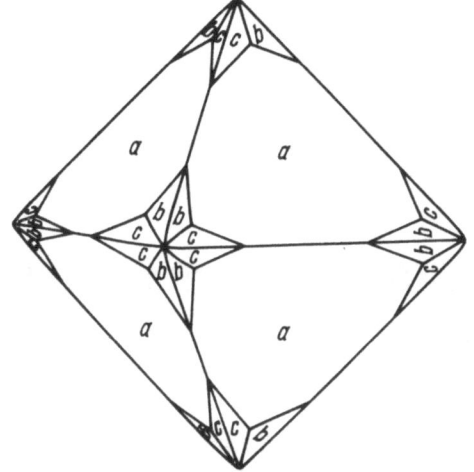

Fig. 1. Faces on a crystal made from a sphere
24 mm in diameter in a solution containing a
dye. a) Octahedron; b) one didodecahedron; c)
another didodecahedron.

Fig. 2. Faces on an alum crystal grown from
a sphere 21 mm in diameter in a solution containing a dye. a) Octahedron; b, c) two didodecahedra; d) tetragon-trioctahedron.

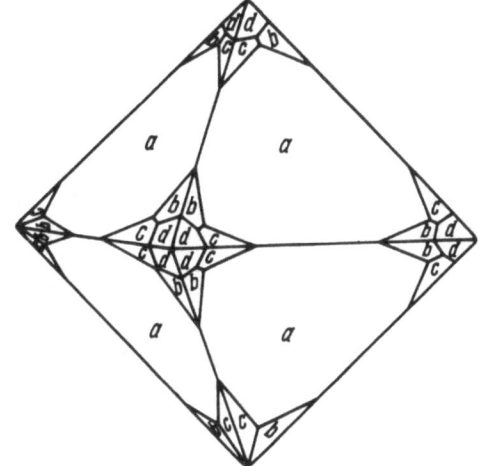

The spheres in the pure alum solution acquired the shape of octahedra with various degrees of perfection; the crystals made from spheres of diameter 21 and 24 mm took the form
of octahedra with very small cube faces, and they were optically nonuniform, but the weight
of them increased somewhat. Crystals grown from the spheres 15 and 18 mm in diameter
took the form of ideal octahedra, and they decreased in weight considerably, and the homo-
geneity they had initially was retained. There is no doubt that the equilibrium form of an alum
crystal found in a pure solution is an octahedron.

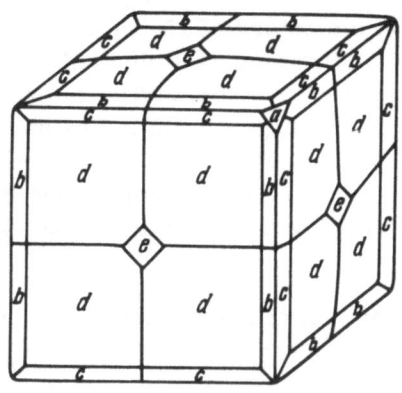

Fig. 3. Faces on a crystal obtained from a sphere
18 mm in diameter in a solution containing a dye.
a) Octahedron; b, c) two didodecahedra; d) tetragon-
octahedron; e) cube.

The crystals in the dye solution differed from one another and require more detailed description. It was impossible to take good photographs, so drawings are given and projections together with the descriptions.

The first crystal (Fig. 1) was produced from a sphere 24 mm in diameter and represented a combination of the faces of the octahedron and two didodecahedra. The octahedron faces were more developed, while there were almost no cube faces. The crystal lost weight in the experiment.

The second crystal was derived from a sphere 21 mm in diameter (Fig. 2), and it had not only the faces of the first crystal but also tetragon trioctahedron faces; the cube faces again were almost absent, while the octahedron faces remained large. This crystal also lost weight. Figure 3 shows the third crystal, which was made from a sphere 18 mm in diameter, and this shows that the octahedron faces are absent from this; there are didodecahedron faces, but these play a subordinate part, and the tetragon trioctahedron faces are the most developed. This crystal increased in weight. The fourth crystal was a model of the third on a reduced scale, and it also increased in weight.

We had previously done similar experiments over long periods, and these indicated that further testing in the presence of the dye would leave only tetragon trioctahedron faces. The following conclusions are drawn from the study.

1. The equilibrium form of an alum crystal is dependent not on the size of the crystal but on the duration of the experiment.

2. The octahedron is the equilibrium for an alum crystal obtained from a pure solution.

3. The dye substantially affects the final form of the crystal and produces tetragon trioctahedron faces, which are usually absent from alum crystals.

4. The smaller crystals increased in weight in the presence of the dye, while the large ones lose weight.

5. The growing crystals in both cases are less uniform.

Literature Cited

1. A. V. Shubnikov, How Crystals Grow [in Russian], Izd. AN SSSR, Moscow and Leningrad (1935).
2. V. F. Parvov, Kristallografiya, 11:3 (1966).
3. A. V. Shubnikov, "Effects of temperature variations on crystals," Zh. Russ. Fiz.-Khim. Obshch., 50(1-3):39 (1918).

THE LIMITING SHAPES OF SINGLE CRYSTALS OF ROCHELLE SALT AND POTASH ALUM

N. N. Sheftal', I. A. Shpil'ko, and G. F. Dobrzhanskii

It has been reported [1] that a uniform freely growing crystal tends to a very simple form; it was also stated that this tendency does not always appear. For instance, alum crystals grown at small supersaturations are complicated in form and contain combinations of octahedra, cube, and rhombododecahedron. If the supersaturation is increased, the form simplifies to the limiting one of a pure octahedron [2]. This octahedron form occurs also for inhomogeneous alum crystals.

Single crystals of Rochelle salt have a fairly simple form but not the limiting one; they are formed by two prisms of almost equal development, namely {110} and {210}. There are also often present faces of the first and second pinacoids, as well as of the third {001} one.

A description has been given [3] of the growth of uniform Rochelle salt crystals over the temperature range 52.5–51.5°C by the use of highly concentrated solutions, which contain only about 10% of water. The crystals grown under such conditions were of much simpler shape; the largest ones had two prisms and the third pinacoid (Fig. 1), but the {210} prism was only weakly developed, while the {100} and {010} pinacoids were very rare and incomplete. Also, the shapes of numerous single crystals of smaller size were actually of limiting simplicity; they were bounded by the {110} prism and third {001} pinacoid (Fig. 2). Although the crystals were grown from point seeds, they all deviated greatly from isometric, being very much elongated on the z axis.

On the other hand, the single crystals grown in this way were highly perfect, which was seen as the absence of signs of the effect known as rain, which consists of a system of minute straight parallel channels running through the crystal in several directions and filled with parent solution. Rain appears in an ordinary crystal not at once but after the growth has been proceeding for some days [4]. It is seen only by careful examination as a system of lines similar to light rays. These defects are so small as to not interfere with technical uses of the crystals.

The crystals made from the highly concentrated solution were more perfect, which agrees with the fact that single crystals grown from melts are more perfect than those grown from solutions in that they contain no trapped solvent. Rochelle salt crystals grown at high temperatures naturally contain very little solvent, so they confirm the tendency [1] for uniform single crystals to take up the limiting form, and they also demonstrate the relations between the approach to this form and the degree of perfection.

One can explain the complicated form of alum crystals grown at low supersaturations by reference to Rochelle salt containing traces of solvent, which is facilitated by the large amount of water of crystallization in alum. To test this assumption we made a series of experiments

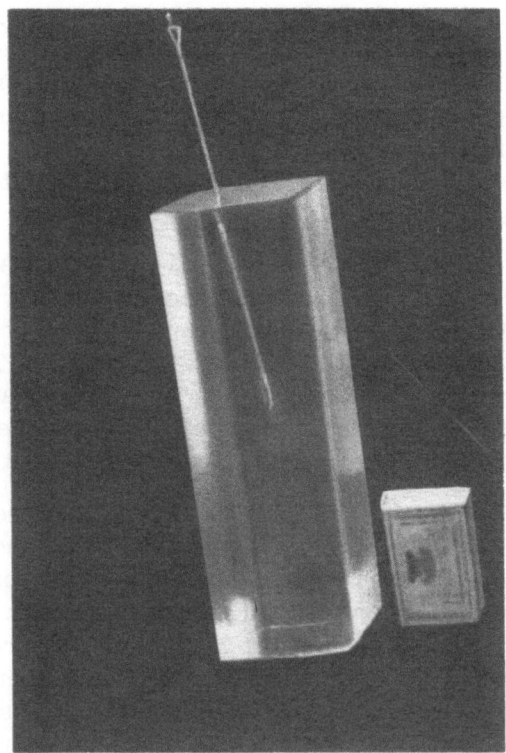

Fig. 1. Homogeneous Rochelle salt crystal (mass about 1 kg) of simple shape grown in a strong solution from a small seed.

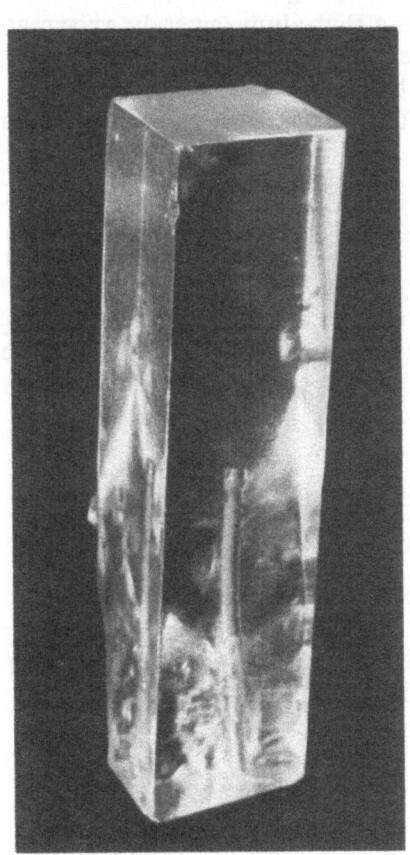

Fig. 2. Very simple Rochelle salt crystal grown in a strong solution.

on alum crystals at elevated temperatures; it was assumed that the results would be close to those that would be obtained with crystals of Rochelle salt, but it was impossible to grow the crystals at 60°C and above. The solution partly decomposed and was very cloudy, and so we were forced to restrict the growth to temperatures not higher than 45°C. We did runs at the following initial saturation temperatures: 44.7, 33.8, and 23°C. The 23°C solution was used in order to compare the shapes of the crystals at elevated temperatures with those grown by Shubnikov in his early studies [2].

The following conditions were used. At the bottom of a crystallizer containing 1.5 liters of solution we placed 3 or 4 seed crystals 3 mm in size; the crystals were grown under static conditions for 12 or 13 days, and the temperature during this time was reduced by 0.9° from 44.7, by 1° from 33.7, and by 2.3° from 23. The larger temperature reduction in the latter case was necessary because the crystals grown at 23°C did not grow at first on account of bubbles forming on their surfaces, and only after the temperature had been reduced by 2.3° for a month did the growth begin, which continued for 12 days.

All the crystals had faces of the octahedron, cubed, and rhombododecahedron; the areas of all the faces were measured carefully, and the proportion of the faces of a given simple form were determined (excluding the part that was in contact with the bottom). The proportion of the octahedron faces in crystals grown at the highest temperature was 80% (with an average of 73 and a lowest value of 79%), while the proportion of cube faces was 10% (as against 12 and 18%, respectively), and the proportion of rhombododecahedron faces was 10% (12 and 13%). There was thus a tendency for the shape to simplify on elevating the temperature, which confirms the view that the cause of complex crystal shapes in alum is inclusions of excess solvent.

It has been stated [5] that the most uniform crystals should grow not at the lowest supersaturation, when the crystallization pressure is very small, but at higher ones, where the self-purification effect is more marked. On this basis one assumes that alum crystals grown at very low supersaturations and having complex shapes are less uniform than ones grown at somewhat higher supersaturations with the limiting simple shape. Buckley [6] describes hundreds of large and uniform alum crystals weighing up to 110 kg as being grown in the form of pure octahedra.

Shubnikov and Shaskol'skaya [4] equilibrium forms for alum crystals also were pure octahedra; the degree of uniformity is related to the shape, as is clear from a paper in this volume [7].

We have thus found that uniform crystals of Rochelle salts and alum tend to become simpler in shape as the uniformity increases, and they do not provide examples of deviation from this tendency, as has previously been supposed [1].

Literature Cited

1. N. N. Sheftal', in: Growth of Crystals, Vol. 1, Consultants Bureau, New York (1959), p. 5.
2. A. V. Shubnikov, Izv. Ros. Akad. Nauk, 819 (1913).
3. N. V. Alyardin, N. N. Sheftal', and Z. I. Zhmurova, Kirstallografiya, 2:193 (1957).
4. A. V. Shubnikov, How Crystals Grow [in Russian], Izd. AN SSSR (1935).
5. N. N. Sheftal', in: Growth of Crystals, Vol. 3, Consultants Bureau, New York (1962), p.3.
6. H. Buckley, Crystal Growth [Russian translation], IL (1954), p. 56.
7. N. M. Shchagina, this volume, p. 317.

80TH BIRTHDAY OF NIKOLAI VASIL'EVICH BELOV *

N. N. Sheftal'

The sum of human knowledge includes a considerable group of sciences concerned with stones; these are general geology, historical geology, dynamic geology, tectonics, paleontology, hydrogeology, engineering geochemistry, geophysics, petrography, mineralogy, and — somewhere near the end — crystallography.

It might be queried whether crystallography is a geological science, for it has long been reckoned amongst the physical sciences, and only those such as Belov have placed it at the center of the triangle whose vertices are physics, chemistry, and mineralogy, probably on account of traditions that have attracted it toward the last vertex.

It seems to me that a paper dealing with this anniversary of and published in the geological series of the Vestnik of Moscow University should deal with this question and give an indication of the place of crystallography in the advances of geological sciences.

In spite of the great advances in physics, chemistry, biology and technical sciences, the earth sciences have not lost their value, and they deal with the history of the earth as a heavenly body, and its history in a period when life was only first appearing, as well as with the transformations occurring in parallel with the change in living organisms, which have left marks of many features in the layers of the earth's crust. Also, there is still value in the study of the basic material of which the earth's crust is made up.

Belov's studies have been concerned to a very considerable extent with precisely this aspect.

Consider the mineralogical textbooks of the 1930s; in these we find only a chemical classification of minerals, while the origin and main parageneses have become the dry descriptive mineralogy of the 19th century, while the living and interesting chemistry of the earth's crust and the history of its minerals appear only as illustrations for the structural data.

Between 1940 and 1970, mineralogy was transformed from the chemistry of the earth's crust into the chemical crystallography of the same, and Belov made a large contribution to this transformation.

With the exception of 5%, the earth's crust consists entirely of silicates; their structures are very important but also are extremely complicated. Belov and his students have elucidated the structures of nearly 50 minerals, which were primarily silicates but also some borates and sulfides; this is a vast contribution to our knowledge of nature, especially for geology, mineralogy, and inorganic chemistry, and the evidence has provided much assistance in the technology of processing mineral materials, whether it be the cement industry or the production of glass or molecular sieves (zeolites).

---

* The Russian text was published in Vestnik Moskovskogo Universiteta, Seriya Geologiya, No. 6, 1971.

N. V. BELOV

This extensive work on structures gave rise to a deeper understanding of the processes of mineral formation, the geoenergetics and radiation features involved, as well as on the evolution of the material in the earth's crust, the systematization involved in the periodic system, the crystallochemical relationships involved in all this.

Belov in his time has been closely associated with some hundreds of people as students and coworkers, and he has impressed them all in the assistance he has been able to provide in his vital approach, with the provision of the best scope for creative activity and advancing one's own understanding of the subject. We may survey briefly this work extending over many years with these numerous people, which form the Belov school. First of all we must note that there is a close relation between theory and practice in his work. The practice of the chemical crystallographer is to elucidate structure, while the theory appears in making the preliminary model, and elucidating the relationships and transitions. Belov encompasses both these aspects, and does not draw an arbitrary line between the two sides of a unifying study, since such division is always accompanied by weaknesses and failures.

Belov has said that in geology I follow Fersman; in fact, he was an active participant in the chemical study of the Khibina problem, and he has been one of the most productive authors of original papers in the journal Priroda, where he acquired in mineralogical circles the reputation of being one of the most highly qualified students in applied and theoretical chemistry. However chemistry, particularly in application to stones, could develop at that time only from the viewpoint of chemical crystallography.

Belov in 1936 translated Hassel's Chemical Crystallography at Fersman's suggestion, and he supplemented it with a number of detailed comments, which provided amongst other things an extensive literature survey of value to the student of the crystallochemical aspect of the subject. His knowledge of structures led him to hold seminars, at which it was simultaneously teacher and student, and this raised many hundreds of problems in relation to the complex relationships between polyhedra and their structure. The work was extended in his Blue Book called "The structure of ionic crystals and metallic phases," which was published in 1947.

This dealt with the ideal structures of crystals, and these are islands of temporary equilibrium in the continuous evolution of material, which is subject to three principles: minimum free energy in its geometrical expression (symmetry, which provides equal location of equal atoms and satisfaction of all bonds), the close-packing principle, which expresses the tendency to attain this minimum at any instant by the shortest path, i.e., with the minimum number of unsatisfied bonds, which means also minimum crystallographic surface, and thirdly the principle of mutually balanced location of interacting large close-packed anions and small cations in the holes between them.

In this book Belov developed the theory of close-packing, which became the basis for complete elucidation of complex structures. In the last chapter of the book and in the conclusions he devotes five or six pages to the application of his theory of the method of delineated space, in which he compared systems of polyhedra for close-packed anion rays, in which he then places the cations; this was done with numerous references to experiment. The possible forms of close-packed structures were reduced greatly in numbers by establishing the space group for the symmetry and morphology of the crystal. He examined the scope for placing the cations in polyhedra, and the final choice among the surviving variants was made on the basis of interatomic distances established by x-ray methods.

Five complicated structures were elucidated by these methods, which showed that they provide the key to structural analysis, and it is precisely on this basis that about 50 very complicated structures have been elucidated under Belov's direction. From 1950 onwards he began to publish in the Mineralogical Symposia of L'vov University his outlines of structural mineralogy, of which the 143rd item appeared in 1971.

His theory and elucidated structures allowed him to consider systematically various features, including morphologic ones, of individual minerals and groups, in conjunction with habit, predominant faces, cleavage, parting, twin planes, the relationship of these to close-packed planes, and the configurations of atoms around the strong cations, especially silicon. He stated that elongation often macroscopically reflects microscopic chains of octahedra, and that the form is sometimes determined not directly by the structure but by a protective layer of anions in planes of high charge, and that parting arises as a result of reagents occurring in close-packed planes, which are trapped by the growing crystal. He thus gave an explanation of the layering and perfect cleavage of long close-packed planes found for some hydroxides and sulfides, together with the analog of type AX_2, which also explains the stability and great abundance of rutile amongst the three forms of titanium dioxide in nature. It is clear here that the instability in this case is related to a number of separated edges for the anion octahedra.

In this respect he did not neglect the relation between the mineralogical features and the technical applications of the minerals.

He considered in detail the scope for cation exchange in zeolites and the performance of these as molecular sieves on the basis of the large holes in the structure, which can be populated reversibly by various molecules entering from outside.

In these ways he laid the foundations of structural mineralogy. One encounters his characteristic condensed style in structure descriptions such as the following: "In neptunite," he writes, "the skeleton is composed of baroque columns of Ti and Mn octahedra; these cross at angles of 80° and form a type of trellis structure, which is linked together by Li octahedra at the points where the columns come together. The translationally identical rods are related by vertical transverse ones, which are produced by pairs of large Na octahedra, between which run the ends of the rods in the second direction."

The numerous silicate structures elucidated under Belov's direction gave rise to indications of new ways of constructing them; these did not fit into Bragg's silicate chemical crystallography, and they won him world-wide recognition in 1956 in all quartzes on chemical crystallography, geochemistry, and mineralogy. Belov himself called this new development the second chapter in the chemical crystallography of silicate (1961). This is the structure of silicates with large cations, mainly Ca, Na, K, and the rare earths, i.e., the structures of the lighter pale leucocratic rock-forming minerals, as distinct from Bragg's dark melanocratic ones.

The basis of the Belov silicates is not the silicon tetrahedra, which is incompatible with the edges of the octahedra formed around the large cations by the oxygen atoms, but a doubled tetrahedron, the Si_2O_7 diortho group, which forms a somewhat distorted trigonal prism, whose edges are compatible with a doubled edge of a Ca octahedron. Belov now considered as the structural base of silicates not silica and not anionic silicon—oxygen radicals, but cations that form into rods of oxygen octahedra. The silicon inactive in chemical analyses is also inert in the structures; it adapts itself to the disposition of the leading cation and forms readily deformed chains, strips, grids, and even rings.

From this viewpoint, one can easily understand the chemical crystallography of the most important minerals in the silicate shell of the earth. The main lines of the statics of chemical crystallography have been completed, but Belov's thoughts went further and deeper, and he considered the structural dynamics of silicates and alumina silicate formation from the parent magma, as well as the subsequent transformations.

An outstanding article of Belov's* characterizes the main structural features of this process; in it he shows that the main mobile and reacting structural units in the primary mineral-producing process is the neutral SiO_2 molecule with its angular configuration, which contains potential bonds to form in situ the static $[SiO_4]^{4-}$ unit. This assertion was confirmed not only by the logic of chemistry and structural studies but also by experimental studies on the silicates of blast furnaces and other ovens.

This position casts a new light on the structural treatment of mineral formation from magma on the base of the Bowen scheme for differentation.

Structural models were given for each step in the process, which were based on the somewhat abstract model for a magma as a molten glass, which approach was discussed vigorously every three or four years at the All-Union Conferences on the Vitreous State.

---

*N. V. Belov, E. N. Belova, G. P. Litvinskaya, and Yu. A. Kharitonov "Silicification, silicosis, polymerization, and aluminosis, in the geocrystallochemistry of silicates and aluminosilicates," Vestnik Mosk. Univ., ser. geol., No. 4 (1970).

Belov considered the entire petrological process involved in Bowen left and right branches, together with the third stage, the weathering process, albertization of plagioclase, and retention of Ca at the earth's surface as limestone and dolomite, which sink and are silicified by rising igneous masses. The Belov diortho group (doubled Si tetrahedron) is the structural basis of the leucocratic silicates. The last work to be mentioned, which is one that represents an important stage in Belov's thought, is "The periodic law and chemical crystallography." * In this he traced the features of chemical crystallography common to the whole periodic system. He illustrated the alternation of the static and dynamic roles, which deal with the explanation of properties and processes, respectively, and he stated that the dynamics of inorganic compounds had derived a great deal from chemical crystallography, in which the first steps were related with mineralogy, which was concerned with the chemistry of stable natural compounds of all elements, together with their isomorphous or other combinations."

A vast amount of evidence was used to show how the introduction of the mineralogical approach to crystallochemistry provided an interpretation of the third basic characteristic of an atom, namely the ionic radius; this explains a large number of otherwise unelucidated features of elements, in particular the very common heterovalent isomorphism along the diagonals, such as the mutual substitution of Na and Ca in all ratios in plagioclases and the absence of isomorphism between chemically similar elements. Belov and others here dealt with the relationships of the periodic system to the coordination numbers of elements in inorganic compounds.

We have noted only the basic trends in the works of Belov, who has written about 600 papers, but the statements are sufficient to show that mineralogical crystallography as represented by Belov's school has made a large contribution to the progress in geology, which can be reckoned as one of the major achievements of Soviet geological science.

A characteristic feature of Belov's work is the continuing growth in its range, which is not surprising, for he has overcome many of the difficulties in the direct elucidation of structures and has provided the necessary organization of materials and people to facilitate a growing interaction with his team, and he has devoted much time to the forwarding of developments, particularly of ideas whose value is undoubted. Major objects for science in the 19th century were gases and solutions, whereas the 20th century has been the age of crystals.

Crystallography traditionally is placed at the center of the triangle mentioned above, but it now appears at the center of a tetrahedron and perhaps towards the fourth vertex, namely technology. Of course this tetrahedron has a tendency to acquire a fifth vertex, namely biology, though this is really a matter for the not very distant future.

Crystal growth is an increasingly important branch of crystallography; a great edifice of novel technology has arisen from synthetic crystals. Crystal growth extends to new ranges of materials that do not occur in nature, although many of them are closely related to minerals, but this means that the Belov museum of structures should increase, and that mineralogy should extend, and in fact is already extending, to artificial crystals. There is no doubt that Belov's structural ideas will be used and extended in such work.

Belov himself understands very well the importance of crystal growth, and he has organized a special course on crystal growth and a laboratory for crystal growth in his department, which has provided a training ground for graduate workers and for the presentation of theses on crystal growth.

Belov's theory is a fairly simple one; he says that structures are to be understood and grasped via models, but at the same time it is very complex. It needs many years of experi-

*In collaboration with Yu. G. Zagal'skaya and E. A. Pobedimskaya.

ence with thinking on models to operate freely with it, but it is clear that this theory of the real structure of substances has become particularly necessary in this century and extends not only to crystals but many other regions of knowledge. It seems to us that much of what is now called theory will in future be given over to machines, but it will still be known by the name of Belov's theory.

Belov's vast experience over many years gives him outstanding authority as a scientist; in a paper in the Compte Rendu he has shown that the current generation of scientists knows only the work of the last decade, all the rest appearing to them as virtually unknown antiquity.

There is no doubt that a knowledge of the history of science is of enormous value to a scientist; it makes his conclusions more reliable, more profound, and more extensive, and which may be the most important of all, more exact. Without a knowledge of the past one cannot properly evaluate the present and imagine the future.

Belov himself is a living history of science; he knows deeply the past and is personally acquainted with what has happened over the past 60 years, and all of this remains with him in his thoughts about the events of today.

It is clear from this the importance of a broad humanitarian education for the scientist. Belov is not only an excellent writer but also a very stimulating speaker, not to say an orator, as he is a master of the living word. I remember his striking address in the small theatre at the Institute of Crystallography on the 80th birthday of W. Bragg. This was an address on a great scientist, which began with a story of the young Bragg appointed to a professorship in Australia after leaving Oxford University, and containing many interesting sketches of his life, which brough us close to him and then to his son, and finally to him at the age of 80. It was a striking illustration of the work that can be done by a scientist over 70 years of age. I looked at my neighbor, a young girl, who listened to Belov enraptured, as one might listen to a piece of beautiful music. A little later, I overheard someone say to listen to Belov is like listening to music.

Scientists are sometimes divided into idealists and materialists, and each of these concepts may be given many different hues; but one can divide them also into idealists and realists, the first being such as Don Quixote, of whose type we do not at present have too many. Belov undoubtedly belongs to the realist, and in that respect his activities are very significant, since he has never been attracted to the Don Quixotes, because their world view appears impossible to him as a scientist.

Belov has used his influence and position to do a great many good deeds, which have assisted a multitude of scientific workers needing support to reach full adult stature. It has been said somewhere that the sole quantity that makes a man great is goodness, and by this one sign alone Belov must be reckoned a very great man.

I finish with a few words on the activities of Belov as the head of the Department of Crystallography and Crystallochemistry at the Geological Faculty of Moscow University.

Since 1963, the Department has taken in 240 specialists on chemical crystallography and crystal growth, and there have been presented fifty candidate's and two doctoral dissertations. The Department has published 450 papers, and it is generally an extremely lively department. Its members have made many discoveries, especially in collaboration with their leader himself, and many of these have dealt with specialized crystallographic disciplines, in particular crystallochemistry. With this I complete my sketch of Belov's work,* but feel that it is somewhat unsatisfactory. I have set on one side as unsuitable a comparison with other leading scientists, since he is in many ways unique and incomparable. What portrait should see him

_____
*See Vestnik Mosk. Univ., ser. geol., No. 6 (1966) for biographic data.

in pictures as a striking humane man of generous and still youthful appearance, with unusually firm gaze and a kindly approach to people. This is how Soviet crystallographers see him, when they encounter him still at many meetings and conferences. One cannot put a price on such features of a man.

Finally we may say that Belov is known not only to Soviet crystallographers but those throughout the world, since for six years he was Vice President of the International Union of Crystallography and three years President, and all those who have known him in this capacity will undoubtedly wish him a long and vitally active life.